Introduction to Time-Frequency and Wavelet Transforms

Shie Qian

National Instruments Corp.

PH PTR

Prentice Hall PTR
Upper Saddle River, New Jersey 07458
www.phptr.com

A CIP catalogue record for this book can be
obtained from the Library of Congress

Editorial/Production Supervision: *Kathleen M. Caren*
Acquisitions Editor: *Bernard Goodwin*
Editorial Assistant: *Michelle Vincenti*
Marketing Manager: *Dan DePasquale*
Manufacturing Buyer: *Alexis R. Heydt-Long*
Cover Design Director: *Jerry Votta*

© 2002 byPrentice Hall PTR
Prentice-Hall, Inc.
Upper Saddle River, NJ 07458

Prentice Hall books are widely used by corporations and government agencies for training, marketing, and resale. The publisher offers discounts on this book when ordered in bulk quantities.
For more information, contact Corporate Sales Department, Phone: 800-382-3419;
Fax: 201-236-7141; Email: corpsales@prenhall.com
Or write: Prentice Hall PTR, Corp. Sales Dept., One Lake Street, Upper Saddle River, NJ 07458

All products or services mentioned in this book are the trademarks or service marks of their respective companies or organizations.

All rights reserved. No part of this book may be reproduced, in any form or by any means, without permission in writing from the publisher.

Printed in the United States of America

10 9 8 7 6 5 4 3 2 1

ISBN 0-13-030360-7

Pearson Education LTD.
Pearson Education Australia PTY, Limited
Pearson Education Singapore, Pte. Ltd.
Pearson Education North Asia Ltd.
Pearson Education Canada, Ltd.
Pearson Educación de Mexico, S.A. de C.V.
Pearson Education—Japan
Pearson Education Malaysia, Pte. Ltd.
Pearson Education, Upper Saddle River, New Jersey

In memory of my father,
Tomy Tsien,
who led me into science and engineering but left us
long before he could see his dream come true.
With deepest appreciation for his sacrifice,
the dream is here.

Preface

For a long time, I wondered if the recently popularized time-frequency and wavelet transforms were merely academic exercises. Do applied engineers and scientists really need signal processing tools other than the FFT? After 10 years of working with engineers and scientists from a wide variety of disciplines, I have finally come to the conclusion that, so far, neither the time-frequency nor wavelet transform appear to have had the revolutionary impact upon physics and pure mathematics that the Fourier transform has had. Nevertheless, they can be used to solve many real-world problems that the classical Fourier transform cannot.

As James Kaiser once said, "*The most widely used signal processing tool is the FFT; the most widely misused signal processing tool is also the FFT.*" Fourier transform-based techniques are effective as long as the frequency contents of the signal do not change with time. However, when the frequency contents of the data samples evolve over an observation period, time-frequency or wavelet transforms should be considered. Specifically, the time-frequency transform is suited for signals with slow frequency changes (narrow instantaneous bandwidth), such as sounds heard during an engine run-up or run-down, whereas the wavelet transform is suited for signals with rapid changes (wide instantaneous frequency bandwidth), such as sounds associated with engine knocking. The success of applications of the time-frequency and wavelet transforms largely hinges on understanding their fundamentals. It is the goal of this book to provide a brief introduction to time-frequency and wavelet transforms for those engineers and scientists who want to use these techniques in their applications, and for students who are new to these topics.

Keeping this goal in mind, I have included the two related subjects, time-frequency and wavelet transforms, under a single cover so that readers can grasp the necessary information and come up to speed in a short time. Professors can cover these topics in a single semester. The co-existence of the time-frequency and wavelet approaches in one book, I believe, will help comparative understanding and make complementary use easier.

This book can be viewed in two parts. While Chapters Two through Six focus on linear transforms, mainly the Gabor expansion and the wavelet transform, Chapters Seven through Nine are dedicated to bilinear time-frequency representations. Chapter Ten can be thought of as a combination of time-frequency and time-scale (that is, wavelets) decomposition. The presentation of the wavelet transform in this book is aimed at readers who need to know only the basics and perhaps apply these new techniques to solve problems with existing commercial software. It may not be sufficient for academic researchers interested in creating their own set of basic functions by techniques other than the elementary filter banks introduced here.

All chapters start with the discussion of basic concepts and motivation, then provide theoretical analysis and, finally, numerical implementation. Most algorithms introduced in this book are a part of the software package, Signal Processing Toolset, a National Instruments product. Visit `www.ni.com` for more information about this software.

This book is neither a research monograph nor an encyclopedia, and the materials presented here are believed to be the most basic fundamentals of time-frequency and wavelet analysis. Many theoretically excellent results, which are not practical for digital implementation, have been omitted. The contents of this book should provide a strong foundation for the time-frequency and wavelet analysis neophyte, as well as a good review tutorial for the more experienced signal-processing reader.

I wrote this book to appeal to the reader's intuition rather than to rely on abstract mathematical equations and wanted the material to be easily understood by a reader with an engineering or science undergraduate education. To achieve this, mathematical rigor and lengthy derivation have been sacrificed in many places. Hopefully, this style will not unduly offend purists.

On the other hand, "*Formulas were not invented simply as weapons of intimidation*" [22]. In many cases, mathematical language, I feel, is much more effective than plain English. Words are sometimes clumsy and ambiguous. For me, it is always a joy to refresh my knowledge of what I learned in school but have not used since.

Some of the material presented in this book is the result of collaborative work which so greatly profited from the contributions of friends and colleagues that I must mention them. It was my graduate advisor, Professor Joel M. Morris, who led me into such a fascinating field. Motivated by the suppression of cross-term interference, in the early 90s the idea of the decomposition of the Wigner-Ville distribution emerged, which led to a series of interesting results. With Shidong Li and Kai Chen, the relationship of the most similar dual and the pseudo inverse was discovered. Based on Wexler and Raz's periodic discrete Gabor expansion [225], Dapang Chen and I obtained its infinite counterpart which resulted in an interesting time-dependent spectrum, the time-frequency distribution series, also known as the Gabor spectrogram. To improve the time-frequency resolution, we also proposed the adaptive Gabor expansion which turned out to be the same scheme as that employed by the matching pursuit method independently developed by Stéphane Mallat and Zhifeng Zhang during the same time period [172]. With Qinye Yin, such an adaptive decomposition scheme was generalized into the Gaussian chirplet cases. The fast-refinement algorithm initially appeared when Qinye Yin visited Austin, Texas. As a result of his insightful observation, the computation of the adaptive Gaussian chirplet approximation has been significantly improved. All these years later, I clearly remember a discussion at Xiang-Geng Xia's office in Malibu, California, in front of the magnificent beach there. The subject was the Gabor expansion-based time-varying filter. As a result of that discussion, a few days later Xiang-Geng called me and said, "With a tight frame, the iteration of the time-varying filter converges!" That memory is indelible.

I would also like to thank Professors Xiang-Geng Xia and Richard G. Baraniuk for their contributions in Chapter 5 and Section 8.3, respectively.

In a larger sense, this book is the result of the enthusiasm and support from numerous customers, colleagues, and friends. I want to take this opportunity to express my sincere thanks to all of them. Particularly, I would like to thank Dr. James Truchard and Jeff Kodosky, the founders of National Instruments Corporation. It is their great enthusiasm and continuous support that keep such a "non-profitable" project evolving and making all those interesting applications take place.

This book has been an on-off project for almost three years and I extend my thanks to Bernard Goodwin at Prentice Hall for his endless patience and generous assistance. There were so many errors in the original draft that I dare not look at it again. Mahesh Chugani carefully read the entire manuscript. His numerous comments and suggestions improved the book significantly.

My deepest thanks are reserved for my mother, Yuzhen Wu, and my family: my wife, Jun, and daughter, Nancy. I am very grateful for their understanding, support, and patience during this formidable project.

錢世鍔
Shie Qian
Austin, Texas

Contents

CHAPTER 1

Introduction

*D*ue to physical limitations, usually we are only able to study a system through signals associated with the system rather than physically opening up the system. For example, physicists and chemists use the spectrum generated by the prism, without breaking up molecules, to distinguish different types of matter.[1] Astronomers apply spectra as well as the Doppler effect, discovered by an Austrian physicist *Christian Johann Doppler* in 1842, to determine distances to planets that a human being may never be able to reach. Doctors utilize the electrocardiograph (ECG), without opening up the body, to trace the electrical activity of the heart and diagnose whether or not a patient suffers from heart problems. Indeed, signal processing has played a fundamental role in the history of civilization. Prior to World War II, however, signal processing was primarily a part of physics. Signals that scientists and engineers dealt with were mainly analog by nature. It was the *sampling theorem*, proved by the mathematician *J. Whittaker* in 1935 [229] and applied to communication by *Claude Shannon* in 1949 [46], that led to a new era of signal processing. Modern signal processing can be thought of as the combination of physics as well as statistics. Because of the discovery of the sampling theorem and the advance of the digital computer over the last couple of decades, we are now able to employ elegant mathematical approaches, such as the virtual prism — Fourier transform, to process all different kinds of signals that our ancestors never would have been able to imagine. Applications of modern signal processing range from the control of the *Mars Pathfinder* more than twenty million miles away to the discovery of abnormal cells inside the body.

1. Spectrum analysis was jointly discovered by the German chemist *Robert Wilhelm Bunsen* (1811 - 1899) and the German physicist *Gustar Robert Kirchhoff* (1824 - 1887). Contrary to popular belief, Bunsen had little to do with the invention of the Bunsen burner, a gas burner used in scientific laboratories. Although *Bunsen* popularized the device, credit for its design should go to the British chemist and physicist *Michael Faraday* (1791 - 1867).

A fundamental mathematical tool employed in signal processing is a *transform*. When we are asked to multiply the Roman numerals LXIV and XXXII, only a few of us will be able to give the correct answer right away. However, if the Roman numerals are first translated into Arabic numerals, 64 and 32, then all of us can get 2048 immediately. The process of converting the unfamiliar Roman numerals into common Arabic numerals is a typical example of transforms [22]. By properly applying transforms, we can simplify calculations or make certain attributes of the signal explicit.

One of the most popular transforms known to scientists and engineers is the Fourier transform that converts a signal from the time domain to the frequency domain. Two hundred years ago, during the study of heat propagation and diffusion, *Jean Baptiste Joseph Fourier* found a series of harmonically related sinusoids to be useful in representing the temperature distribution through a body. The method of computing the weight of each sinusoidal function is now known as the *Fourier transform*. The Fourier transform can not only benefit the study of heat distribution, but is also extremely useful for many other mathematical operations, such as solutions to differential equations. The application of the Fourier transform with which scientists and engineers are most familiar may be convolution theory. By applying the Fourier transform, one can convert time-consuming convolutions into more efficient multiplications.

In fact, the Fourier transform is not simply a mathematical trick to make calculations easier; it also acts as a mathematical *prism* to beak down a signal into a group of waveforms (different frequencies), as a prism breaks up light into a color spectrum. With the help of the Fourier transform, we can interpret radiation from distant galaxies, diagnose a developing fetus, and make inexpensive cellular phone calls. With the establishment of quantum mechanics, the significance of Fourier's discovery becomes even more obvious. By using the Fourier transform, for instance, we can quantitatively describe a fundamental and inescapable property of the world – the *Heisenberg uncertainty principle*. That is, in certain pairs of quantities, such as the position and velocity of a particle, cannot both be predicted with complete accuracy.

The Fourier transform is so powerful that people tend to apply it everywhere without noticing one fundamental difference between the mathematical prism and a real prism. The spectrum produced by the prism in the morning is different from that in the evening. Using a fancy word, we may say that the prism gives *instantaneous spectra*. Using a prism to examine spectra of light, there is no need for the information about light that existed a million years ago and the light that will be there tomorrow. However, this is not the case for the Fourier transform. To compute the Fourier transform, we not only need previous information, but also information that has not yet occurred. The spectrum computed by the Fourier transform is the spectrum averaged over an infinitely long time before the present to an infinitely long time after the present!

Figure 1-1 illustrates two linear chirp signals. Each is a time reversed version of the other. While frequencies of the signal on the left plot increase with time, frequencies of the signal on the right decrease with time. Although the frequency behavior of the two signals is obviously different, their frequency spectra computed by the Fourier transform, as shown in Figure 1-2, are identical. The Fourier transform preserves all information about the time waveform (if it did not, we could not reconstruct the signal from the transform), but information about time or space is buried deep within the phases, which is beyond our comprehension.

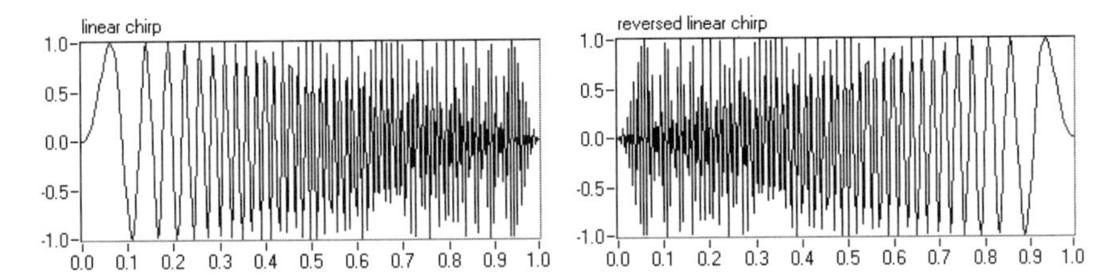

Figure 1-1 The linear chirp signal in the plot on the right is a time reversed version of the signal in the plot on the left. While frequencies of the chirp signal on the left increase with time, frequencies of the signal on the right decrease with time.

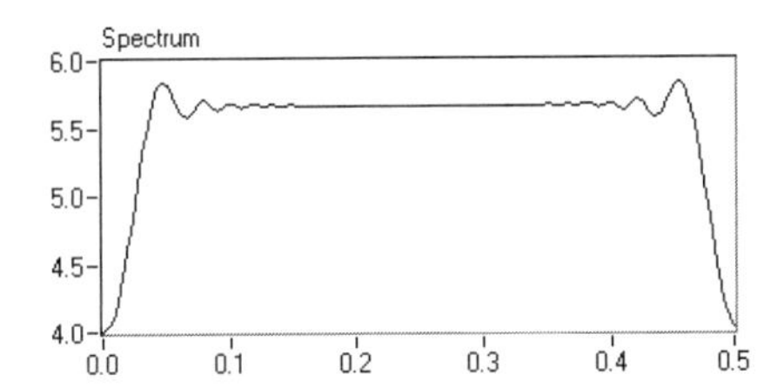

Figure 1-2 Although the two linear chirp signals in Figure 1-1 have completely different time waveforms, their frequency spectra are identical. The Fourier transform preserves all information about the time waveform (if it did not, we could not reconstruct the signal from the transform), but information about time or space is buried deep within the phases, which is beyond our comprehension.

Figure 1-3 depicts the spectrum of an engine sound (the corresponding time waveform is illustrated in the top plot of Figure 1-4). When listening to this signal, we can clearly identify several knocking sounds caused by out of phase firing inside the engine. As indicated by the wavelet transform, the second plot in Figure 1-4, the knocking sound is actually quite strong. To compute the Fourier transform, we have to include the signal before knocking takes place and also the signal after the knocking ends. What the spectrum computed by using the Fourier transform tells us are the frequencies contained in the entire time waveform, not the frequencies at a particular time instant. The Fourier transform provides the signal's average characteristics. Although the amplitude of engine knock sounds could be rather large in a very short time period, the energy of the sound, compared to the entire background noise, is negligible. Consequently, there will be no obvious signatures in the spectrum to show the presence of engine knocking. The Fourier transform smears the signal's local behavior globally. "The Fourier transform is poorly suited to very brief signals, or signals that change suddenly and unpredictably; yet in signal processing, brief changes often carry the most interesting information" [22].

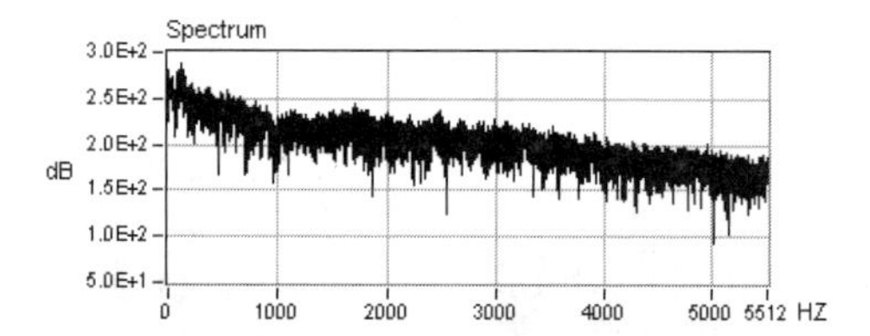

Figure 1-3 Because the energy of the engine knocking sound is relatively small, the presence of engine knocks is completely overwhelmed in the averaged spectra computed by the Fourier transform.

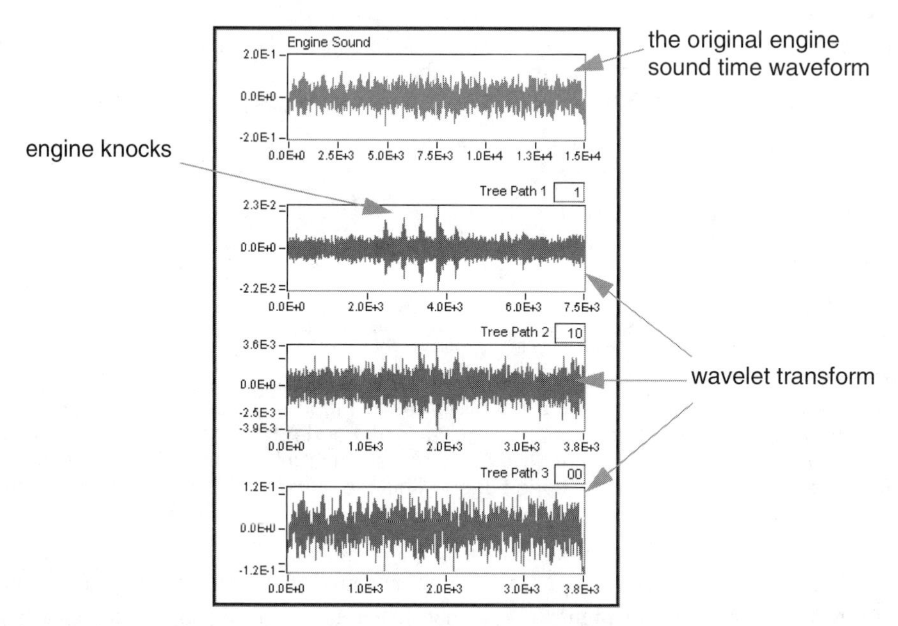

Figure 1-4 While engine knock sounds are completely concealed in the background noise in the time waveform plot (top), the wavelet transform (second from the top) clearly indicates the existence of engine defects.

Although most natural spectra are time dependent (for example, the light during the morning is different than that during the evening), the Fourier transform makes "changing frequency" unthinkable. As Gabor wrote, "even experts could not at times conceal an uneasy feeling when it came to the physical interpretation of results obtained by the Fourier method" [113].

The shortcoming of the Fourier transform has been recognized for a long time. The development of Fourier's alternatives, involving a great many individuals, started at least a half century ago. The first two important articles, dealing with the limitation of the Fourier transform, appeared right after World War II: one by *Dennis Gabor* in 1946 (who later received the Nobel Prize for the invention of holography) [113] and the other by *J. Ville* in 1948 [219]. Since the result obtained by Ville was similar to the one introduced by *Eugene Wigner* (who received the Nobel Prize for the discoveries concerning the theory of the atomic nucleus and elementary particles) in the area of quantum mechanics in 1932 [226], traditionally Ville's method is named the *Wigner-Ville distribution*.

The initial reaction to neither Gabor nor Ville's work was enthusiastic. The difficulty associated with the *Gabor expansion* was that the sets of elementary functions that are suitable for time-frequency analysis in general do not form orthogonal bases. The problem with the Wigner-Ville distribution has been the co-called cross-term interference that makes the resulting presentation difficult to be interpreted. It was two sets of papers, *Claasen* and *Mecklenbräuker* [91] and *Bastiaans* [72], which both appeared in the early 1980's, that triggered great interest in revisiting Gabor and Ville's pioneer work. Since then, there has been a tremendous amount of activities and numerous developments. Some of them appear to be reaching a level of maturity for real applications.

The recognition of the wavelets transform is much more recent, though a similar methodology can be traced as early as the beginning of the twentieth century [305]. *Wavelets* are not a "bright new idea" but concepts that have existed in other forms in many different fields. For instance, the numerical implementation of the wavelet transform is nothing more than the well-established filter banks. As *Stéphane Mallat* wrote, "This wavelet theory is truly the result of a dialogue between scientists who often met by chance, and were ready to listen... this is a particularly sensitive task (mentioning who did what), risking aggressive replies from forgotten scientific tribes" [27]. "Tracing the history of wavelets is almost a job for an archaeologist" [22].

The set of basis functions employed by Fourier, sine and cosine functions, not only is the mathematical model of the most fundamental natural phenomena – the *wave*, but also is a solution of differential equations.[2] Unfortunately, this is not the case for time-frequency and wavelet transforms. Neither time-frequency nor wavelet transforms will likely have the revolutionary impact upon science and engineering that the Fourier transform has had. However, the time-frequency and wavelet transforms do offer many interesting features that the Fourier transform does not possess.

In addition to detecting engine knocks, for example, the wavelet transform is also successfully used for train wheel diagnosis. It has been found that one of the main causes of train accidents was the result of defective wheels and bearings. Hence, on-line train wheel and bearing diagnoses are exceptionally important for avoiding potential catastrophes. The parameter that engineers believe can be used to effectively detect hidden flaws in wheels and bearings is variations of railroad track. Defective wheels and bearings usually will generate an impulse like noise

2. It would be hard to exaggerate the significance of the differential equation to the history of civilization. It was the differential equation that enabled *Sir Issac Newton* to use the current information, such as position, velocity, and acceleration, to predict the future. Newton's discovery led the French mathematician and astronomer *Pierre Simon Laplace* to believe, "Nothing would be uncertain to it, and the future, like the past, would be present before its eyes." Because of such successful application of the differential equation, Laplace imagined a single formula that would describe the motion of every object in the universe for all time.

as the train moves on the track, making abnormal track variations. Such noise could be effectively filtered out by the wavelet transform.

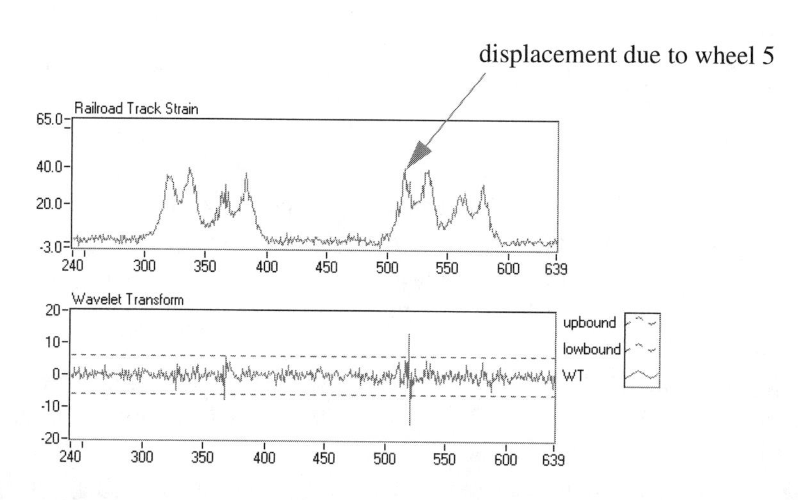

Figure 1-5 Displacement of the railroad track during the time that eight wheels pass over a strain gauge, and the corresponding wavelet transform. There is no clear signature between the normal and abnormal wheels in the time waveform (the upper plot). However, in the wavelet transform domain (the lower plot), we can readily identify a potential problem at the fifth wheel (between $x = 500$ and $x = 550$).

Figure 1-5 illustrates a typical train wheel on-line testing result. When a wheel is far away from the strain gauge mounted beneath the track, the corresponding track displacement is small. It increases as the train wheel approaches the strain gauge. The displacement reaches a maximum when a wheel is right above the strain gauge. The plot on the top of Figure 1-5 shows the displacement history during the time that eight wheels pass the strain gauge. While the *X*-axis describes the time index, the *Y*-axis indicates the magnitude of track displacement. Each bump corresponds to one wheel passing over the strain gauge mounted underneath the railroad track. Obviously, there is no clear signature between the normal and abnormal wheels in the time waveform. However, in the wavelet transform domain, the plot on the bottom of Figure 1-5, we can readily identify a potential problem at the fifth wheel (between $x = 500$ and $x = 550$). The wavelet transform-based on-line diagnosis system is expected to substantially reduce potential train accidents caused by defective wheels or bearings.

Another interesting application of the wavelet transform is for detecting oil leakage. One of the most challenging tasks in an oil field is on-line pipeline leakage monitoring. This is particularly true concerning incidents directly caused by organized oil theft. This is not only an economic loss for the oil company, but also environmental pollution, a public issue.

When a leakage incident occurs, the oil pressure in the vicinity of the leakage point drops rapidly. Such a drop is presumably propagated in all directions along the pipeline. Consequently:

1. Oil pressure decreases at both the inlet and outlet
2. The oil flow rate at the outlet decreases, while the oil flow rate at the inlet increases

Based on the time difference of the pressure drops observed at the inlet and outlet, conceptually, the leakage location can then be determined by

$$\frac{length\ of\ pileline + pressure\ wave\ velocity \times time\ difference}{2} \tag{1.1}$$

Figure 1-6 depicts the layout of the pressure and flow meters. Note that when both conditions 1 and 2 mentioned previously are simultaneously satisfied, all other combinations can be excluded from the leakage. For instance, the decrease of pressure and flow rate at both ends can be considered as the result of the inlet pump slowing down. Conversely, increased pressure and flow rate indicates the pump is speeding up.

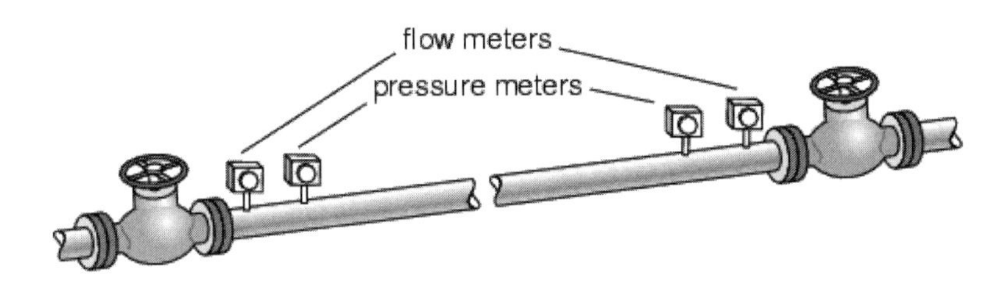

Figure 1-6 Oil leakage will cause pressures at both the outlet and inlet to decrease, the flow rate at the outlet to decrease, and the flow rate at the inlet to increase.

It is said that one invents with intuition and one proves with logic. This is certainly true in this application. The idea is straightforward but the implementation is very challenging. The main difficulties include:

1. Synchronization of all pressure and flow meters that typically are 60 km apart.
2. Variation of pressure wave velocity. The pressure wave velocity is related to temperature, the density of the medium, as well as the elasticity of the pipe material. To facilitate oil movement, the raw oil is often heated at each station, especially in cold weather. Due to the non-uniform temperature distribution, the pressure wave velocity is not constant. Consequently, the actual formula for estimating the location of the leakage is much more involved than that which one may anticipate (e.g., Eq. (1.1)).
3. Background noise. Compared to the main oil flow, the leakage usually is negligible. Therefore, the rapid changes of pressure caused by incidents of leakage often are not noticed.

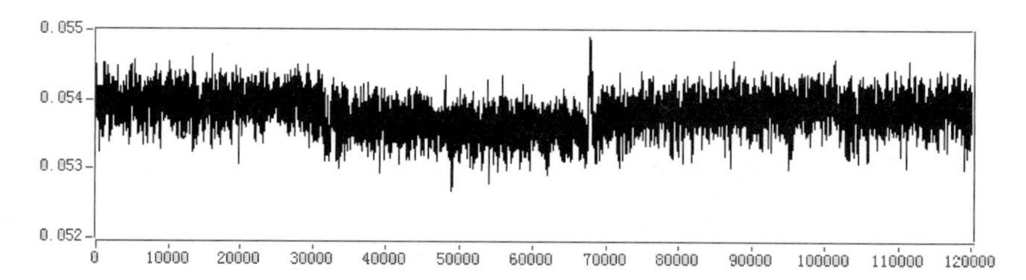

Figure 1-7 A typical pressure signal. Although there is a drop caused by a leakage in the vicinity of 31,000, there is no obvious indication in the time waveform (Data provided by Zhuang Li, College of Engineering, Tianjin University, China).

Figure 1-7 illustrates a typical oil pressure signal. Although there is a drop caused by leakage in the vicinity of 31,000, there is no obvious indication in the time waveform. However, by applying the wavelet transform,[3] one can accurately determine the time instant of the pressure drop.

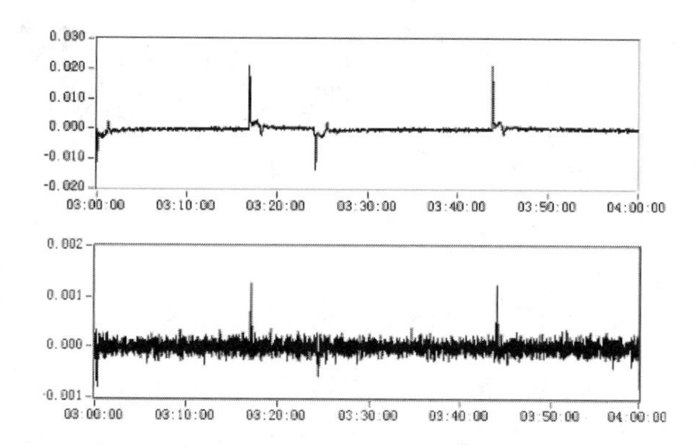

Figure 1-8 The upper plot is the wavelet transform of the signal at the pipeline inlet, whereas the lower plot shows the wavelet transform of the signal at the pipeline outlet. There were two incidents of leakage from the pipeline. The first one occurred between 3:00 and 3:18 am and the second was between 3:24 and 3:44 am (Data provided by Zhuang Li, College of Engineering, Tianjin University, China).

Figure 1-8 depicts the wavelet transforms of signals recorded between 3:00 to 4:00 am on April 13, 2001. The upper plot shows the wavelet transform of the signal at the pipeline inlet, whereas the lower plot is the wavelet transform of the signal at the pipeline outlet. As indicated in the wavelet transform domain, there were two incidents of leakage from the pipeline. The first one occurred between 3:00 and 3:18 am and the second was between 3:24 and 3:44 am. For the

3. Due to the time accuracy required in this application, the wavelet transform must be time-shift invariant.

second leakage, the time difference between the pressure drop at the pipeline inlet and the drop at the other end of the pipeline outlet was computed as 14.95 seconds. The leakage location was identified as 39.34km south of the Changzhou Station in the Victoria Oil Field, China. The actual point of theft was 39.29 km south of the station (less than 50 meters away!). Since the length of the pipeline is 61.48 km, the relative error was 0.26%.

Figure 1-9 Captured oil thieves' truck. For a veteran oil thief, the entire process of filling an eight to ten ton tank only takes 15 to 20 minutes (Image courtesy of Zhuang Li, College of Engineering, Tianjin University, China).

For a veteran oil thief, the entire process of filling an eight to ten ton tank only takes 15 to 20 minutes. Given the extent of the oil pipeline networks and road conditions, the system must respond quickly so that security personnel can catch oil thieves in the act. For a sixty-kilometer pipeline, the wavelet transform-based detector can discover and pinpoint the leakage location within two minutes. During two months of preliminary testing, this system successfully detected 79 leakages, resulting in the capture of 25 illegal oil trucks.

The Fourier transform compares a signal with a set of sine and cosine functions. Each sine and cosine function oscillates at a different frequency. Hence, the Fourier transform indicates magnitudes of the signal at each individual frequency. On the other hand, the wavelet transform compares a signal with a set of short waveforms (called *wavelets*). Each wavelet has a different time duration (or *scale*). The shorter the time duration, the wider the frequency bandwidth, and vice versa. In mathematical jargon, the process of stretching or compressing the fundamental wavelet (usually called the *mother wavelet*) is named *dilation*. As wavelets get narrower and narrower, eventually they become impulse-like functions (equivalent to wide frequency band). Consequently, the wavelet transform is very powerful for detecting impulse-like signals, such as engine knocks, noise created by defective train wheels and bearings, and the pressure drop caused by oil leakages. In those examples, the wavelet transform is obviously superior to its counterpart – the Fourier transform.

Besides wideband (short time duration) signal detection, the wavelet transform is also widely used for 2D image processing. Figure 1-10 is the 2D wavelet transform of an image of the fingerprint. In this example, we break up one full image (on the left) into four sub-images. Obviously, the major features of the image of the fingerprint are contained in the upper-left plot which is one the quarter of the original image. The remaining three sub-images have relatively less important "information" and thereby have less influence on the reconstruction. This suggests that instead of storing or transferring the entire big image, we only need to store or transfer most of the upper-left image and the number of prominent pixels in the other three sub-images. Because the number of pixels in each sub-image is a quarter of the number of pixels contained in the original image, by applying a 2D wavelets transform and proper coding schemes, we are able to save a great amount of memory and tremendous communication bandwidth.

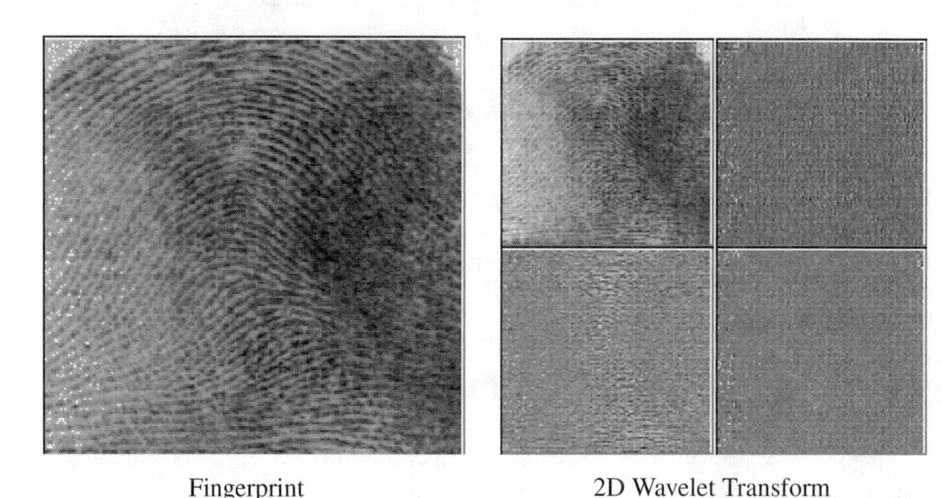

Fingerprint 2D Wavelet Transform

Figure 1-10 The major features of the image of the fingerprint are contained in the upper-left plot within an image a quarter of the size of the original image. The remaining three sub-images have relatively less important "information" and thereby have less influence on the reconstruction.

While the wavelet transform is suited to highly nonstationary signals with sudden peaks or discontinuities, the time-frequency transform is very powerful for analyzing signals whose frequency changes slowly with time (i.e., narrowband instantaneous frequency bandwidth).

One example is an application for time-varying harmonic analysis. Here, the term harmonic refers to frequencies that are integer (or fractional) multiples of a fundamental frequency, such as the vibration observed from rotating machinery. For a running engine, the causes of vibrations are very different; some may be associated with bearings and others may be associated with the cooling fan, but the vibration frequencies are all functions of a fundamental frequency — the engine rotation speed. For instance, the vibration frequency related to the bearing may be equal to the fundamental frequency multiplied by the number of balls inside the bearing housing. The vibration frequency related to the fan may be equal to the fundamental frequency multiplied by the number of blades. By applying the Fourier transform to the time waveform of the vibration signal, we will obtain a group of harmonics in the frequency domain. Because all these harmonics have explicit physical interpretations, by analyzing the amplitudes and phases of different harmonics, engineers can often determine whether the engine is running normally.

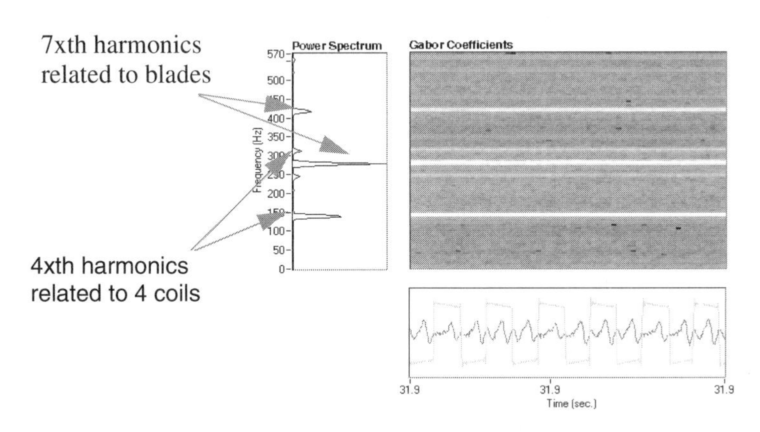

Figure 1-11 The physical characteristics of rotating machines can often be reflected in the harmonics. This figure shows the signal obtained from a four-phase electrical motor, with a seven-blade cooling fan, running at constant speed. While the 4xth harmonics correspond to vibrations related to the number of coils, the 7xth harmonics are related to vibrations associated with the number of blades of the cooling fan. At a constant speed, all these harmonics can be well identified from both the conventional power spectrum and the joint time-frequency plot.

Figure 1-11 illustrates a sound waveform recorded from an electrical motor running at a constant speed. Intuitively, not only will the motor rotation generate the sound, but also all the other parts that vibrate due to the motor rotation will make noise. The sound waveform plotted at the bottom, in fact, is a combination of all the kinds of vibrations caused by motor rotation. Moreover, the vibration frequencies are multiples of the fundamental frequency – the motor rotation speed. In addition to the sound waveform, the bottom plot also depicts the tachometer pulses. Every two tachometer pulses indicate one revolution. The plot on the left shows the Fourier transform-based power spectrum. The plot in the middle is the magnitude of the corresponding Gabor coefficients. Since the electrical motor contains four coils, we expect to observe harmonics with frequencies at four, eight, and twelve times the rotational speed. In addition, its cooling fan has seven blades. So, we will also expect to observe harmonics at seven, fourteen, and twenty-one times the rotational speed. As shown in Figure 1-11, when the rotational speed is constant, we can clearly see the fourth and seventh orders from both the Fourier transform-based power spectrum and the Gabor coefficients (in which they appear as horizontal white lines).

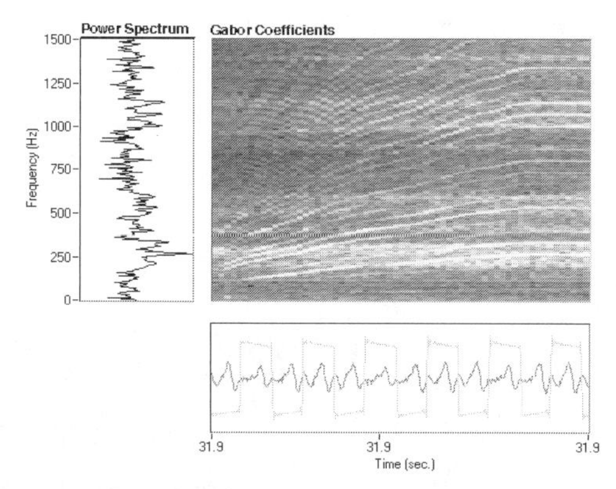

Figure 1-12 During the electrical motor run-up, the fundamental frequency bandwidth becomes wide and its harmonics overlap. Consequently, we are no longer able, from the conventional power spectrum, to distinguish the harmonics caused by different vibrating components. But this is not the case for the Gabor coefficients. Whether or not the electrical motor runs at constant speed, the signatures of different harmonics (white lines), in terms of the Gabor coefficients, are always obvious.

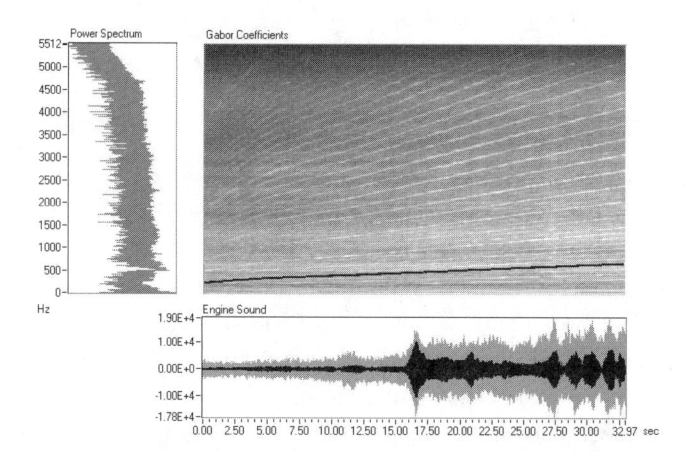

Figure 1-13 Gabor expansion-based time-varying filter. The bottom plot illustrates the electrical motor run-up sound (light color) vs. the extracted seventh harmonic (dark color) time waveforms. The dark line in the Gabor coefficients plot marks the Gabor coefficients corresponding to the seventh harmonic.

However, the Fourier transform-based classical harmonic analysis only works when the fundamental frequency is constant (e.g., a motor running at a constant speed). It is not suitable for harmonics when the fundamental frequency evolves over time (e.g., motor speed changes). As we shall see in Section 2.4, when the frequency changes, the frequency bandwidth becomes wide. The frequency bandwidth is proportional to the rate of change of frequency. The faster the change of frequency, the wider the corresponding frequency bandwidth. As the fundamental frequency bandwidth becomes wide, the bandwidth corresponding to the harmonics will become wide too. Finally, the harmonics will overlap in the frequency domain.

Figure 1-12 plots a sound time waveform recorded when the motor runs up. Because the harmonics overlap, the vibrations caused by different sources can no longer be distinguished from the conventional power spectrum. But this is not the case for the Gabor coefficients. Whether or not the motor runs at constant speed, the signatures of different harmonics, in terms of the Gabor coefficients, are always obvious. This observation suggests that we can use the Gabor transform to study such time-varying harmonics. For example, we can extract the Gabor coefficients that are associated with the desired time-varying harmonic. Then, we can take the inverse of the Gabor transform to obtain a corresponding time waveform. Figure 1-13 depicts the results of the Gabor expansion-based time-varying filtering, which shows how the seventh order is extracted. Such a method has been successfully used in the automobile industry for engine run-up/down testing.

The major advantage of time-frequency transforms is that they describe how the spectrum of a signal changes with time. Such an instantaneous spectrum has been found to be very useful in many applications, such as aneurysm research.[4] The results obtained from the Gabor spectrogram may eventually lead to an economical aneurysm screening approach.

As one study indicated, the occurrence of aneurysms is common in the world population [209]. It is estimated that about 5% of the adult population have aneurysms. Statistics also show that about 28,000 people suffer a subarachnoid hemorrhage from a ruptured aneurysm annually in North America. The mortality and mobility from a ruptured aneurysm are very high. Often there are no warning signs other than vague non-specific symptoms prior to the catastrophic rupture. Except for Computer Tomography (CT) and Magnetic Resonance Imaging (MRI), at present, there is no simple and economical method for screening.

It has been determined that some aneurysms emit specific resonant sounds of varying frequencies. So, the frequency of the sound created by blood flow is a potential feature in diagnosing the aneurysm. However, the sound associated with an aneurysm is generated by a complicated dynamic fluid system involving the wall, chamber, surrounding blood vessels, and moving blood, under varying pressure. It has been shown that the sound recorded from the aneurysm is due to vibration stimulated by the blood flow inside the aneurysm and nearby blood vessels. This vibrational system is non-linear and time-varying. In addition, the sound emitted by an aneurysm is non-stationary and is usually combined with the biological noises generated by the heart, respiratory system, and eye movements.

Figure 1-14 shows an aneurysm signal recorded directly from an antercranial aneurysm during surgery. This record has 512 data samples corresponding to 300 milliseconds in time. The

4. An aneurysm is a bulge or sac formed by the ballooning of the wall of an artery or a vein. It may become the site of a blood clot that breaks away and lodges in the tissues of such vital organs as the heart and the brain, causing serious, even fatal, heart failure or brain damage.

spectrum of the signal is illustrated on the left. It shows the range of resonant frequencies (450 to 550 Hz), but provides no other useful information. The center plot depicts the corresponding Gabor spectrogram that describes the signal's instantaneous spectra.

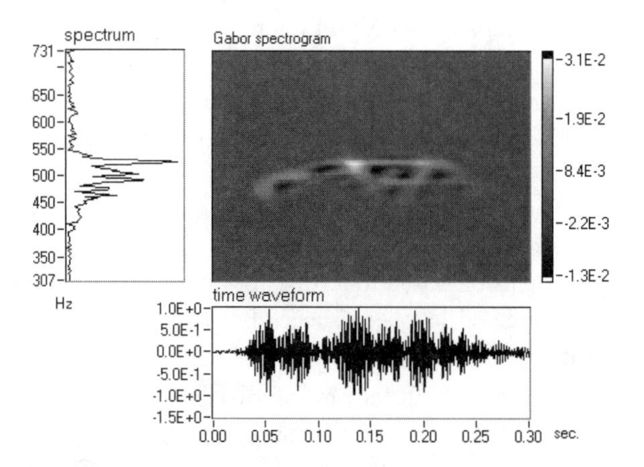

Figure 1-14 A Gabor spectrogram of bloodflow sound, recorded directly from an antercranial aneurysm during surgery, serves as an X-ray picture, helping researchers get a better understanding of the mechanism of an aneurysm system (Data provided by Mingui Sun of Medical School at the University of Pittsburgh).

With the help of the Gabor spectrogram, the physical system that produces this signal may be hypothetically explained as follows. The main arc may be produced by the variation of blood pressure which changes the physical parameters of the resonant system. More specifically, at the very beginning of the vibration (in the vicinity of .05 second), the spectral component is relatively wide, corresponding to a low Q (center frequency vs. frequency bandwidth) value of the vibrational system due to the existence of a large damping effect. As the stimulation increases, the pressure-induced vibration becomes stronger, and the spectral component narrows, indicating an increasing Q value. As a result, the vibration tends to concentrate on a single frequency (between 0.1 to 0.15 seconds), yielding a sinusoidal-like waveform. As the vibration continues another interesting phenomenon occurs. It can be observed that several branches, with harmonic-like patterns, deviate from the main arc when the vibrational magnitude becomes considerably larger (between 0.15 to 0.25 second). However, after a short period of time these branches again merge back to the main arc. This interesting behavior may indicate that, at the branching points, the vibrational system may have reached the upper limit of its linear range. However, the stimulation is still applied providing the system with additional energy. This over-stimulation causes the vibration to enter a nonlinear stage. Conversely, as the stimulation decreases, the vibrational system loses energy. Finally, the system stops vibrating when the damping effect becomes dominant again.

In this example, the Gabor spectrogram is used as an x-ray picture, helping researchers to get a better understanding of the mechanism of an aneurysm system. As shown in Figure 1-14, information provided by the Gabor spectrogram is not available in either the time waveform or the conventional frequency spectrum. The resulting observation may eventually lead to a simple and efficient aneurysm diagnostic approach that will not only be less expensive than CT and MRI, but also pain-free.

Figure 1-15 Sinking of shipment containers due to soil liquefaction at Port Island, Kobe City, after the 1995 Hyogoken-Nanbu (Kobe) earthquake (Image courtesy of Prof. F. Yamazaki of the Institute of Industrial Science at University of Tokyo).

Besides applications in the area of the life sciences, the Gabor spectrogram is also reported to have been successfully applied in the area of earthquake engineering, such as the detection of soil liquefaction from seismic records conducted at the University of Tokyo [156]. Soil liquefaction is an earthquake-related phenomenon that takes place in saturated cohesionless soils in the sub-surface layers. The cause of liquefaction is the rise of the water pore-pressure under undrained condition during shaking of the ground. The increase of the pore pressure reduces the soil shear resistance to almost zero, causing the soil to behave as a liquid. Consequently, the energy of horizontal vibrations (seismic shear waves from depth) transferred by the soil to the ground surface will be substantially reduced, particularly, the high frequency contents. Soil liquefaction has been recognized as the main reason for the collapse of earth dams and slopes, and the failure of foundations, superstructures, and lifelines, such as gas and electrical power supplies (see Figure 1-15).

Recent earthquakes in the US and Japan have shown the need of advanced earthquake disaster mitigation management in order to prevent or minimize damage, especially in urban areas. The basic idea is to monitor the seismic motion and perform actions based on the information obtained during an earthquake. The main tool of (means in) advanced earthquake disaster mitigation management are the early warning and preliminary damage assessment systems, the backbone of which is a ground-motion monitoring network consisting of three-axial accelerometers. Such systems are used to stop the high-speed bullet-trains in Japan before the arrival of the destructive seismic waves, to shut off the gas supply in the areas where the ground shaking is too severe, and to estimate the damages and casualties immediately after an earthquake when the exact data are still not collected. Early localization of the liquefied areas is also of interest, particularly for the lifelines, and efforts have been made to determine liquefaction occurrence shortly after an event.

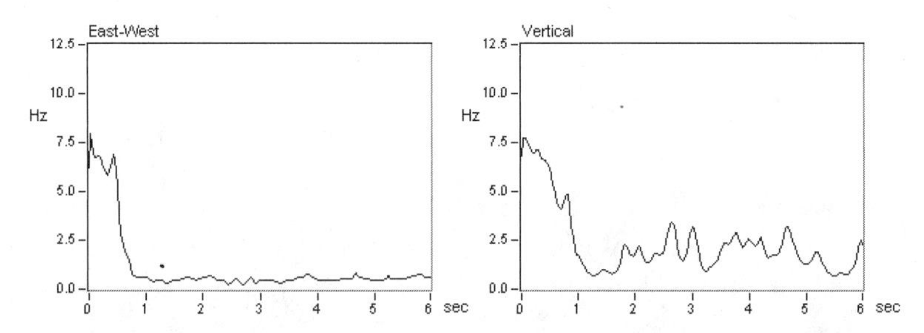

Figure 1-16 Mean instantaneous frequencies for the East-West component at the site where extensive liquefaction took place are obviously lower than that for a non-liquefaction case in Figure 1-17 (Data from Higashi-Kobe Bridge record, Strong Motion Array Observation Record Database CD-ROM Vol.5, Association for Earthquake Disaster Prevention, Tokyo, Japan, 1997).

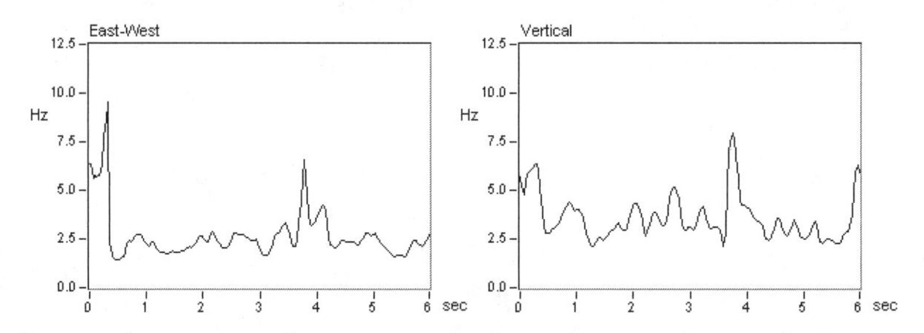

Figure 1-17 Mean instantaneous frequencies for the East-West component at the site where no liquefaction occurred are obviously higher than that for the liquefaction case in Figure 1-16 (Data from JMA-Kobe record, JMA Kobe Observatory, Japan Weather Association CD-ROM).

Proposals for detectors of soil liquefaction include downhole piesometers and sensors for measuring the rise of the water level in a hollow pipe inserted of in the ground. Such approaches, however, require costly placing work or are unreliable for long-term measurements.

An alternative is to use the seismic records, collected from the ground-motion monitoring networks, for detection of liquefaction. Since the 1964 Niigata earthquake, a number of seismic records from liquefied-soil sites have been obtained and studied. The records show that the horizontal ground acceleration alternates uniquely after the onset of liquefaction – its frequency abruptly drops off towards the 0.5 - 1 Hz range and its amplitude decreases – while the vertical acceleration is rather stable. Methods, directly based on the earthquake waveforms, for liquefaction detection have been developed. Recently, researchers from the Institute of Industrial Science at the University of Tokyo have employed the seismic signal instantaneous spectrum to quantify the alternation of the horizontal ground acceleration. The mean instantaneous frequency computed from that spectrum can well characterize the frequency changes, as shown in Figure 1-16 and Figure 1-17. Occurrence of liquefaction is judged, based on the relative difference in the mean instantaneous frequency of the horizontal and vertical acceleration.

In addition to the applications presented above, readers can find other successful stories in references [86], [87], [88], [89], [171], [200], [207], [208], [209], [210], [223], [395], and [396]. There is no doubt that the time-frequency and wavelet transforms have begun to pervade modern technology as well as our everyday life. It is the goal of this book to give a brief introduction to such exciting techniques.

This book was primarily written as a technical reference book for scientists and engineers who don't have the opportunity to learn this topic in their schools, but it can also be used as a one-semester advanced undergraduate course on time-frequency and wavelet transforms. The first nine chapters can be roughly partitioned into three parts; a brief review of classical signal analysis, linear transforms (including both the Gabor expansion and the wavelet transform), and time-dependent spectra. All topics start with ideas, concepts, and motivations and are then followed by theoretical developments and, finally, numerical implementations. The presentation focuses on basic concepts and the motivations behind those methods, rather than abstract mathematical formulas. To me, the evolution of time-frequency and wavelet transforms is as interesting as the result; I see them not only as tools but also as a story to tell of ideas and people.

Chapter 10 is dedicated to the adaptive representation and adaptive spectrogram. Although the development of the adaptive representation initially was motivated by the limitation – a fixed window – possessed by the Gabor expansion, the result turns out to be a much more general and powerful signal decomposition scheme. It has been found that the adaptive representation (also known as *matching pursuit*) is very useful for signals whose elementary model cannot form a set of orthogonal functions, such as most seismic signals.

The concepts of time-frequency and wavelet transforms are rather straightforward, but the rigorous treatments require a certain level of mathematical preparation. Because this book aims to serve practicing scientists and engineers and university students who are only expected to understand elementary calculus and linear algebra, I have tried to write this book in a style that appeals to the reader's intuition. I am careful to explain the physical events, while avoiding abstract mathematics and artificial abstractions. In some cases, mathematical rigor and lengthy derivations have been sacrificed to get to the practical point.

However, "formulas were not invented simply as weapons of intimidation"[22]. Plain English is often clumsy and sometimes ambiguous. While having attempted to do everything with the simple mathematics, in a few places I have to introduce "sophisticated" methods, due to their overwhelming advantages in terms of either mathematical manipulation or physical explanation.

In short, this book is neither a research monograph nor an encyclopedia of time-frequency and wavelet transforms. There may be occasions for academia's giants to write such books, and even for the rest of us to buy and, conceivably, read them. The material presented in this book is mainly based on what I have found in my experiences to be relatively popular, simple, and successful methods. Due to the underlying scope of the book, many interesting proposals, particularly those that are mathematically "arrogant," have been omitted. If readers are interested in the encyclopedia-type of accounts, they must search elsewhere. I would not wish a reader to start this book under false pretenses.

Finally, I am pleased to be able to post all computer examples introduced in this book on the National Instruments web site, www.ni.com, so that readers can get first-hand experience with time-frequency and wavelet transforms and share my enthusiasm about this topic!

CHAPTER 2

Fourier Transform — A Mathematical Prism

This chapter provides a brief review of the fundamentals of signal processing, which serves as a quick reference throughout this book. In most cases, we will only present the facts and results without detailed justifications. The reader can find more rigorous treatments, with an engineering flavor, from *Papoulis* [36] and *Oppenheim* et al. ([35] and [34]). The concepts and examples introduced in this chapter will be extensively used for future developments.

As most scientists and engineers know, one of the most important signal processing tools is the Fourier transform. Although it has mainly been credited to *Jean Baptiste Joseph Fourier*, the development of Fourier analysis has a long history involving a great many individuals, such as *L. Euler* in 1748 and *D. Bernoulli* in 1753. During the study of heat propagation and diffusion, Fourier found series of harmonically related sinusoids to be useful in representing the temperature distribution through a body. In an 1807 memoir he claimed that "any" periodic signal could be represented by a series of harmonically related sine and cosine functions. The reaction of Fourier's contemporaries, however, was less enthusiastic than he might have hoped. His memoir was awarded the grand prize for mathematics in 1812, but with the comment that "his analysis... leaves something to be desired in regards to its generality, and even from the point of view of rigor."[1] The judgment of posterity has been more generous. The English physicist *Lord Kelvin* found it "difficult to say whether (Fourier's results') uniquely original quality, or their transcendently intense mathematical interest, or their perennially important instructiveness for physical science, is most to be praised."[2]

1. H. S. Carslaw, *Introduction to the Theory of Fourier's Series and Integrals*, p. 7, Macmillan and Co., Ltd., London, 1930.

The Institut de France appointed four distinguished mathematicians and scientists to examine Fourier's paper. Three of the four – *S. F. Lacroix*, *G. Monge*, and *P. S. de Laplace* – were in favor of publication of the paper, but the fourth, *J. L. Lagrange*, remained adamant in rejecting trigonometric series. Lagrange strongly believed that it was impossible to represent signals with corners (i.e., with discontinuous slopes) using trigonometric series. Because of Lagrange's vehement objections, Fourier's paper never appeared. After several other attempts to have his work accepted and published, Fourier finally undertook the writing of his own book, *The Applied Theory of Heat*, in 1822, 15 years after he first presented his results. In that book, Fourier further obtained a representation for aperiodic signals, not as weighted sums of harmonically related sinusoids, but as weighted integrals of sinusoids that are not all harmonically related ([35] and [34]).

While Fourier's treatment of this topic was significant, many of the basic ideas behind it had been discovered by others. Also, Fourier's mathematical arguments were still imprecise, and it remained for *P. L. Dirichlet* in 1829 to provide precise conditions under which a periodic signal could be represented by a Fourier series. As a matter of fact, Fourier did not actually contribute to the mathematical theory of Fourier series. However, he did have the clear insight to see the potential for this series representation, and it was to a great extent his work and his claims that sped the subsequent work on Fourier series.

The set of basis functions that Fourier chose, sine and cosine functions, not only is the solution of differential equations, but also is the mathematical model of a *wave* — a most fundamental natural phenomenon. Although this was not known in Fourier's time – radio, infrared, visible light, and x-rays are all waves differing only in frequency. Applying the Fourier transform, one can mathematically translate a signal that varies with time (or in some cases, in space) into a set of waves with different frequencies, as the prism breaks light into different colors. Being able to break down signals into frequencies has myriad uses, including tuning your radio to your favorite station, interpreting radiation from distant galaxies, using ultrasound to check the health of a developing fetus, and making cheap long distance telephone calls.

With the discovery of quantum mechanics, the significance of Fourier's discovery became even more obvious. On the "position space" side of the Fourier transform, one can talk about an elementary particle's position; on the other side, in "Fourier space," one can talk about its momentum or think of it as a wave. By the Fourier transform, one can explain the *uncertainty principle*. That is, matter at very small scales behaves differently from matter on a human scale – an elementary particle does not simultaneously have a precise position and a precise momentum [22]. The Fourier transform provides a perfect mathematical model for the uncertainty principle, one of the most fundamental phenomena in nature.

The Fourier transform has myriad applications, but the mathematical tools that Fourier employed were rather basic – nothing more than *inner product* and *expansion*. In Section 2.1,

2. W. Thomson, "Heat," in *Encyclopedia Britannica*, Vol. 11, Ninth Edition, p. 578, Charles Scribner's Sons, New York, 1880.

we briefly review, from the *frame* point of view, concepts related to signal decomposition. Frame theory has been found very useful for analyzing the completeness, stability, and redundancy of linear discrete signal representations, but the general frame theory has a wider scope for deep mathematical issues, which is beyond the scope of this book. Since this book is mainly written for practical scientists/engineers and senior college students, the presentation is limited to a rather superficial level. However, materials presented in Section 2.1, I strongly believe, should serve as a good start for both practical scientists/engineers, who only want to know what it is, and graduate students, who will further pursue the mathematical details.

Following the brief introduction of the frame theorem, the development of the Fourier transform is reviewed in Section 2.2. From the mathematical point of view, there are an infinite number of ways to decompose a given signal. Why is Fourier's discovery so important? This is because the set of basis functions employed by Fourier is a mathematical model of the most common natural phenomenon – a *wave*. It is the Fourier transform that links the two most important quantities in the universe – *time* and *frequency*. Applying the Fourier transform, we can break up a time waveform into a group of frequencies, just as a prism breaks light up into a spectrum of different colors. Consequently, we can accomplish many miracles that are beyond any imagination of our predecessor.

Because of the Fourier transform, the signal now has two faces – time waveform and frequency spectrum. Section 2.3 is dedicated to a brief review of the basic relationships between these two representations of a signal. Taking into account that the Fourier transform is extensively covered in most science and engineering schools, this section only presents those properties which are closely related to the future development.

Having obtained a vehicle for transforming a signal from time domain into frequency domain, or vice versa, the next challenge is creating the mathematical language to characterize a signal's time waveform and frequency representations. It is the mathematical language that enables us to apply the machine, such as the digital computer, to automatically process the signal. This has been traditionally achieved by formulating them as a signal's energy density function. By doing so, we can directly borrow the concepts from probability theory to quantitatively characterize a signal's time and frequency behavior. For example, the variance can be thought of as the measure of a signal's spread in the time or frequency domain. Particularly, we introduce a method of evaluating the mean frequency and the frequency bandwidth from the time waveform directly, without computing the Fourier transform first as is usually suggested. Such representations not only simplify the computation, but also help us to better understand the relationship between the time and frequency behaviors of a signal.

The time and frequency representations of a signal are linked by the Fourier transform. Hence, the signal's time and frequency behaviors are not independent. They are related to each other. Such a dependency can be best characterized by the product of time duration and frequency bandwidth, or uncertainty principle, which is the central topic of Section 2.5.

The materials reviewed in this chapter are fundamental for future development, though many of them may not be new for some readers. The presentation and organization of this chapter is completely motivated by time-frequency analysis. Hopefully, the reader will find that reading this chapter is both interesting and enlightening.

2.1 Frame

The term *signal* generally refers to a function of one or more independent variables that contains information about the behavior or nature of some phenomenon. Common examples of signals include electrical current, image, speech signals, stock indexes, etc., which are all produced by some time-varying process. While electrical current, speech signals, and stock indexes are functions of time, the image signal $s(x,y)$ is a 2D (two-dimensional) light intensity function, where x and y denote spatial coordinates. The function $s(x,y)$ is proportional to the brightness of an image at the point (x,y).

A primary task of signal analysis is to understand what information can be drawn from a signal. How can we extract this information? A mathematical tool for signal analysis is to compare the signal with a set of known functions $\{\psi_n\}_{n\in Z}$, i.e.,

$$c_n = \langle s, \psi_n \rangle = \int_{-\infty}^{\infty} s(t)\psi^*{}_n(t)dt \tag{2.1}$$

for a continuous-time signal $s(t)$, or

$$c_n = \langle s, \psi_n \rangle = \sum_m s[m]\psi^*{}_n[m] \tag{2.2}$$

for a discrete-time signal $s[m]$. The operation described by (2.1) and (2.2) is named as the *inner product* or *dot product*. The process of inner product may be thought of as a mathematical scale (or balance). While the signal $\mathbf{s}$ corresponds to the object to be measured, the set of pre-selected functions $\{\psi_n\}_{n\in Z}$ can be considered as the set of standard weights. The inner product is the measure of similarity between the signal $\mathbf{s}$ and the function ψ_n, and is proportional to the degree of closeness between $\mathbf{s}$ and ψ_n. The more the signal $\mathbf{s}$ is similar to the function ψ_n, the bigger the inner product. When $\mathbf{s}$ and ψ_n are parallel (either direct or reverse), the magnitude (absolute value of the amplitude) of the inner product reaches its maximum.[3] When $\mathbf{s}$ and ψ_n are nearly perpendicular, the inner product is close to zero.

Naturally, we would expect the following properties from inner product operations.

- If $\mathbf{s}_i = \mathbf{s}_k$, then $<s_i,\psi_n> = <s_k,\psi_n>$ for all n.
- If $\mathbf{s}_i$ is close to $\mathbf{s}_k$, then the corresponding inner products $<s_i,\psi_n>$ and $<s_k,\psi_n>$ should be close to each other too, for all n. Mathematically, this property can be written as

$$\sum_n |\langle s_i, \psi_n\rangle - \langle s_k, \psi_n\rangle|^2 \le B\|s_i - s_k\|^2 \qquad 0 < B < \infty \tag{2.3}$$

3. It can be simply proved by the *Schwarz* inequality.

where B is a non-negative real number. Let $\boldsymbol{s} = \boldsymbol{s}_i - \boldsymbol{s}_k$. Eq. (2.3) reduces to

$$\sum_n |\langle s, \psi_n \rangle|^2 \leq B\|s\|^2 \qquad 0 < B < \infty \tag{2.4}$$

- Conversely, if $<\boldsymbol{s}_i, \psi_n>$ is close to $<\boldsymbol{s}_k, \psi_n>$, then $\boldsymbol{s}_i$ and $\boldsymbol{s}_k$ should be close too, for all n. Analogous to (2.4), this property can be described by

$$A\|s\|^2 \leq \sum_n |\langle s, \psi_n \rangle|^2 \qquad 0 < A < \infty \tag{2.5}$$

where A is a non-negative real number.

Combining (2.4) and (2.5) yields

$$A\|s\|^2 \leq \sum_n |\langle s, \psi_n \rangle|^2 \leq B\|s\|^2 \qquad 0 < A \leq B < \infty \tag{2.6}$$

In this case, we say that the set of sequences $\{\psi_n\}_{n \in Z}$ form a *frame*, a family of vectors that characterizes any signal $\boldsymbol{s}$ from its inner products $\{\langle s, \psi_n \rangle\}_{n \in Z}$. For $A = B$, $\{\psi_n\}_{n \in Z}$ forms a *tight frame*. Obviously, the condition of frame in (2.6) implies the following.

- If $\|s\| > 0$, the signal $\boldsymbol{s}$ has at least one point not equal to zero.
- If $\|s\|^2 < \infty$, then $\sum_n |\langle s, \psi_n \rangle|^2 < \infty$.

If there is a set of dual functions $\{\hat{\psi}_n\}_{n \in Z}$ for the frame $\{\psi_n\}_{n \in Z}$, then we can further recover the original signal s from the set of inner products $\{c_n\}$, i.e.,

$$s = \sum_n \langle s, \psi_n \rangle \hat{\psi}_n = \sum_n c_n \hat{\psi}_n \tag{2.7}$$

The dual functions form a *dual frame*, i.e.,

$$\frac{1}{B}\|s\|^2 \leq \sum_n |\langle s, \hat{\psi}_n \rangle|^2 \leq \frac{1}{A}\|s\|^2 \qquad 0 < A \leq B < \infty \tag{2.8}$$

Usually, we also name the inner product, (2.1) and (2.2), as *analysis* or *transform*, and conversely, the reconstruction (2.7) as *synthesis*, *inverse transform*, or *decomposition*.

For a given frame $\{\psi_n\}_{n \in Z}$, its dual frame $\{\hat{\psi}_n\}_{n \in Z}$ in general is not unique. Moreover, the computation of the corresponding dual frame is not trivial unless $A = B$ (that is, a tight frame). The reader can find a comprehensive treatment about this topic in *Daubechies* [13]. For a tight frame, we can use $\{\psi_n\}_{n \in Z}$ as a dual function, i.e.,

$$\hat{\psi}_n = \frac{1}{A}\psi_n \tag{2.9}$$

Consequently, the reconstruction (2.7) becomes

$$s = \frac{1}{A}\sum_n \langle s, \psi_n \rangle \psi_n = \frac{1}{A}\sum_n c_n \psi_n \tag{2.10}$$

where the parameter A can be used as a measure of the redundancy in the expansion set. When $A = B = 1$, $\{\psi_n\}_{n \in Z}$ forms an *orthogonal basis.* In this case, there is no redundancy in the resulting expansion.

We say that the set of $\{\psi_n\}_{n \in Z}$ is a *basis* for a given space ς if the set of $\{\langle s, \psi_n \rangle\}_{n \in Z}$ is unique for any particular $s \in \varsigma$. In this case, the sets of $\{\psi_n\}_{n \in Z}$ and $\{\hat{\psi}_n\}_{n \in Z}$ form a *biorthogonal basis.* That is, $\langle \psi_n, \hat{\psi}_m \rangle = k\delta(m-n)$, where

$$\delta(m-n) = \begin{cases} 0 & m \neq n \\ 1 & m = n \end{cases} \tag{2.11}$$

and k is an arbitrary constant. When $\{\psi_n\}_{n \in Z}$ is the same as the set of dual functions, the set $\{\psi_n\}_{n \in Z}$ is called an *orthogonal basis.* While the biorthogonal basis is a special case of the general frame, the orthogonal basis is a special tight frame. Frames are over-complete versions of non-orthogonal bases and tight frames are over-complete versions of orthogonal bases. Non-orthogonal bases do not reproduce the signal energy exactly, and reconstructing a signal from these coefficients may amplify any error introduced on the coefficients.

In finite dimensions, the analysis and synthesis operations are simply matrix-vector multiplications. For example, assume that $\vec{x}$ is an M-by-*1* row vector and G is an N-by-M analysis matrix, where $N \geq M$. If there is an N-by-M matrix H such that

$$H^T G\vec{x} = \vec{x} \tag{2.12}$$

then G (as well as H) forms a frame. If $H = G$, then G forms a tight frame.

If $N = M$ in (2.12), then G (as well as H) forms a basis. When the expansion vectors form a basis, the matrix must be square and non-singular. In this case, the expansion vectors in G and H are biorthogonal to each other. For the biorthogonal basis, the synthesis matrix is the inverse of the analysis matrix. If $N = M$ and $H = G$ in (2.12), then the vectors in G are orthogonal to each other. For the orthogonal basis, the synthesis matrix is simply the transpose of the analysis matrix. In finite dimensions, vectors can always be removed from a frame to get a basis, but in infinite dimensions that is not always possible.

The frame theory was originally developed by *Duffin* and *Schaeffer* to reconstruct band-limited signals from irregularly spaced samples [100]. It has been found to be very useful for analyzing the completeness, stability, and redundancy of linear discrete signal representations ([267] and [13]).

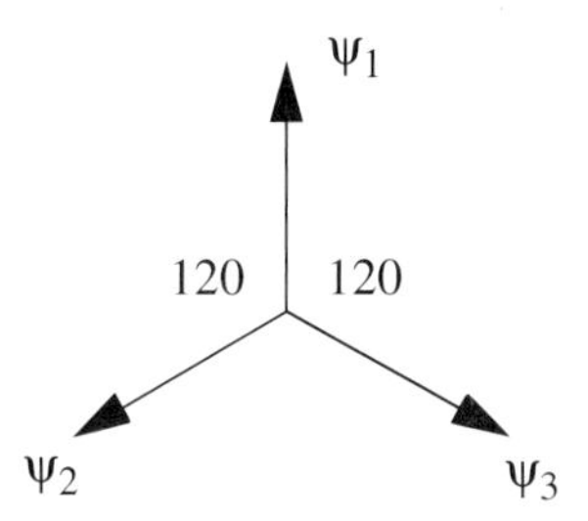

Figure 2-1 A 3D tight frame in 2D space. The set of $\{\psi_n\}_{n \in Z}$ is redundant because we can remove any one of the three vectors, using the remaining two vectors to represent arbitrary 2D functions.

Example 2.1 A 3D tight frame in 2D space

$$H = \begin{bmatrix} \psi_1 \\ \psi_2 \\ \psi_3 \end{bmatrix} = \begin{bmatrix} 0 & 1 \\ -\frac{\sqrt{3}}{2} & -\frac{1}{2} \\ \frac{\sqrt{3}}{2} & -\frac{1}{2} \end{bmatrix} \tag{2.13}$$

A 3D tight frame in 2D space is illustrated in Figure 2-1. Apparently, any vector is the sum of the remaining two vectors. So the set of $\{\psi_n\}_{n \in Z}$ is linearly dependent. For any two dimensional signals s,

$$\sum_{n=1}^{3} |\langle s, \psi_n \rangle|^2 = \frac{3}{2}[s(1)^2 + s(2)^2] = \frac{3}{2}\|s\|^2 \tag{2.14}$$

The frame bound $A = 3/2 > 1$, which implies that the frame (2.13) contains redundancy; three vectors represent two dimensional signals. We can remove any one of them to form a basis. Since (2.13) forms a tight frame, we can recover the original signal according to (2.10). Note that in this case, there are an infinite number of sets of coefficients, $\{c_n\}$, and an infinite number of dual functions, which can lead to the same sequence s.

For instance, since $\sum \psi_n = 0$ (rows are linearly dependent), we can always have

$$\sum_{n=1}^{3} (\langle s, \psi_n \rangle + k)\psi_n = s \tag{2.15}$$

where k is an arbitrary constant. Equation (2.15) shows that adding an arbitrary constant to inner products will not alter the result of reconstruction. However, it is interesting to note that the sum of the squares of $\langle s, \psi_n \rangle + k$,

$$\sum_{n=1}^{3} |\langle s, \psi_n \rangle + k|^2 = \frac{3}{2}\|s\|^2 + 3k^2 \geq \frac{3}{2}\|s\|^2 \tag{2.16}$$

is bigger than the norm of $\{\langle s, \psi_n\rangle\}_{n \in Z}$. The relationship (2.16) exhibits an important fact that when the sets of analysis and synthesis functions are identical, the inner product has a minimum L^2-norm (or minimum length).

There are two operations commonly used in signal processing that are closely related to inner product operation. One is *correlation* given by

$$R_n(\tau) = \int_{-\infty}^{\infty} s(t)\psi^*_n(t-\tau)dt \qquad or \qquad R_n[\tau] = \sum_m s[m]\psi^*_n[m-\tau] \tag{2.17}$$

which reflects the similarity between the signal s and a time shifted version of the function ψ_n. The other is known as *convolution* or *filtering*,

$$y_n(t) = \int_{-\infty}^{\infty} s(\tau)\psi^*_n(t-\tau)d\tau \qquad or \qquad y_n[m] = \sum_k s[k]\psi^*_n[m-k] \tag{2.18}$$

which indicates the similarity between the signal s and the time reversed and shifted function ψ_n.

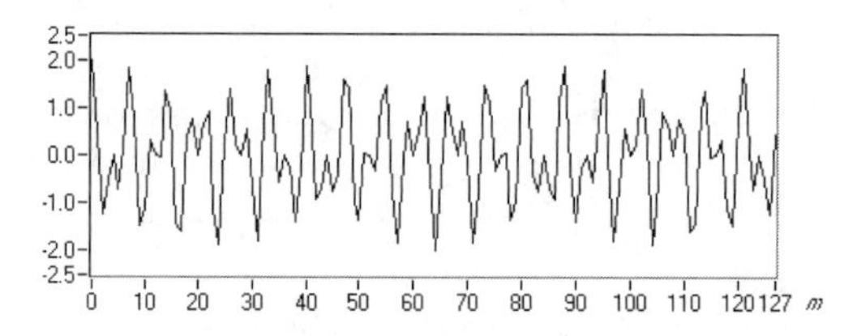

Figure 2-2 The sum (real part) of two complex sinusoidal functions

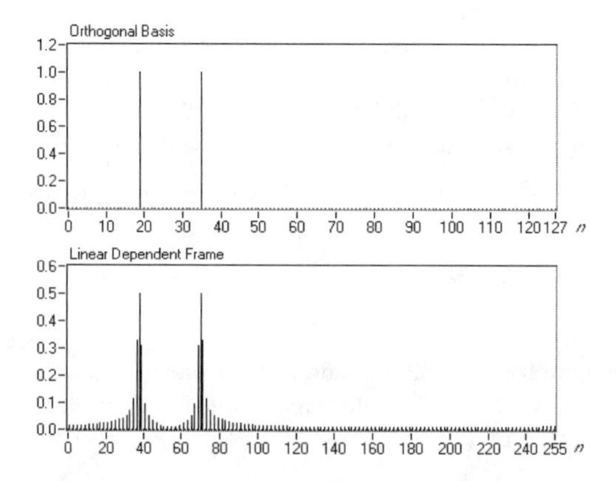

Figure 2-3 The amplitudes of resulting transforms computed by an orthogonal basis ($N = M$) and a linearly dependent tight frame ($N = 2M$). The orthogonal basis gives two non-zero coefficients $c_{19} = c_{35} = 1$, as one would anticipate. Two frequency components, however, have projections almost everywhere in the case of the linearly dependent tight frame.

Example 2.2 Linearly dependent vs. orthogonal

A set of harmonically related complex sinusoidal functions

$$\left\{\frac{N}{M}e^{j2\pi mn/N}\right\} \qquad m = 0, 1, \ldots M-1 \qquad n = 0, 1, 2, \ldots N-1 \tag{2.19}$$

constitutes a tight frame in M-dimensional space. The term *harmonically related complex sinusoidal functions* refers to the sets of periodic sinusoidal functions with fundamental frequencies that are all multiples of a single positive frequency $2\pi/N$. For $N \geq M$, the set (2.19) is linearly dependent with the frame bound $A = N/M$. When $N = M$, the set (2.19) forms an orthogonal basis. For $N < M$, the set (2.19) is not complete in M-dimensional space (the number of rows is less than the number of columns).

Let's consider a sum of two complex sinusoidal functions given by

$$s[m] = \exp\left(j19\frac{2\pi}{128}m\right) + \exp\left(j35\frac{2\pi}{128}m\right) \tag{2.20}$$

where $m = 0,1,2,\ldots 127$ (128 points). The real part of (2.20) is plotted in Figure 2-2. Figure 2-3 illustrates the amplitudes of the resulting transforms computed by an orthogonal basis ($N = M$) and a linearly dependent tight frame ($N = 2M$), respectively.

As shown in the top plot of Figure 2-3, the orthogonal basis gives two non-zero coefficients $c_{19} = 1$ and $c_{35} = 1$, as one anticipates. The resulting decomposition has the exact same form as (2.20). In the case of the linearly dependent tight frame (the bottom plot of Figure 2-3), in addition to $c_{38} = 0.5$ and $c_{70} = 0.5$, all odd index coefficients c_n are not equal to zero. Although the original signal (2.20) only contains two frequencies, $38\pi/128$ and $70\pi/128$, the linearly dependent transform shows as if the original signal contains frequencies $n\pi/128$ too, for $n = 1,3,5,\ldots 255$. The resulting decomposition is

$$s[m] = 0.5\exp\left(j19\frac{2\pi}{128}m\right) + 0.5\exp\left(j35\frac{2\pi}{128}m\right) + \sum_{n = 1, 3, 5, \ldots 255} a_n e^{j2\pi mn/256} \tag{2.21}$$

which is different from (2.20), though $s[m]$ in (2.20) and (2.21) actually are identical. Unlike the orthogonal basis, where each vector encodes information that is encoded nowhere else, one component could have projections everywhere for a linearly dependent frame.

Let's now remove c_{19} from the orthogonal case and c_{38} from the linearly dependent frame (same frequencies), and then compute the corresponding reconstructions. While the reconstruction error (mean square error) in the case of the linearly dependent frame is 0.0313, the error for the orthogonal basis is 0.5, sixteen times larger than its counterpart! In general, the reconstruction by the orthogonal basis is more sensitive to coefficient disturbance than that by the linearly dependent frame.

The tight frame plays an important role in signal processing. First of all, it is easy to compute the expansion coefficients in the case of the tight frame. As one can imagine, the computation of the expansion coefficients hinges on the existence of the dual function $\{\hat{\psi}_n\}_{n \in Z}$. For the tight frame, we can directly choose the set of elementary functions as the dual functions. Consequently, the expansion coefficients can be readily obtained, through (2.1) or (2.2), once the elementary functions are determined. The resulting coefficients possess a minimum norm. However, when the set of elementary functions does not form a tight frame, such as the set of

Gabor elementary functions $\{h(t-mT)e^{jn\Omega t}\}$ with the product $T\Omega$ near to 2π, we have to first compute the dual function $\{\hat{\psi}_n\}_{n\in Z}$ and then compute the Gabor expansion coefficients $\{c_n\}$ through (2.1) or (2.2). Unfortunately, in most cases the computation of the dual frame is rather challenging.

Secondly, if $\hat{\psi}_n = \psi_n$, then the expansion coefficient c_n is exactly the signal's projection on the function $\psi_n(t)$. When $\hat{\psi}_n \neq \psi_n$, the expansion coefficient c_n, the inner product of $s(t)$ and $\hat{\psi}_n(t)$, reflects the similarity between the analyzed signal and the dual function rather than the similarity between the signal $s(t)$ and the elementary function $\psi_n(t)$. If $\{\hat{\psi}_n\}_{n\in Z}$ and $\{\psi_n\}_{n\in Z}$ are significantly different, then $\{c_n\}$ may not reflect the signal's behavior regarding the prudently selected elementary functions $\{\psi_n\}_{n\in Z}$ at all. In time-frequency analysis, such as the Gabor expansion, having the set of elementary functions $\{\psi_n\}_{n\in Z}$ optimally localized does not guarantee that the coefficient $\{c_n\}$ also reflects the signal's local behavior. When $\{\hat{\psi}_n\}_{n\in Z}$ is badly localized in the joint time-frequency domain, $\{c_n\}$ will fail to describe the signal's time-varying nature.

If a tight frame is also linearly independent,[4] then it further forms an orthogonal basis. The transformation computed by an orthogonal basis does not contain redundancy, as shown at the top of Figure 2-3. For an orthogonal basis, each vector encodes information that is encoded nowhere else. If the elementary functions that constitute the signal to be analyzed is a subset of the selected orthogonal basis, then the resulting decomposition will have the exact same form as the original signal (Example 2.2). For a linearly dependent tight frame, as shown at the bottom of Figure 2-3, information encoded by one will also be encoded by its neighbors. Although the original signal (2.20) only contains two frequencies, the resulting decomposition (2.21) shows as if it also contains all frequencies equal to $n\pi/128$, for $n = 1,3,5,...255$.

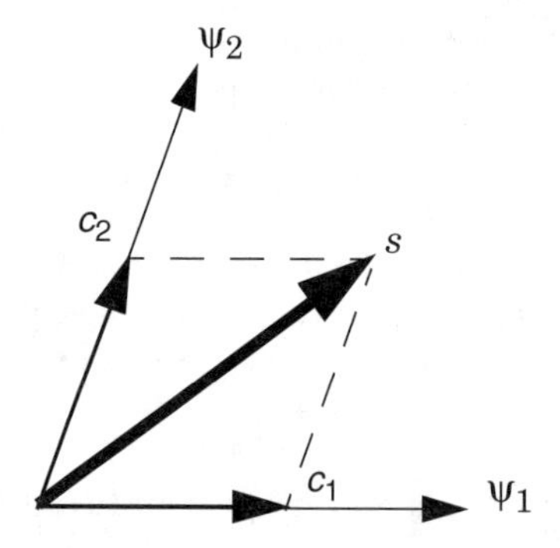

Figure 2-4 For the biorthogonal expansion, scales c_1 and c_2 are not the signal's orthogonal projections on ψ_1 and ψ_2. Consequently, the coefficients $\{c_n\}$ cannot be interpreted as the weights of the set of functions $\{\psi_n\}_{n\in Z}$, as one usually expects.

4. The class of frames that consists of linearly independent vectors are *Riesz* bases (unconditional basis).

Although the biorthogonal decomposition also does not contain redundancy, the resulting coefficients $\{c_n\}$ are not the signal's projections on the set of synthesis functions $\{\psi_n\}_{n \in Z}$, as shown in Figure 2-4. In this case, the coefficients $\{c_n\}$, the inner products of the signal s and the dual functions $\{\hat{\psi}_n\}_{n \in Z}$, do not reflect the weights of the synthesis functions $\{\psi_n\}_{n \in Z}$. Consequently, the physical interpretation of $\{c_n\}$ becomes ambiguous.

However, the redundancy is not necessarily a bad thing, as shown in Example 2.2. Reconstruction of a signal from an orthogonal or biorthogonal basis is usually more sensitive to noise than the scheme based on redundant frames.[5]

The harmonically related complex sinusoidal orthogonal basis introduced in Example 2.2, in fact, is the basis used by the fast Fourier transform. When the dimension N goes to infinity, we can further obtain the *Fourier series*. That is, any periodic time signal $\tilde{s}(t) = \tilde{s}(t + lT)$, where $l = 0, \pm 1, \pm 2, \ldots$, can be decomposed as a linear combination of $\{exp\{j2\pi nt/T\}\}$, i.e.,

$$\tilde{s}(t) = \sum_{n=-\infty}^{\infty} c_n \exp\left\{j\frac{2\pi}{T}nt\right\} \tag{2.22}$$

Because in this case the set of dual functions and the set of elementary functions have the same form, we can readily obtain the expansion coefficients of the Fourier series in (2.22) by the regular inner product operation (2.1), e.g.,

$$c_n = \int_{-T/2}^{T/2} \tilde{s}(t) \exp\left\{-j\frac{2\pi}{T}nt\right\} dt \tag{2.23}$$

which indicates the similarity between the signal and a set of harmonically related complex sinusoidal functions $\{exp\{j2\pi nt/T\}\}$. The expansion coefficients c_n in (2.22) are a signal's orthonormal projections on the complex sinusoidal functions $\exp\{2\pi n/T\}$.[6] They indicate the amount of signal present at the frequency $2\pi n/T$. Because $\exp\{j2\pi nt/T\}$ corresponds to impulses in the frequency domain, the weights used for the frequency measure in the Fourier transform possess the finest frequency tick marks (that is, the finest frequency resolution). The measurements, the Fourier coefficients $\{c_n\}$, precisely describe the signal's behavior at the frequency $2\pi n/T$.

5. My wife and I used to take down information from phone messages, such as addresses or phone numbers, onto a scratch pad. Then, I would copy it into my address book and trash the scratch paper right way. My wife, however, liked to keep all her notes. Once in a while, her notes could be seen almost everywhere in the house: on the kitchen counter, wall, and refrigerator. I used to complain about it many times, but without success. It seemed to me that all those scratch papers were redundant, since she already had copied them into her address book. One day, I lost my address book at the airport. Consequently, I was at the risk of losing almost all of my connections. It was then that I realized that redundancy is not necessarily a bad thing. If my wife were me, she might have been able to recover much of the information from the scratch papers that she preserved. (Luckily for me, a few weeks later the airport authorities mailed the address book back to me.)

6. The reader should bear in mind that the expansion coefficients c_n, in general, are not orthogonal projections of the given signal $s(t)$ on elementary functions $\psi_n(t)$ unless $\psi_n(t)$ are equal to their dual functions.

Another well-known orthonormal basis is the sinc function given by

$$\mathrm{sinc}(t) = \frac{\sin(\pi t)}{\pi t} \tag{2.24}$$

which is plotted in Figure 2-5. The set $\{\mathrm{sinc}(t\text{-}n\mathrm{T})\}_{n\in Z}$ is complete and orthonormal, i.e.,

$$\langle \mathrm{sinc}(t-nT), \mathrm{sinc}(t-n'T)\rangle = T\delta[n-n'] \tag{2.25}$$

If the signal is band limited, such as $|S(\omega)| = 0$ for $\omega > \pi/\Delta t$, where Δt denotes a sampling interval, then we have

$$s(t) = \sum_{n=-\infty}^{\infty} c_n \,\mathrm{sinc}(t-nT) \tag{2.26}$$

where

$$c_n = \frac{1}{T}\int_{-\infty}^{\infty} s(t)\mathrm{sinc}(t-nT)dt \tag{2.27}$$

which is the sampling of the continuous-time signal $s(t)$ at time instant nT. Formula (2.26) is usually referred to as the *sampling theory.*

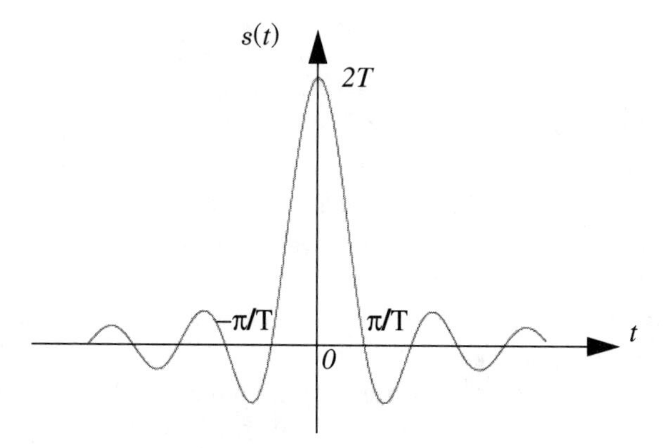

Figure 2-5 Sinc function

The concepts of inner product and expansion are fundamental for signal analysis. When the expansion is invoked, the first problem is the selection of the desired elementary functions. The word *desired* here implies:

- The elementary functions should have desirable physical interpretations.
- The form of the set of elementary functions $\{\psi_n\}_{n \in Z}$ should be simple to build.

To discover a signal's property at different frequencies, we hope that the elementary functions are optimally concentrated in the frequency domain, such as $\{\exp\{j2\pi nt/T\}\}$, that corresponds to the frequency impulses at $2\pi n/T$ and constitutes an orthonormal space. In this case, the dual function and elementary function have the same form. The resulting inner product $<s(t),\exp\{j2\pi nt/T\}>$, the Fourier transform, precisely indicates the signal's behavior at the frequency $2\pi n/T$.

To characterize the signal's behavior in the time and frequency domains simultaneously, the elementary functions need to be localized in both time and frequency domains, such as the windowed complex sinusoidal function – Gabor elementary function – or short duration waveform – *wavelet*.

Moreover, it is highly desired that the set of elementary functions should be easy to build. In addition to the Fourier series (multiples of a single positive frequency), good examples include the Gabor expansion and the wavelet transform. For the Gabor expansion, the set of elementary functions is constituted of time-shifted and frequency-modulated single prototype functions $h(t-mT)\exp\{jn\Omega t\}$. In the wavelet transform, different time and frequency tick marks are accomplished by translating and dilating a single *mother wavelet* $\psi((t-b)/a)$. In all cases, once we decide the mother function, the entire set of elementary functions can be readily obtained.

Once the elementary functions are selected, the remaining question is how to compute the dual functions. Usually, elementary functions motivated by physical interpretation do not form tight frames. Consequently, implementations of desired expansions are not as simple as those for the Fourier series. To obtain the best measurement, we usually pursue a near tight frame if the redundancy is acceptable. When the set of desirable elementary functions is not orthogonal and redundancy is not allowed (for instance, we do not want information encoded everywhere as in Figure 2-3), then we can apply an adaptive representation (also known as *matching pursuit*), as described in Chapter 10.

Finally, note that the position of $\{\psi_n\}_{n\in Z}$ and its dual $\{\hat{\psi}_n\}_{n\in Z}$ is exchangeable. That is,

$$s(t) = \sum_n \langle s, \hat{\psi}_n \rangle \psi_n(t) = \sum_n \langle s, \psi_n \rangle \hat{\psi}_n(t) \tag{2.28}$$

Which one, $\{\psi_n\}_{n\in Z}$ or $\{\hat{\psi}_n\}_{n\in Z}$, is used for the analysis function to compute the expansion coefficients depends on the application at hand. If we are mainly interested in the expansion coefficients, such as is the case for most applications of the short-time Fourier transform and wavelet transform, then we may use $\psi_n(t)$ for the analysis function because it is selected first and, thereby, makes it easier to meet our requirements.

The general expansion has a wider scope for deep mathematical issues. In this book, we have limited our discussions to the cases where expansions can be realized with the help of a fair amount of elementary linear algebra.

2.2 Fourier Transform

The Fourier transform that we now know has, in fact, two different cases: one applied for a periodic signal and the other for an aperiodic signal with finite energy, that is

$$\int_{-\infty}^{\infty} |s(t)|^2 dt < \infty \tag{2.29}$$

In an 1807 memoir Fourier claimed that "any" periodic signal could be represented by a series of harmonically related sine and cosine functions. The resulting representation is now known as the *Fourier series* (2.22). Features of the Fourier series, which are important to signal processing, include:

- The set of elementary functions in the Fourier series, harmonically related complex sinusoidal functions, is the mathematical model of the most important natural phenomena — *waves* — with different oscillations for different frequencies. Therefore, the Fourier series and the Fourier transform bridge the time and frequency domains. By using the Fourier transform, we are able to examine the frequency behavior for a given time domain function (just as a prism breaks up light into a spectrum of different colors), or vice versa. Note that there are an infinite number of ways to perform signal decomposition. However, only those mathematical transforms whose elementary functions have meaningful physical interpretations would be of significance to scientists and engineers.
- In the frequency domain, a complex sinusoidal function is equivalent to a pulse. So, the Fourier transform possesses the finest frequency resolution for measuring a signal's frequency content.
- The structure of the set of elementary functions is simple. They are all multiples of a single fundamental frequency $2\pi/T$, where T denotes the period of the periodic signal.
- The set of elementary functions is orthogonal. Consequently, the Fourier coefficients are nothing more than inner products of the analyzed signal and harmonically related complex sinusoidal functions. Moreover, each vector (one frequency) encodes information that is encoded nowhere else.

Fifteen years after Fourier discovered the Fourier series, in 1822, he introduced a representation for aperiodic signals as weighted integrals of complex sinusoids that are not all harmonically related. To obtain this result, we could start with the Fourier series. Since an aperiodic signal can be viewed as a periodic signal with an infinitely long period, we can extend the Fourier series to an aperiodic signal. More precisely, in the Fourier series representation (2.22), as the period T increases, the fundamental frequency $2\pi/T$ decreases, and harmonically related components become closer in frequency. When the period goes to infinity, the frequency components form a continuum and the Fourier series sum becomes an integral. Consequently, we can have

$$s(t) = \frac{1}{2\pi}\int_{-\infty}^{\infty} S(\omega)e^{jt\omega}d\omega \tag{2.30}$$

where

$$S(\omega) = \int_{-\infty}^{\infty} s(t)e^{-j\omega t}dt \tag{2.31}$$

in which the aperiodic signal $s(t)$ has to be of finite energy, as shown in (2.29). $S(\omega)$ is the measure of the similarity between the signal $s(t)$ and the complex sinusoidal functions. For periodic signals the complex exponential building blocks are harmonically related; for aperiodic signals they are infinitesimally close in frequency. The representation (2.31), in terms of a linear combination, takes the form of an integral rather than a sum. The formula (2.31) is called the *continuous Fourier transform* and (2.30) is known as the *inverse Fourier transform.*

For applications in digital signal processing, it is necessary to extend the continuous-time Fourier transform to discrete-time signals. Let $s[m\Delta t] = s(t)$, where Δt denotes the sampling interval. Without loss of generality, let $\Delta t = 1$. Then the discrete-time Fourier transform can be derived as

$$\tilde{S}(\theta) = \sum_{m} s[m]e^{-j\theta m} \tag{2.32}$$

where $\theta = \omega\Delta t$. $\theta/2\pi$ is the normalized frequency. The inverse discrete-time Fourier transform is

$$s[m] = \frac{1}{2\pi}\int_{-\pi}^{\pi} \tilde{S}(\theta)e^{j\theta m}d\theta \tag{2.33}$$

Because the time variable of $s(t)$ is digitized, its frequency counterpart becomes a periodic function in the frequency domain; that is, $\tilde{S}(\theta) = \tilde{S}(\theta + 2\pi l)$, for $l = 0, \pm1, \pm2,...$. Note that the frequency variable θ in (2.32) and (2.33) is a continuous variable.

When digital computers are used, $\tilde{S}(\theta)$ can only be evaluated at discrete points, such as $\theta = 2\pi n/N$, where $0 \leq n < N$. In this case, (2.32) and (2.33) have to be further modified. The resulting transform is in discrete-time and discrete-frequency, i.e.,

$$\tilde{S}[n] = \sum_{m=0}^{M-1} \tilde{s}[m]W_N^{-nm} \qquad n = 0, 1, 2, \ldots N-1 \tag{2.34}$$

where $W_N = e^{2\pi/N}$. Because we sample the frequency variable θ, the time samples in (2.34) further reduce to being periodic; that is,

$$\tilde{s}[m] = s[m + lM] \qquad m = 0, 1, 2 \ldots M-1 \qquad l = 0, \pm1, \pm2, \ldots \tag{2.35}$$

In this book, we reserve the name, *discrete Fourier transform* (DFT), to the discrete-time and discrete-frequency Fourier transform (2.34) rather than the discrete-time but continuous-frequency Fourier transform (2.32). The inverse of the *DFT* is

$$\tilde{s}[m] = \frac{1}{M}\sum_{n=0}^{N-1} \tilde{S}[n]W_M^{nm} \qquad m = 0, 1, 2, \ldots M-1 \tag{2.36}$$

When $M = N$ in (2.34) and (2.36) (the set of harmonically related complex sinusoidal functions forms an orthogonal basis, as introduced in Example 2.2), the corresponding DFT can be efficiently computed by a fast algorithm named the fast Fourier transform (FFT).

The FFT employs a mathematical trick, which could go back to Gauss in about 1805 (predating the Fourier series!), to evaluate the discrete Fourier transform. By cutting the number of computations from n^2 to $n\log n$, the FFT "catapulted the calculation of Fourier coefficients out of horse-and-buggy days into supersonic travel" [22]. In what follows, we shall investigate the Fourier transforms of several important signals in time-frequency analysis.

Example 2.3 Fourier transform of a complex sinusoidal function $s(t) = e^{j\omega_0 t}$

$$S(\omega) = \int_{-\infty}^{\infty} e^{j\omega_0 t} e^{-j\omega t} dt = 2\pi\delta(\omega - \omega_0) \tag{2.37}$$

For a discrete-time signal with N-points

$$s[m] = e^{j2\pi n_0 m/N} = W_N^{-n_0 m} \tag{2.38}$$

the corresponding FFT is[7]

$$\tilde{S}[n] = \sum_{m=0}^{M-1} e^{j2\pi n_0 m/N} W_N^{-nm} = \frac{1}{N}\delta(n - n_0 - lN) \tag{2.39}$$

in which $l = 0, \pm 1, \pm 2, \ldots$. In the frequency domain, the sinusoidal-type functions have only one point. Hence, they are perfectly localized. In the time domain, however, they are not localized. Therefore, the sinusoidal-type functions are not suitable for analyzing or synthesizing signals presenting fast local variations, such as transients or abrupt changes.

Example 2.4 Fourier transform of a real-valued cosine function $s(t) = \cos(\omega_0 t)$

$$\begin{aligned} S(\omega) &= \int_{-\infty}^{\infty} \cos(\omega_0 t) e^{-j\omega t} dt \\ &= \frac{1}{2}\int_{-\infty}^{\infty} (e^{j\omega_0 t} + e^{-j\omega_0 t}) e^{-j\omega t} dt = \pi\delta(\omega - \omega_0) + \pi\delta(\omega + \omega_0) \end{aligned} \tag{2.40}$$

Unlike its complex counterpart in Example 2.3, the real-valued cosine function corresponds to two pulses at frequencies $-\omega_0$ and ω_0.

Example 2.5 Fourier transform of a rectangular pulse signal

$$s(t) = \begin{cases} |A| & |t| \le T \\ 0 & |t| > T \end{cases}$$

7. In practice, we can always make the number of sample points M equal to the number of frequency bins N by zero padding.

as shown in Figure 2-6. The corresponding Fourier transform is

$$S(\omega) = A\int_{-T}^{T} e^{-j\omega t}dt = -\frac{A}{j\omega}e^{-j\omega t}\Big|_{-T}^{T} = 2A\frac{\sin\omega T}{\omega} \tag{2.41}$$

which is plotted in Figure 2-7. The rate of oscillation of $S(\omega)$ in the frequency domain is inversely proportional to the width of the rectangular pulse T. The narrower the width of $s(t)$, the higher the oscillation of $S(\omega)$. In contrast to the sinusoidal functions, the rectangular pulse is finitely supported in the time domain but badly localized in the frequency domain.

If we write the Fourier transform of the rectangular pulse in terms of the sinc function in (2.24), then

$$S(\omega) = 2A\frac{\sin\omega T}{\omega} = 2AT\frac{\sin\omega T}{\omega T} = 2AT\,\mathrm{sinc}\left(\frac{\omega T}{\pi}\right) \tag{2.42}$$

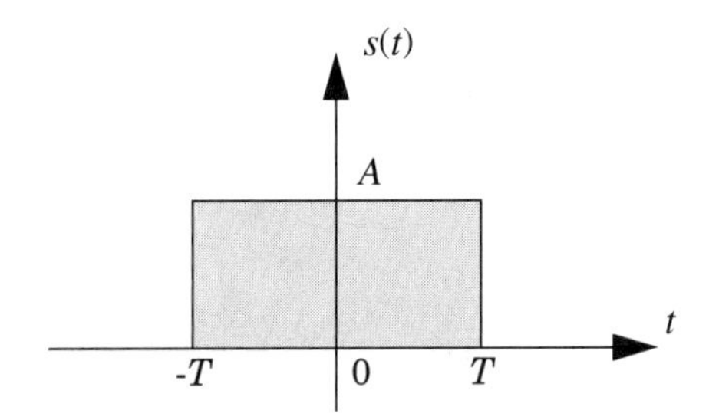

Figure 2-6 Rectangular pulse

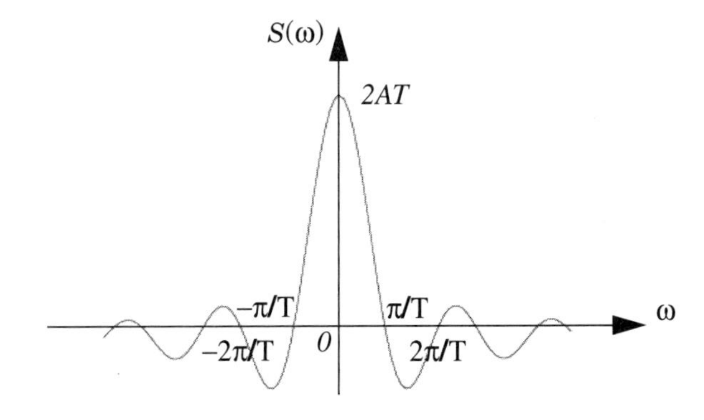

Figure 2-7 Fourier transform of rectangular pulse

Example 2.6 Fourier transform of the Gaussian function

$$s(t) = \sqrt{\frac{\alpha}{2\pi}}\exp\left\{-\frac{\alpha}{2}(t-t_{\mu})^2\right\} \tag{2.43}$$

which is the standard Gaussian function with the mean time t_{μ}, as shown in Figure 2-8. The quantity α^{-1} characterizes the variance of the Gaussian function.

The Fourier transform of the Gaussian function is

$$S(\omega) = \int_{-\infty}^{\infty} \sqrt{\frac{\alpha}{2\pi}} \exp\left\{-\frac{\alpha}{2}(t-t_{\mu})^2\right\} e^{-j\omega t} dt = \exp\left\{-\frac{1}{2\alpha}\omega^2 + j\omega t_{\mu}\right\} \tag{2.44}$$

Note that the Fourier transform of the Gaussian function is also Gaussian. The variance of the frequency representation in (2.44) is the reciprocal of the time variance in (2.43). The narrower the spread in the time domain, the wider the spread in the frequency domain, and vice versa. The time shift in (2.43) corresponds to a phase shift in (2.44). Unlike the sinusoidal or rectangular pulse functions discussed earlier, the Gaussian function is localized in both time and frequency. Moreover, the Gaussian function is infinitely differentiable. It plays an important role in almost all science and engineering disciplines.

Figure 2-8 *Carl Friedrich Gauss*' portrait and the Gaussian curve in a ten deutsche mark bill. The Gaussian curve is infinitely differentiable and plays a unique role in many science and engineering disciplines. Gauss contributions to astronomy, math, physics, statistics, and engineering can hardly be exaggerated. Germany is among a few countries in the world today that admire great scientists, such as Gauss (1777-1855), in everyday life.

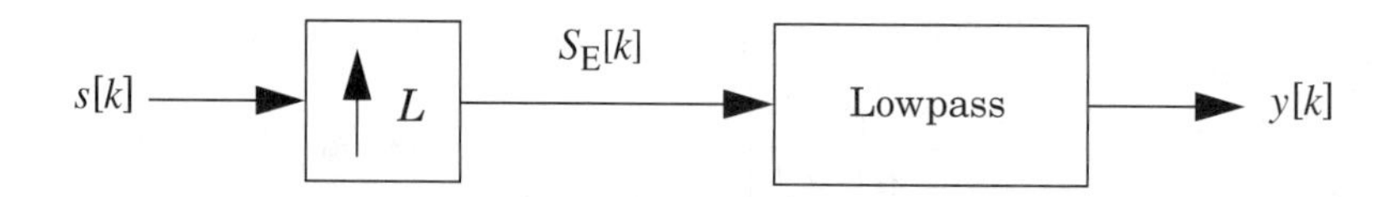

Figure 2-9 Interpolation filter

Example 2.7 Interpolation filter

Figure 2-9 is a block diagram of an interpolation filter. The left block is commonly called an L-fold expander. The right block is a conventional lowpass filter. The expander inserts L zeros between each sample $s[m]$ and produces the output $s_E[m]$ as

$$s_E[m] = \begin{cases} s[m/L] & m = 0, \pm L, \pm 2L \\ 0 & otherwise \end{cases} \tag{2.45}$$

Then,

$$\tilde{S}_E(\theta) = \sum_m s_E[m]e^{-j\theta m} = \sum_m s_E[mL]e^{-j\theta m} = \sum_m s[m]e^{-jL\theta m} = \tilde{S}(L\theta) \tag{2.46}$$

which implies that $\tilde{S}_E(\theta)$ is an L-fold compressed version of $\tilde{S}(\theta)$, as demonstrated in Figure 2-10. The multiple copies of the compressed spectrum are usually called *images*. By applying a lowpass filter after the expander, as shown in Figure 2-10, we can obtain upsampled signals. The operation of the interpolation filtering can be achieved by lowpass filtering, i.e.,

$$y[m] = \sum_{k=0}^{N} s[k]\gamma[m-kL] \tag{2.47}$$

where $\gamma[m]$ denotes a lowpass filter with the cut-off frequency at π.

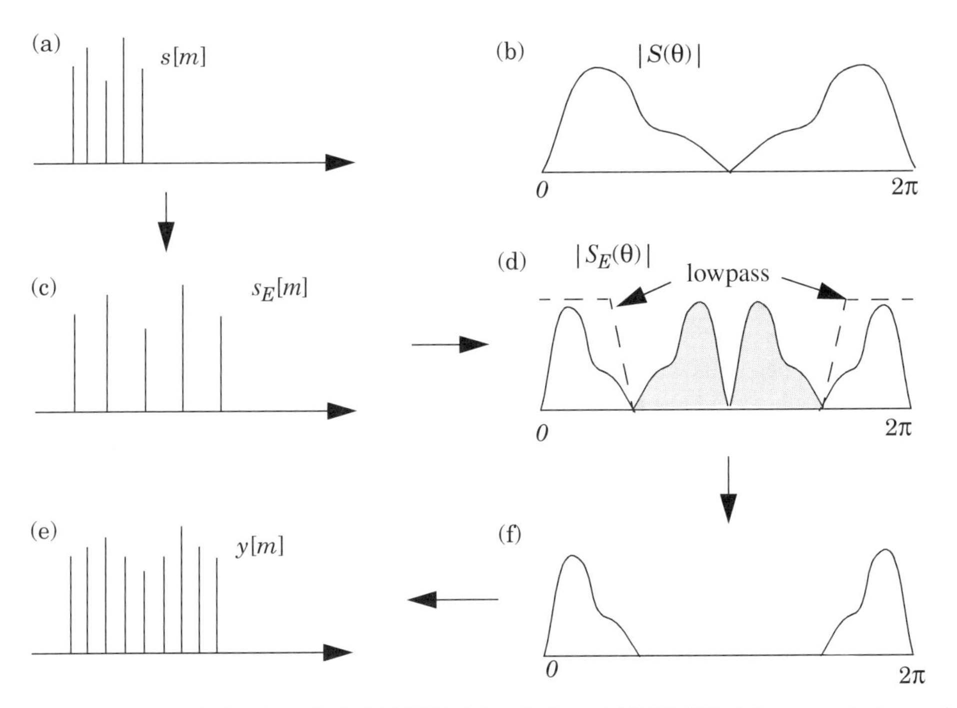

Figure 2-10 Interpolation (a) $s[m]$, (b) $|S(\theta)|$, (c) $s_E[m]$, and (d) $|S_E(\theta)|$, (e) upsampled samples $y[m]$, and (f) image removed spectrum.

The interpolation filter is a very important filter, which can be used for resampling the existing digital samples. As will be introduced in Chapter 7, the sampling rate of the discrete Wigner-Ville distribution is usually four times higher than the signal's bandwidth, which is twice the conventional sampling frequency. Applying the interpolation filter discussed in Example 2.7, we can resolve aliasing without altering the existing hardware structure.

The square of the Fourier transform $|S(\omega)|^2$ is called the *power spectrum*, which describes how the signal energy is distributed in the frequency domain. The Fourier transform $S(\omega)$ in general is complex-valued linear function of the analyzed signal, whereas the power spectrum $|S(\omega)|^2$ is a real-valued quadratic of the Fourier transform $S(\omega)$.

According to the *Wiener-Khinchin* theorem, the power spectrum can also be written as the Fourier transform of the signal's auto-correlation function, i.e.,

$$|S(\omega)|^2 = \int_{-\infty}^{\infty} R(\tau)e^{-j\omega\tau}d\tau \tag{2.48}$$

where the *auto-correlation function* $R(\tau)$ is computed by

$$R(\tau) = \int_{-\infty}^{\infty} s(t)s^*(t-\tau)dt \tag{2.49}$$

The equation (2.48) is a general form for both the classical power spectrum and the time-dependent spectrum. For example, if we make $R(\tau)$ time dependent, such as $R_t(\tau)$, then the resulting Fourier transform is a function of time and frequency, i.e.,

$$P(t, \omega) = \int_{-\infty}^{\infty} R_t(\tau)e^{-j\omega\tau}d\tau \tag{2.50}$$

The deficiency of classical Fourier analysis is that the complex sinusoidal basis functions used for the Fourier transform are not concentrated in the time domain (Example 2.3). Consequently, in the Fourier transform the inner products between the analyzed signal and the complex sinusoidal functions are not explicitly associated with time. Based on the Fourier transform alone, it is not clear whether the signal's frequency contents have changed in time. The time information is hidden in the phase of the Fourier transform, which is often beyond our comprehension. On the other hand, the frequency contents of the majority of signals encountered in our everyday life are time dependent, such as the economic index, medical data, the monthly rainfall record, sunlight, and speech signals. It has been recognized for a long time that the time or frequency representations alone are not adequate for many applications. As *Dennis Gabor* wrote, "even experts could not at times conceal an uneasy feeling when it came to the physical interpretation of results obtained by the Fourier method" [113].

2.3 Relationship between Time and Frequency Representations

In the preceding section, we briefly reviewed the concepts of signal expansion from the frame point of view. In particular, we discussed the Fourier transform. In what follows we shall further investigate the relationship between a signal's time and frequency representations, which will serve as the foundation for later development. Since the subject of Fourier analysis is presumably taught in all science and engineering schools, in this section we will only review the concepts that are directly related to time-frequency and wavelet analysis. To have the reader focus on the concepts rather than mathematical details, only the continuous-time cases are discussed in this section, though their discrete-time counterparts are equally important. The reader can find the corresponding properties for discrete-time signals from related references, such as [36], [35], and [34].

- **Shifting** in time by t_0 results in multiplication by a phase factor in the frequency domain, i.e.,

$$s(t - t_0) \leftrightarrow e^{-j\omega t_0} S(\omega) \tag{2.51}$$

Conversely, a shift in frequency by ω_0 results in modulation by a complex exponential in the time domain, i.e.,

$$S(\omega - \omega_0) \leftrightarrow e^{j\omega_0 t} s(t) \tag{2.52}$$

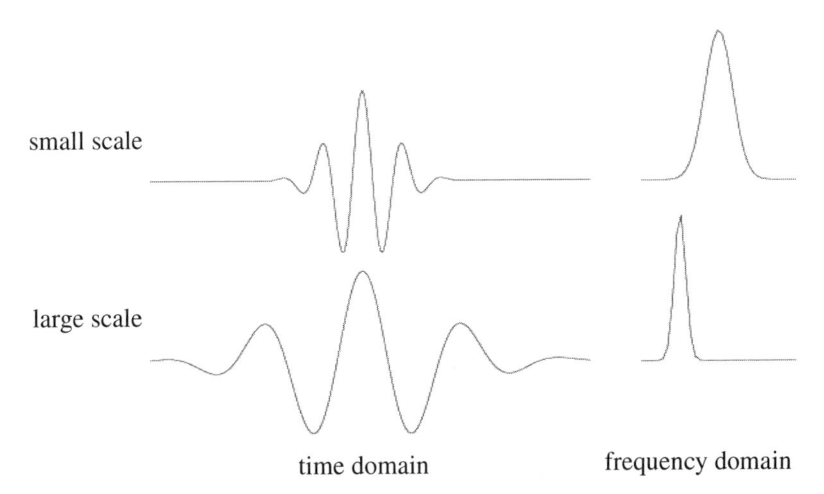

Figure 2-11 The plots on the left are the time waveforms. The plots on the right are the magnitudes of the corresponding Fourier transforms. As the time waveform is compressed (top), both its center frequency and bandwidth, as shown in the corresponding Fourier transform, increase accordingly. This shows that the signal's time and frequency representations cannot have short duration simultaneously.

- **Scaling** in the time domain leads to inverse scaling in the frequency domain, i.e.,

$$s\left(\frac{t}{\alpha}\right) = S(a\omega) \tag{2.53}$$

where a is a real constant. Figure 2-11 illustrates $s(t/\alpha)$ and the corresponding Fourier transform. It is interesting to note that as the time waveform is compressed, both the center frequency and bandwidth of the signal will increase, which illustrates two important results. First, we can scale the time variables to simply adjust the signal's frequency centers. Consequently, instead of using the harmonically related complex sinusoidal functions, such as those in the Fourier transform, we may employ the scaled functions to build tick marks to measure a signal's frequency behavior. The resulting representation is now known as the wavelet transform.

Second, the signal's time and frequency representations cannot have short duration simultaneously. When the time duration gets larger, the frequency bandwidth must get smaller, and vice versa. This assertion, known as the *uncertainty principle*, can be given various interpretations, depending on the meaning of the term "duration." Because the uncertainty principle plays an extremely important role in time-frequency analysis, we will treat it separately in Section 2.5.

- **Conjugate Function**

$$s^*(t) \leftrightarrow S^*(-\omega) \tag{2.54}$$

If $s(t)$ is real, then $s(t) = s^*(t)$. Hence, $S(\omega) = S^*(-\omega)$, which is named the *Hermitian function*.

- **Derivatives**

$$\frac{d^n}{dt^n}s(t) \leftrightarrow (j\omega)^n S(\omega) \tag{2.55}$$

This is a very useful relationship, which will be used extensively in this book to convert the function from the time domain to the frequency domain, and vice versa.

- **Convolution Theorem** For signals $s(t)$ and $\gamma(t)$, convolution is defined as

$$s(t) \otimes \gamma(t) = \int_{-\infty}^{\infty} s(\tau)\gamma(t-\tau)d\tau \tag{2.56}$$

If the Fourier transform of $s(t)$ and $\gamma(t)$ are $S(\omega)$ and $G(\omega)$, respectively, then the convolution of $s(t)$ and $\gamma(t)$ is equal to the inverse Fourier transform of $S(\omega)G(\omega)$; that is,

$$\int_{-\infty}^{\infty} s(\tau)\gamma(t-\tau)d\tau = \frac{1}{2\pi}\int_{-\infty}^{\infty} S(\omega)G(\omega)e^{j\omega t}d\omega \tag{2.57}$$

Conversely,

$$\frac{1}{2\pi}\int_{-\infty}^{\infty} S(\Omega)G(\Omega-\omega)d\Omega = \int_{-\infty}^{\infty} s(t)\gamma(t)e^{-j\omega t}dt \tag{2.58}$$

- **Parseval's Formula** Setting $t = 0$ in (2.57), we obtain

$$\int_{-\infty}^{\infty} s(\tau)\gamma(-\tau)d\tau = \frac{1}{2\pi}\int_{-\infty}^{\infty} S(\omega)G(\omega)d\omega \tag{2.59}$$

With $\gamma(-t) = h^*(t)$, we have $G(\omega) = H^*(\omega)$. Then, (2.59) reduces to

$$\int_{-\infty}^{\infty} s(\tau)h^*(\tau)d\tau = \frac{1}{2\pi}\int_{-\infty}^{\infty} S(\omega)H^*(\omega)d\omega \tag{2.60}$$

which is known as *Parseval's formula*. If $s(t) = h(t)$, we have $S(\omega) = H(\omega)$, and (2.60) becomes

$$\int_{-\infty}^{\infty} |s(t)|^2 dt = \frac{1}{2\pi}\int_{-\infty}^{\infty} |S(\omega)|^2 d\omega \tag{2.61}$$

which implies that during Fourier transformation energy is conserved.

2.4 Characterization of Time Waveform and Power Spectrum

Since the invention of the Fourier transform, any given signal now has two faces; one shows how the signal's amplitude changes with time, and the other indicates what frequencies the signal has. Time and frequency representations have greatly helped us to better understand numerous natural phenomena that our ancestors didn't comprehend. In the preceding section we reviewed the relationship between a signal's time and frequency representations. In what follows we shall further discuss how to describe a signal's time and frequency behaviors mathematically. It is the mathematical description that enables us to use machines, such as digital computers, to automatically process the signal in which we are interested.

Based on the Parseval's formula discussed in Section 2.3, we have

$$\int_{-\infty}^{\infty} \frac{|s(t)|^2}{E} dt = \int_{-\infty}^{\infty} \frac{|S(\omega)|^2}{2\pi E} d\omega \tag{2.62}$$

where E denotes the signal energy, i.e.,

$$E = \|s(t)\|^2 = \int_{-\infty}^{\infty} |s(t)|^2 dt = \frac{1}{2\pi}\int_{-\infty}^{\infty} |S(\omega)|^2 d\omega \tag{2.63}$$

It is easy to verify that $|s(t)|^2/E$ and $|S(\omega)|^2/2\pi E$ are valid density functions. While $|s(t)|^2/E$ can be considered as a signal's time domain energy density, $|S(\omega)|^2/2\pi E$ can be thought of as a signal's frequency domain energy density. We can employ the concepts developed in probability theory, such as moments, to quantitatively characterize a signal's time and frequency behaviors.

The first moment of $|s(t)|^2/E$ and $|S(\omega)|^2/2\pi E$ denote a signal's *mean time* and *mean frequency*, i.e.,

$$\langle t \rangle = \frac{1}{E}\int_{-\infty}^{\infty} t|s(t)|^2 dt \tag{2.64}$$

and

$$\langle \omega \rangle = \frac{1}{2\pi E}\int_{-\infty}^{\infty} \omega|S(\omega)|^2 d\omega \tag{2.65}$$

which describe a signal's gravitational centers in the time and frequency domains, respectively.

Obviously, the mean time $\langle t \rangle$ can be directly computed from the time domain, whereas to compute the mean frequency $\langle \omega \rangle$ (2.65) it seems that we need to first calculate the corresponding Fourier transform. In what follows, we will derive a method to evaluate the mean frequency $\langle \omega \rangle$ without computing the Fourier transform. The result not only is simple, but will also help us to comprehend a signal's frequency behavior based on its time representation.

Replacing $\omega S(\omega) = H(\omega)$ in (2.65) yields

$$\langle \omega \rangle = \frac{1}{2\pi E}\int_{-\infty}^{\infty} H(\omega)S^*(\omega)d\omega \tag{2.66}$$

Applying Parseval's formula (2.60) on (2.66), we obtain

$$\langle \omega \rangle = \frac{1}{E}\int_{-\infty}^{\infty} h(t)s^*(t)dt \tag{2.67}$$

where $h(t)$ and $s(t)$ denote the inverse Fourier transform of $H(\omega)$ and $S(\omega)$, respectively. Note that according to the derivative property (2.55), we can write the function $h(t)$ in terms of $s(t)$, i.e.,

$$h(t) = -j\frac{d}{dt}s(t) \tag{2.68}$$

Without loss of generality, let $s(t) = a(t)\exp\{j\varphi(t)\}$, where the real-valued functions $a(t)$ and $\varphi(t)$ denote amplitude and phase, respectively. Then (2.67) becomes

$$\begin{aligned}\langle \omega \rangle &= \frac{1}{E}\int_{-\infty}^{\infty} -j\frac{d}{dt}s(t)s^*(t)dt \\ &= \frac{1}{E}\int_{-\infty}^{\infty} -j\{a'(t)e^{j\varphi(t)} + ja(t)\varphi'(t)e^{j\varphi(t)}\}s^*(t)dt \\ &= \frac{1}{E}\int_{-\infty}^{\infty} -j\{a'(t) + ja(t)\varphi'(t)\}a(t)dt \\ &= \frac{1}{E}\int_{-\infty}^{\infty} \varphi'(t)|a(t)|^2 dt - \frac{j}{E}\int_{-\infty}^{\infty} a'(t)a(t)dt\end{aligned} \tag{2.69}$$

Because the left side of (2.65), the mean frequency, is real, the second term of the right side in (2.69) must be zero. Consequently, (2.69) reduces to

$$\langle \omega \rangle = \int_{-\infty}^{\infty} \varphi'(t)\frac{|s(t)|^2}{E}dt \tag{2.70}$$

which says that the mean frequency $\langle \omega \rangle$ is the weighted average of the instantaneous quantities $\varphi'(t)$ over the entire time domain.

In most literature $\varphi'(t)$, the first derivative of the phase, is named as the instantaneous frequency. In fact, this is incorrect. As in the speech signal shown in Figure 2-12, for most signals there is more than one frequency at any time instant t. In other words, the instantaneous frequency is a multi-valued function, whereas $\varphi'(t)$ is a single value function. It is misleading to define the first derivative of the phase as a signal's instantaneous frequency. What does $\varphi'(t)$ really stand for? The answer, unfortunately, is not simple and remains an open research topic.

As we will see later, a certain type of time-dependent power spectrum $P(t,\omega)$, such as the Wigner-Ville distribution, was motivated by a signal's time-frequency density function. For these functions, their conditional mean frequency is equal to the first derivative of a signal's phase, i.e.,

$$\langle \omega \rangle_t = \frac{\int_{-\infty}^{\infty} \omega P(t, \omega) d\omega}{\int_{-\infty}^{\infty} P(t, \omega) d\omega} = \varphi'(t) \tag{2.71}$$

which suggests that $\varphi'(t)$, the first derivative of a signal's phase, is the mean instantaneous frequency. So, in this book we refer to the first derivative of a signal's phase $\varphi'(t)$ as the *mean instantaneous frequency.*

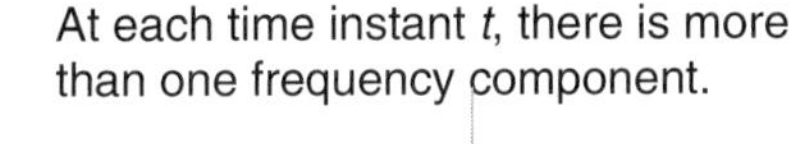

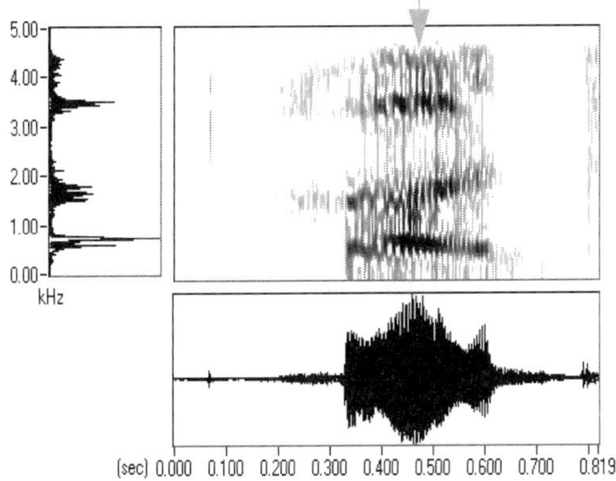

Figure 2-12 Joint time-frequency representations of the word "hood." Obviously, at each time instant t there is more than one frequency component. The instantaneous frequency is a multi-valued function, whereas $\varphi'(t)$ is a single value function. Hence, it is misleading to define the first derivative of the phase as a signal's instantaneous frequency (Data provided by *Y. Zhao* of the University of Missouri).

However, due to the quadratic nature of the time-dependent power spectra, the so-called mean instantaneous frequency of multi-component signals, in general, is not simply the sum of the weighted individual frequencies. That is, suppose that

$$s(t) = a(t)e^{j\varphi(t)} = \sum_{n=1}^{N} a_n(t)e^{j\varphi_n(t)} \tag{2.72}$$

Except for a few very simple cases,[8] in general,

$$\frac{d}{dt}\varphi(t) \neq \sum_{n=1}^{N} a_n(t)\varphi_n'(t) \tag{2.73}$$

where $a_n(t)$ denote real-valued weighting functions. This is due to the fact that the power spectrum of the sum of signals $s_i(t)$, $i = 1, 2, 3, \ldots$, is not equal to the sum of each individual signal's power spectrum. The reader can find a comprehensive treatment on instantaneous frequency from *Wei* and *Bovik*'s paper [224].

By applying the concepts of the variance and standard deviation, we can further quantitatively characterize a signal's energy spread in the time and frequency domains. Usually, we use the time domain standard deviation Δ_t and the frequency domain standard deviation Δ_ω to measure a signal's *time duration* and *frequency bandwidth*, respectively.

$$\Delta_t^2 = \frac{1}{E}\int_{-\infty}^{\infty} (t - \langle t\rangle)^2 |s(t)|^2 dt = \int_{-\infty}^{\infty} t^2 \frac{|s(t)|^2}{E} dt - \langle t\rangle^2 \tag{2.74}$$

and

$$\Delta_\omega^2 = \frac{1}{2\pi}\int_{-\infty}^{\infty} (\omega - \langle\omega\rangle)^2 \frac{|S(\omega)|^2}{E} d\omega = \frac{1}{2\pi}\int_{-\infty}^{\infty} \omega^2 \frac{|S(\omega)|^2}{E} d\omega - \langle\omega\rangle^2 \tag{2.75}$$

Just as in the case of computing the mean frequency, we can also express the frequency variance (2.75) as a function of time.

Based on Parseval's relationship, we can rewrite (2.75) as

$$\Delta_\omega^2 = \frac{1}{2\pi E}\int_{-\infty}^{\infty} (\omega - \langle\omega\rangle)^2 |S(\omega)|^2 d\omega = \frac{1}{2\pi E}\int_{-\infty}^{\infty} H(\omega)H^*(\omega)d\omega = \frac{1}{E}\int_{-\infty}^{\infty} h(t)h^*(t)dt \tag{2.76}$$

where $H(\omega) = (\omega - \langle\omega\rangle)S(\omega)$.

According to the shift property (2.52),

$$\frac{1}{2\pi}\int_{-\infty}^{\infty} S(\omega)e^{j(\omega - \langle\omega\rangle)t} d\omega = s(t)e^{-j\langle\omega\rangle t} \tag{2.77}$$

Taking the derivative of both sides of (2.77) yields

$$\frac{1}{2\pi}\int_{-\infty}^{\infty} j(\omega - \langle\omega\rangle)S(\omega)e^{j(\omega - \langle\omega\rangle)t} d\omega = \frac{d}{dt}[s(t)e^{-j\langle\omega\rangle t}] \tag{2.78}$$

8. One exception is a sum of two monotone signals with identical amplitudes.

Therefore,

$$h(t) = \frac{1}{2\pi}\int_{-\infty}^{\infty} H(\omega)e^{-j\omega t}d\omega \tag{2.79}$$
$$= \frac{1}{2\pi}\int_{-\infty}^{\infty} (\omega - \langle\omega\rangle)S(\omega)e^{j\omega t}d\omega$$
$$= -je^{j\langle\omega\rangle t}\frac{d}{dt}s(t)e^{j\langle\omega\rangle t}$$
$$= -je^{j\langle\omega\rangle t}[j(\varphi'(t) - \langle\omega\rangle)a(t)e^{j(\varphi(t) - \langle\omega\rangle)t} + a'(t)e^{j(\varphi(t) - \langle\omega\rangle)t}]$$
$$= e^{j\varphi(t)}[(\varphi'(t) - \langle\omega\rangle)a(t) - ja'(t)]$$

Substituting $h(t)$ into (2.76), one obtains

$$\Delta_\omega^2 = \frac{1}{E}\int_{-\infty}^{\infty} (\varphi'(t) - \langle\omega\rangle)^2 a^2(t)dt + \frac{1}{E}\int_{-\infty}^{\infty} (a'(t))^2 dt \tag{2.80}$$

which says that the frequency bandwidth is completely determined by the amplitude variation $a'(t)$ as well as the phase variation $\varphi'(t)$. The smoother the magnitude and phase, the narrower the frequency bandwidth. When both the magnitude and phase are constant, such as for a complex sinusoidal signal $\exp\{j\omega_0 t\}$, the frequency bandwidth reduces to zero. To better understand this important concept, let's look at some examples.

Example 2.8 Normalized Gaussian function

$$s(t) = g(t) = \sqrt[4]{\frac{\alpha}{\pi}}e^{-\frac{\alpha}{2}t^2} \tag{2.81}$$

Obviously,

$$E = \int_{-\infty}^{\infty} |s(t)|^2 dt = 1 \tag{2.82}$$

and $\varphi'(t) = 0$. Based on equation (2.70), the mean frequency $\langle\omega\rangle = 0$. From equation (2.80), the variance is

$$\Delta_\omega^2 = \alpha^2\sqrt{\frac{\alpha}{\pi}}\int_{-\infty}^{\infty} t^2 e^{-\alpha t^2}dt = \frac{\alpha}{2} \tag{2.83}$$

Example 2.9 Frequency modulated Gaussian function

$$s(t) = g(t)e^{j\omega_0 t} \tag{2.84}$$

where $g(t)$ is a normalized Gaussian function given by (2.81). Then,

$$\varphi'(t) = \omega_0 \tag{2.85}$$

$$\langle\omega\rangle = \sqrt{\frac{\alpha}{\pi}}\int_{-\infty}^{\infty} \omega_0 e^{-\alpha t^2} dt = \omega_0$$

$$\Delta_\omega^2 = \alpha^2\sqrt{\frac{\alpha}{\pi}}\int_{-\infty}^{\infty} t^2 e^{-\alpha t^2} dt = \frac{\alpha}{2}$$

Compared to Example 2.8, although the phase in this example is not equal to zero, the variation of the phase is constant; that is, $\varphi'(t) = \omega_0$. Consequently, the frequency bandwidth is unchanged.

Example 2.10 Gaussian chirplet

$$s(t) = g(t)e^{j\beta t^2} \tag{2.86}$$

where $g(t)$ is a normalized Gaussian function given by (2.81). Then,

$$\varphi'(t) = 2\beta t \tag{2.87}$$

$$\langle\omega\rangle = \sqrt{\frac{\alpha}{\pi}}\int_{-\infty}^{\infty} 2\beta t e^{-\alpha t^2} dt = 0$$

$$\Delta_\omega^2 = \alpha^2\sqrt{\frac{\alpha}{\pi}}\int_{-\infty}^{\infty} t^2 e^{-\alpha t^2} dt + \sqrt{\frac{\alpha}{\pi}}\int_{-\infty}^{\infty} (2\beta t)^2 e^{-\alpha t^2} dt = \frac{\alpha}{2} + \frac{4\beta^2}{\alpha}$$

Because the derivative of the phase in (2.86) is not constant (that is, $\varphi'(t) = 2\beta t$) the frequency bandwidth in this example is larger than that in either Example 2.8 or Example 2.9. The extra term $4\beta^2/\alpha$ is proportional to the sweep rate β.

In Examples 2.8, 2.9, and 2.10, we investigated three signals that have identical amplitudes but different phases. We can see that the more rapid the phase change, the wider the frequency bandwidth.

Example 2.11 Scaling function $s(t/\alpha)$

The signal energy is

$$\int_{-\infty}^{\infty} \left|s\left(\frac{t}{\alpha}\right)\right|^2 dt = a\int_{-\infty}^{\infty} \left|s\left(\frac{t}{\alpha}\right)\right|^2 d\left(\frac{t}{\alpha}\right) = aE \tag{2.88}$$

The first derivatives of the amplitude and phase are $\alpha^{-1}a'(t/\alpha)$ and $\alpha^{-1}\varphi'(t/\alpha)$, respectively. If the mean frequency and frequency bandwidth of $s(t)$ are $<\omega>$ and $2\Delta_\omega$, then for the scaled function $s(t/\alpha)$,

$$\langle\omega(\alpha)\rangle = \frac{1}{\alpha E}\int_{-\infty}^{\infty} a^{-1}\varphi'\left(\frac{t}{\alpha}\right)a^2\left(\frac{t}{\alpha}\right)dt = \frac{1}{\alpha E}\int_{-\infty}^{\infty} \varphi'\left(\frac{t}{\alpha}\right)a^2\left(\frac{t}{\alpha}\right)d\left(\frac{t}{\alpha}\right) = \frac{\langle\omega\rangle}{\alpha} \tag{2.89}$$

and

$$\Delta_\omega^2(\alpha) = \frac{1}{\alpha E}\int_{-\infty}^{\infty}\left\{\left[\alpha^{-1}a\left(\frac{t}{\alpha}\right)\right]^2 + \left(\alpha^{-1}\varphi'\left(\frac{t}{\alpha}\right) - \langle\omega(\alpha)\rangle\right)^2 a^2\left(\frac{t}{\alpha}\right)\right\}dt \tag{2.90}$$

$$= \frac{1}{\alpha^2 E}\int_{-\infty}^{\infty}\left\{a^2\left(\frac{t}{\alpha}\right) + \left(\varphi'\left(\frac{t}{\alpha}\right) - a\langle\omega\rangle\right)^2 a^2\left(\frac{t}{\alpha}\right)\right\}d\left(\frac{t}{\alpha}\right)$$

$$= \frac{1}{\alpha^2}\Delta_\omega^2$$

which indicates that time scaling leads to the mean frequency shifting and the frequency bandwidth scaling, as illustrated in Figure 2-11. It is interesting to note, however, that the ratio between the bandwidth and the mean frequency is independent of the scaling factor α; that is

$$\frac{2\Delta_\omega(\alpha)}{\langle\omega(\alpha)\rangle} = \frac{2\Delta_\omega}{\langle\omega\rangle} = Q \tag{2.91}$$

In other words, the scaling does not change the ratio between the frequency bandwidth and the mean frequency. This property is commonly referred to as *constant Q*.

2.5 Uncertainty Principle

Because the time and frequency representations are linked by the Fourier transform, a signal's time and frequency behaviors are not independent. Particularly, when the signal has finite time support, its frequency bandwidth must be unlimited, and vice versa. There is no signal that has finite time duration and finite frequency bandwidth simultaneously. Such relationships form a central topic in time-frequency analysis.

To prove this fact, we may start from the converse hypothesis. That is, assume that a signal $s(t)$ has finite time duration T and bandwidth B, i.e.,

$$s(t) = \frac{1}{2\pi}\int_{-B/2}^{B/2} S(\omega)e^{jt\omega}d\omega = 0 \qquad \text{for } |t| \ge T/2 \tag{2.92}$$

We should be able to find at least one point t_0 inside the time duration T such that $s(t_0)$ is nonzero, i.e.,

$$s(t_0) = \frac{1}{2\pi}\int_{-B/2}^{B/2} S(\omega)e^{j(t_0-t)\omega}e^{jt\omega}d\omega \neq 0 \qquad \text{for } |t| \ge T/2 \qquad |t_0| < T/2 \tag{2.93}$$

By replacing the first complex exponential with its power series

$$e^{j\omega(t_0-t)} = \sum_{n=0}^{\infty}\frac{[j(t_0-t)]^n}{n!}\omega^n \tag{2.94}$$

$s(t_0)$ in (2.93) becomes

$$s(t_0) = \frac{1}{2\pi}\sum_{n=0}^{\infty}\frac{(t_0-t)^n}{n!}\int_{-B/2}^{B/2}(j\omega)^n S(\omega)e^{jt\omega}d\omega \qquad \text{for } |t| \ge T/2 \qquad |t_0| < T/2 \tag{2.95}$$

Note that from the relationship (2.55), the integration in (2.95) is nothing more than the signal's n^{th} derivative; that is,

$$\frac{1}{2\pi}\int_{-B/2}^{B/2}(j\omega)^n S(\omega)e^{jt\omega}d\omega = \frac{d^n}{dt^n}s(t) = 0 \qquad \text{for } |t| \ge T/2 \text{ and } \forall n \ge 0 \tag{2.96}$$

Consequently, $s(t_0)$ in (2.95) is always equal to zero, which contradicts our hypothesis (2.93). Hence, there is no such non-zero signal that has finite time duration and frequency bandwidth.

If we adopt the second moment as the measure of time duration and frequency bandwidth, then we can further obtain the *uncertainty principle*.

Theorem

If

$$\sqrt{t}s(t) \to 0 \tag{2.97}$$

for $|t| \to \infty$, then

$$\Delta_t\Delta_\omega \ge \frac{1}{2} \tag{2.98}$$

The equality only holds when $s(t)$ is the Gaussian function, i.e.,

$$s(t) = Ae^{-\alpha t^2} \tag{2.99}$$

Proof

For the sake of simplicity, let's assume that $\langle t\rangle = 0$ and $\langle\omega\rangle = 0$. Consequently, (2.74) and (2.75) become

$$\Delta_t^2 = \int_{-\infty}^{\infty} t^2|s(t)|^2 dt \tag{2.100}$$

and

$$\Delta_\omega^2 = \frac{1}{2\pi}\int_{-\infty}^{\infty}\omega^2|S(\omega)|^2 d\omega \tag{2.101}$$

Then,

$$\Delta_t^2\Delta_\omega^2 = \int_{-\infty}^{\infty} t^2|s(t)|^2 dt\frac{1}{2\pi}\int_{-\infty}^{\infty}\omega^2|S(\omega)|^2 d\omega \tag{2.102}$$

Replacing $\omega S(\omega) = H(\omega)$ in (2.102) yields

$$\Delta_t^2\Delta_\omega^2 = \int_{-\infty}^{\infty} t^2|s(t)|^2 dt\int_{-\infty}^{\infty} h(t)h^*(t)dt \tag{2.103}$$

where we applied Parseval's relationship. Because of the derivative property in (2.55),

$$\omega S(\omega) \leftrightarrow -j\frac{d}{dt}s(t) \tag{2.104}$$

Applying (2.104) to (2.103) yields

$$\Delta_t^2\Delta_\omega^2 = \int_{-\infty}^{\infty} t^2|s(t)|^2 dt \int_{-\infty}^{\infty} \left|\frac{d}{dt}s(t)\right|^2 dt \tag{2.105}$$

From Schwarz' inequality, it follows that

$$\int_{-\infty}^{\infty} t^2|s(t)|^2 dt \int_{-\infty}^{\infty} \left|\frac{d}{dt}s(t)\right|^2 dt \geq \left|\int_{-\infty}^{\infty} ts(t)\frac{d}{dt}s(t)dt\right|^2 \tag{2.106}$$

because

$$\int_{-\infty}^{\infty} ts(t)\frac{d}{dt}s(t)dt = \frac{1}{2}\int_{-\infty}^{\infty} t\frac{d}{dt}s^2(t)dt = \left.\frac{ts^2(t)}{2}\right|_{-\infty}^{\infty} - \frac{1}{2}\int_{-\infty}^{\infty} s^2(t)dt = -\frac{1}{2} \tag{2.107}$$

Inserting (2.107) into (2.106), we obtain the uncertainty inequality relationship (2.98). If (2.98) is an equality, then (2.106) must also be an equality. This is possible only if $s'(t) = kts(t)$; that is, $s(t)$ is as given by (2.99).

2.6 Discrete Poisson-Sum Formula

The Poisson-sum formula is very useful in signal analysis [36]. In this section, we will investigate the discrete version of the Poisson-sum formula, which plays an important role in deriving the discrete Gabor expansion.

Let $\{\tilde{a}[n]\}$ be a periodic sequence of period $L = \Delta M M$. Then we define $\{\tilde{b}[n]\}$ as the periodic extension of $\{\tilde{a}[n]\}$; that is

$$\tilde{b}[n] = \sum_{m=0}^{M-1} \tilde{a}[n - m\Delta M] \tag{2.108}$$

Obviously,

$$\tilde{b}[n] = \tilde{b}[n + \Delta M] \tag{2.109}$$

Let's denote $B[k]$ the DFT of $\{\tilde{b}[n]\}$, that is,

$$B[k] = \sum_{n=0}^{\Delta M-1} \tilde{b}[n] W_{\Delta M}^{-nk} = \sum_{n=0}^{\Delta M-1}\left(\sum_{m=0}^{M-1} \tilde{a}[n - m\Delta M]\right) W_{\Delta M}^{-nk} \tag{2.110}$$

Substituting $i = n - m\Delta M$, (2.110) reduces to a single summation, such as

$$B[k] = \sum_{i=0}^{L-1} \tilde{a}[i] W_{\Delta M}^{-ik} \tag{2.111}$$

Based on (2.110), we have

$$\tilde{b}[n] = \frac{1}{\Delta M}\sum_{k=0}^{\Delta M-1} B[k]W_{\Delta M}^{nk} = \frac{1}{\Delta M}\sum_{k=0}^{\Delta M-1}\left(\sum_{i=0}^{L-1}\tilde{a}[i]W_{\Delta M}^{-ik}\right)W_{\Delta M}^{nk} \tag{2.112}$$

From (2.108), we have

$$\sum_{m=0}^{M-1}\tilde{a}[n-m\Delta M] = \frac{1}{\Delta M}\sum_{k=0}^{\Delta M-1}\left(\sum_{i=0}^{L-1}\tilde{a}[i]W_{\Delta M}^{-ik}\right)W_{\Delta M}^{nk} \tag{2.113}$$

which is known as the *Poisson-sum formula*.

CHAPTER 3

Short-Time Fourier Transform and Gabor Expansion

In the conventional Fourier transform, the signal is compared with complex sinusoidal functions. Because sinusoidal basis functions are spread over the entire time domain and are not concentrated in time, the Fourier transform does not explicitly indicate how a signal's frequency contents evolve in time.

A straightforward approach to overcome this problem is to perform the Fourier transform on a block by block basis rather than to process the entire signal at once, which is named the windowed Fourier transform or the short-time Fourier transform (STFT). In Section 3.1, we briefly introduce the methodology of the STFT. For the continuous-time STFT, essentially we always can use the same function to perform the STFT and its inverse, recovering the original time function through the signal's STFT. In other words, we always can write a signal as the superposition of weighted analysis functions. The weight is equal to the signal's STFT. However, the representation based on the continuous-time STFT is highly redundant. For a compact representation, we often prefer to use the sampled STFT. In this case, the inverse problem is no longer trivial. The inverse of the sampled STFT can be accomplished by the Gabor expansion, though it apparently was not Gabor's original motivation.

Section 3.2 is devoted to the general introduction of the Gabor expansion. Although the idea of the Gabor expansion was rather straightforward, its implementation has been an open research topic. The reader can find the rigorous treatment from *Janssen* ([136], [141], [144], and [145]) as well as *Feirchtinger* and *Stroms* [15]. The central topic is how to design analysis and synthesis window functions. For the continuous-time Gabor expansion, it has been found that, except for a few functions, for a given analysis or synthesis function the analytical solution of its dual function in general does not exist. This has motived researchers to pursue numerical solutions, such as the discrete Gabor expansion. Unlike its continuous time counterpart, the discrete

Gabor expansion is relatively simple and can be realized with the help of elementary linear algebra.

At the critical sampling for a given function, the dual function is unique. In this case, it is difficult to make both functions have desirable properties. Hence, in many applications we impose oversampling rather than the critical sampling. Consequently, the solution of the dual functions is not unique. The natural question then is how to choose the dual functions. In Section 3.4, we introduce the orthogonal-like Gabor expansion. The concept of the orthogonal-like Gabor expansion was introduced by *Qian* et al. [194], which has been found very useful from both a theoretical and application point of view. In Section 3.5, we further develop a fast algorithm for computing dual functions.

Our presentation in this chapter starts with the periodic discrete Gabor expansion. The main limitation of the periodic Gabor expansion is that the window lengths have to be the same as the data length. This will become very inefficient when the number of data samples is large. To overcome this problem, in Section 3.6 we develop a scheme in which the lengths of the window functions and the data samples are independent. The resulting representation is named the discrete Gabor expansion. Since the discrete Gabor expansion is a subset of the periodic discrete Gabor expansion, all results developed from the periodic discrete Gabor expansion, such as the orthogonal-like Gabor expansion and the fast dual algorithm, are also valid for the discrete Gabor expansion.

Over the years, many techniques have been successfully developed to implement the Gabor expansion, such as Zak transform-based algorithms ([75], [76], [237] and [238]), filter bank methods [48], the frame-based method ([170] and [176]), as well as the pseudo-frame approach [166]. In this book we have limited our discussions to an STFT-based method that was first introduced by *Bastiaans* ([71], [73], and [74]) and later extended by *Wexler* and *Raz* [225]. One important feature of this method is that while ensuring perfect reconstruction, the discrete-time dual window functions have identical length.

3.1 Short-Time Fourier Transform

The main deficiency of the classical Fourier transform is that it does not explicitly tell how a signal's frequency contents change over time. Although the phase characteristic of the Fourier transform $S(\omega)$ contains the time information, it is difficult (if not impossible) to establish the point-to-point relationship between $s(t)$ and $S(\omega)$ based upon the classical Fourier transform.

A straightforward approach to overcome this problem is to perform the Fourier transform on a block by block basis rather than to process the entire signal at once. The result of such a windowed Fourier transform can then be thought of as a signal's frequency behavior during the time period covered by a corresponding data block. Figure 3-1 illustrates the corresponding procedure. First, multiply the signal $s(t)$ by a window function $\gamma(t)$. Then, compute the Fourier transform of the product $s(t)\ \gamma^*(t)$. If the function $\gamma(t)$ is centered at $t = 0$ and has a short-time duration, such as the Gaussian function, the Fourier transform of the product $s(t)\ \gamma^*(t)$ presumably describes the signal's frequency behavior in the vicinity of the time instant $t = 0$. By moving

$\gamma(t)$ and repeating the same process, we can obtain a rough idea of how the signal's frequency contents evolve over time. Because of the short-time duration of the function $\gamma(t)$, this method is traditionally known as a short-time Fourier transform.

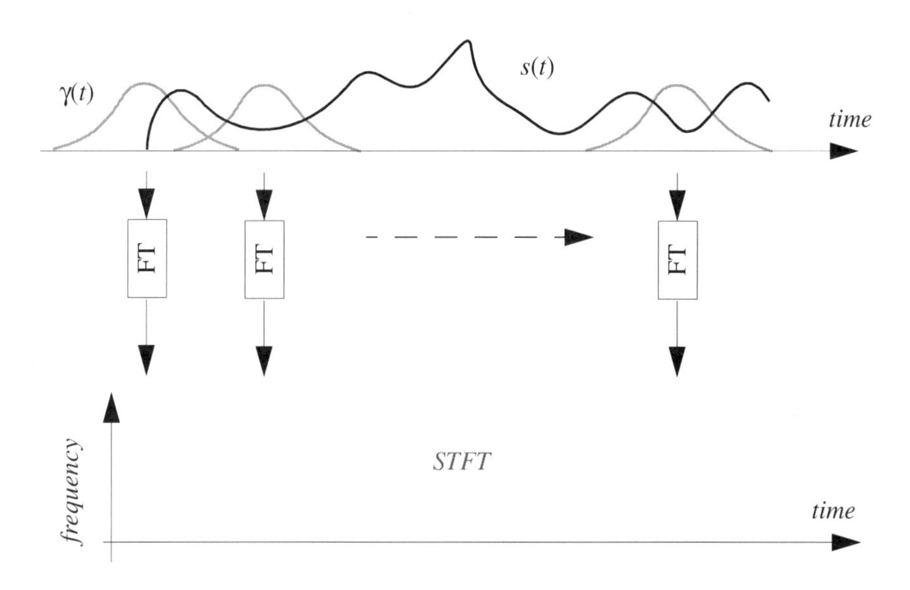

Figure 3-1 Short-time fourier transform

Mathematically, the short-time Fourier transform can be described by[1]

$$STFT(t, \omega) = \int_{-\infty}^{\infty} s(\tau)\gamma^*(\tau - t)e^{-j\omega\tau}d\tau = \int_{-\infty}^{\infty} s(\tau)\gamma^*_{t,\,\omega}(\tau)d\tau \tag{3.1}$$

The right side of (3.1) indicates that $STFT(t,\omega)$, from the concept of the expansion introduced in Chapter 2, is nothing more than a regular inner product of the signal $s(t)$ and a set of elementary functions $\{\gamma_{t,\,\omega}\}_{t,\,\omega \in R}$. Hence, $STFT(t,\omega)$ can also be considered as the measure of the similarity between a signal $s(t)$ and the time-shifted and frequency-modulated window function $\gamma(t)$. Let Δ_t and Δ_ω denote the time duration and frequency bandwidth of $\gamma(\tau)$. Suppose that the function $\gamma(t)$ is centered at $t = 0$ in the time domain and centered at $\omega = 0$ in the frequency domain. Then, $STFT(t,\omega)$ in (3.1) characterizes a signal's behavior in the vicinity of $[t-\Delta_t/2,\ t+\Delta_t/2] \times [\omega-\Delta_\omega/2,\ \omega+\Delta_\omega/2]$.

1. Strictly speaking, Figure 3-1 and (3.1) are not equivalent. The process depicted in Figure 3-1 in fact is

$$\int_{-\infty}^{\infty} s(\tau - t)\gamma^*(\tau)e^{-j\omega\tau}d\tau$$

For an accurate measurement, it is natural to desire that the area of $[t-\Delta_t/2, t+\Delta_t/2] \times [\omega-\Delta_\omega/2, \omega+\Delta_\omega/2]$ is as small as possible; that is, Δ_t and Δ_ω should be as narrow as possible. The resolution of the short-time Fourier transform-based measurement (3.1) is completely determined by the elementary function's time duration Δ_t and frequency bandwidth Δ_ω. Unfortunately, the selections of a signal's time duration Δ_t and frequency bandwidth Δ_ω are not independent, and are related through the Fourier transform. If we let Δ_t and Δ_ω be a signal's standard deviations, as introduced in Chapter 2, then the product $\Delta_t\Delta_\omega$ must satisfy the uncertainty inequality; that is

$$\Delta_t\Delta_\omega \geq \frac{1}{2}$$

Hence, there is a trade-off in the selection of the time and frequency resolution. If $\gamma(t)$ is chosen to have good time resolution (short-time duration or smaller Δ_t), then its frequency resolution must deteriorate (larger Δ_ω), and vice versa. The equality only holds when $\gamma(t)$ is a Gaussian type of function.

If $\{\gamma_{t,\omega}\}_{t,\omega \in R}$ is considered as a ruler to measure a signal's time-frequency behavior, then the different time and frequency tick marks for the STFT are obtained by the time-shifting and frequency-modulating a single prototype window function $\gamma(t)$. Because all elements in the set of $\{\gamma_{t,\omega}\}_{t,\omega \in R}$ are merely time-shifted and frequency-modulated versions of a single prototype function $\gamma(t)$, all elements have the same time duration and frequency bandwidth. Consequently, the pursuit of a set of elementary functions with good time-frequency resolution reduces to the design of a single function $\gamma(t)$. This is a very important feature of the short-time Fourier transform and makes it very easy to implement.

In Chapters 5 and 6, we will discuss another way to construct time and frequency tick marks in which different time and frequency tick marks are achieved by shifting and dilating a single *mother* function. The resulting transform is commonly known as the *wavelet* transform. Although the results of wavelet transform and short-time Fourier transform are very different, in both cases the set of elementary functions comes from a single prototype function.

The square of the STFT is named the STFT spectrogram to distinguish it from the time-dependent spectrum based upon other linear techniques, such as the Gabor expansion and the adaptive representations. The STFT spectrogram is the simplest time-dependent spectrum that can be used to get a rough idea of a signal's energy distribution in the joint time-frequency domain. While the STFT in general is complex, the STFT spectrogram is always real-valued. One central topic in the area of time-frequency analysis is how a signal's energy is distributed in the joint time-frequency domain. In what follows, we will examine in what conditions the STFT spectrogram could be considered as the signal's time-frequency density function.

Example 3.1 Time marginal, energy, and mean instantaneous frequency

For $s(t) = a(t)e^{j\varphi(t)}$ and a real-valued analysis function $\gamma(t)$,

$$STFT(t, \omega) = \int_{-\infty}^{\infty} a(\tau)e^{j\varphi(\tau)}\gamma(\tau - t)e^{-j\omega\tau}d\tau \tag{3.2}$$

The STFT spectrogram is the square of the STFT. Note that $STFT(t,\omega)$ is the Fourier transform of $a(\tau)e^{j\varphi(\tau)}\gamma(\tau-t)$. Applying Parseval's formula, we can obtain the time marginal by

$$\frac{1}{2\pi}\int_{-\infty}^{\infty} |STFT(t,\omega)|^2 d\omega = \int_{-\infty}^{\infty} \left|a(\tau)e^{j\varphi(\tau)}\gamma(\tau-t)\right|^2 d\tau = \int_{-\infty}^{\infty} a^2(\tau)\gamma^2(\tau-t)d\tau \tag{3.3}$$

which implies that when the analysis function reduces to the delta function, the integration of the STFT spectrogram at each time instant t will be equal to the instantaneous power $a^2(t)$. However, in this case, the corresponding $STFT(t,\omega)$ will become $a(t)e^{j\varphi(t)}$ at all frequencies ω. No new information is provided from the short-time Fourier transform!

If the STFT spectrogram is considered as a signal's energy distribution in the joint time-frequency domain, the signal energy could be computed by

$$\frac{1}{2\pi}\int_{-\infty}^{\infty}\int_{-\infty}^{\infty} |STFT(t,\omega)|^2 d\omega dt = \int_{-\infty}^{\infty}\int_{-\infty}^{\infty} a^2(\tau)\gamma^2(\tau-t)d\tau dt = \int_{-\infty}^{\infty} a^2(t)dt\int_{-\infty}^{\infty} \gamma^2(t)dt \tag{3.4}$$

Hence, as long as the window function $\gamma(t)$ is normalized, the average of the STFT spectrogram is equal to the energy contained in the original signal as we anticipate.

We can also compute the mean instantaneous frequency, as follows, if the STFT spectrogram can be thought of as a signal's energy distribution in the joint time-frequency domain.

$$\langle\omega\rangle_t = \frac{\frac{1}{2\pi}\int_{-\infty}^{\infty} \omega|STFT(t,\omega)|^2 d\omega}{\frac{1}{2\pi}\int_{-\infty}^{\infty} |STFT(t,\omega)|^2 d\omega} \tag{3.5}$$

Note that the numerator is the mean frequency of $|STFT(t,\omega)|^2$. And, $STFT(t,\omega)$ is the Fourier transform of a time function $a(\tau)e^{j\varphi(\tau)}\gamma(\tau-t)$. Applying (2.70) leads to

$$\langle\omega\rangle_t = \frac{\int_{-\infty}^{\infty} \varphi'(\tau)a^2(\tau)\gamma^2(\tau-t)d\tau}{\int_{-\infty}^{\infty} a^2(\tau)\gamma^2(\tau-t)d\tau} \tag{3.6}$$

which says that in general the mean instantaneous frequency computed by the STFT spectrogram is not equal to the derivative of the signal's phase, unless $\gamma(t)$ is a delta function. Unfortunately, in this case the short-time Fourier transform won't provide any useful information, as shown earlier.

Although the STFT spectrogram is widely used to characterize a signal's time-frequency energy distribution, its result is subject to the selection of the analysis window function, as shown in Example 3.1. By applying the STFT spectrogram, one can't get an accurate measurement in both the time and frequency domains simultaneously.

Example 3.2 STFT of a Gaussian-type function

$$s(t) = \sqrt[4]{\frac{\beta}{\pi}}e^{-\beta t^2/2} \tag{3.7}$$

Intuitively, in the joint time-frequency domain, $s(t)$ is centered at (0,0). Its time duration and frequency bandwidth are determined by β. Assume that the window function is also a Gaussian-type function, i.e.,

$$\gamma(t) = \sqrt[4]{\frac{\alpha}{\pi}} e^{-\alpha t^2/2} \tag{3.8}$$

Substituting $s(t)$ and $\gamma(t)$ into (3.1) yields

$$STFT(t,\omega) = \sqrt[4]{\frac{\alpha\beta}{\pi^2}} \int_{-\infty}^{\infty} \exp\left\{-\frac{\beta}{2}\tau^2\right\} \exp\left\{-\frac{\alpha}{2}(\tau-t)^2\right\} e^{-j\omega\tau} d\tau \tag{3.9}$$

$$= \sqrt[4]{\frac{\alpha\beta}{\pi^2}} \int_{-\infty}^{\infty} \exp\left\{-\left(\frac{\alpha+\beta}{2}\right)\tau^2 + \alpha t\tau - \frac{\alpha}{2}t^2\right\} e^{-j\omega\tau} d\tau$$

$$= \exp\left\{-\frac{\alpha\beta}{2(\alpha+\beta)}t^2\right\} \sqrt[4]{\frac{\alpha\beta}{\pi^2}} \int_{-\infty}^{\infty} \exp\left\{-\left(\frac{\alpha+\beta}{2}\right)\left[\tau^2 - \frac{2\alpha t}{\alpha+\beta}\tau + \left(\frac{\alpha t}{\alpha+\beta}\right)^2\right]\right\} e^{-j\omega\tau} d\tau$$

$$= \exp\left\{-\frac{\alpha\beta}{2(\alpha+\beta)}t^2\right\} \sqrt[4]{\frac{\alpha\beta}{\pi^2}} \int_{-\infty}^{\infty} \exp\left\{-\left(\frac{\alpha+\beta}{2}\right)\left(\tau - \frac{\alpha}{\alpha+\beta}t\right)^2\right\} e^{-j\omega\tau} d\tau$$

Applying (2.44) from Example 2.6, we have

$$STFT(t,\omega) = \sqrt{\frac{2\sqrt{\alpha\beta}}{\alpha+\beta}} \exp\left\{-\frac{\alpha\beta}{2(\alpha+\beta)}t^2 - \frac{1}{2(\alpha+\beta)}\omega^2 + j\frac{\alpha}{\alpha+\beta}\omega t\right\} \tag{3.10}$$

The corresponding STFT spectrogram is

$$SP(t,\omega) = |STFT(t,\omega)|^2 = \frac{2\sqrt{\alpha\beta}}{\alpha+\beta} \exp\left\{-\frac{\alpha\beta}{\alpha+\beta}t^2 - \frac{1}{\alpha+\beta}\omega^2\right\} \tag{3.11}$$

which shows that the STFT spectrogram is concentrated at (0,0), the center of the signal $s(t)$. The contours of equal height of SP in (3.11) are ellipses. The contour for the case where the levels are down to e^{-1} of their peak value is the ellipse indicated in Figure 3-2. The area of the particular ellipse at this level is

$$A = \frac{\alpha+\beta}{\sqrt{\alpha\beta}}\pi = \frac{1+r}{\sqrt{r}}\pi \tag{3.12}$$

where $r = \beta/\alpha$ is a matching indicator. The area A reflects the concentration of the STFT. Naturally, the smaller the value of A, the better the resolution. The resolution of the STFT is subject to the selection of an analysis function. The minimum of A in (3.12) occurs when $r = 1$. In other words, when the variance of the analysis function α perfectly matches the time duration of the analyzed signal β, $SP(t,\omega)$ in (3.11) will have the best resolution. However, in general, because the signal duration β will likely be unknown, it would be difficult to achieve the optimal resolution. Moreover, one should bear in mind that even the optimal resolution, $A = 2\pi$, is twice as large as that of the Wigner-Ville distribution demonstrated in Example 7.1.

From (3.1), the short-time Fourier transform $STFT(t,\omega)$ can be considered as the Fourier transform of the product $s(\tau)\gamma(\tau-t)$. Hence, given $STFT(t,\omega)$ and the analysis window function $\gamma(t)$, we can recover the original signal simply by performing the inverse Fourier transform, i.e.,

$$s(t)\gamma^*(\tau) = \frac{1}{2\pi}\int_{-\infty}^{\infty} STFT_\gamma(\tau - t, \omega)e^{j\omega\tau}d\omega \tag{3.13}$$

where the subscript of $STFT_\gamma(\tau,\omega)$ emphasizes that the STFT is computed by the window function $\gamma(t)$. As a matter of fact, by applying the window function $\gamma(t)$ we can also recover the original signal $s(t)$ from $STFT_h(\tau,\omega)$ computed by any other window function $h(t)$ as long as

$$\left|\int_{-\infty}^{\infty} h^*(t)\gamma(t)dt\right| < \infty \tag{3.14}$$

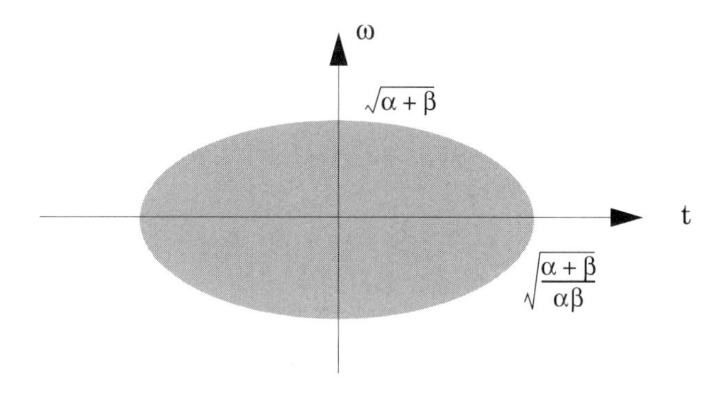

Figure 3-2 The area of the ellipse reaches its minimum when the variance of the analysis function perfectly matches the signal time duration, that is, $\alpha = \beta$. The minimum area is 2π, which is twice as large as that of the Wigner-Ville distribution in Figure 7-1.

For example,

$$\begin{aligned}
&\frac{1}{2\pi}\int_{-\infty}^{\infty}\int_{-\infty}^{\infty} STFT_h(\tau, \omega)\gamma(t-\tau)e^{j\omega t}d\tau d\omega \qquad (3.15)\\
&= \frac{1}{2\pi}\int_{-\infty}^{\infty}\int_{-\infty}^{\infty}\int_{-\infty}^{\infty} s(\alpha)h^*(\alpha-\tau)e^{-j\omega\alpha}\gamma(t-\tau)e^{j\omega t}d\alpha d\tau d\omega\\
&= \int_{-\infty}^{\infty}\int_{-\infty}^{\infty} s(\alpha)h^*(\alpha-\tau)\gamma(t-\tau)\delta(t-\alpha)d\alpha d\tau\\
&= s(t)\int_{-\infty}^{\infty} h^*(t-\tau)\gamma(t-\tau)d\tau
\end{aligned}$$

However, in all possible $STFT_h(\tau,\omega)$, $STFT_\gamma(\tau,\omega)$ possesses the minimum L^2-norm (or minimum length). To see this, let us multiply both sides of (3.15) by $s^*(t)$ and then integrate over t, i.e.,

$$\begin{aligned}
\int_{-\infty}^{\infty}|s(t)|^2dt &= \frac{1}{2\pi}\int_{-\infty}^{\infty}\int_{-\infty}^{\infty} STFT_h(\tau, \omega)\int_{-\infty}^{\infty} s^*(t)\gamma(t-\tau)e^{j\omega t}dtd\tau d\omega \qquad (3.16)\\
&= \frac{1}{2\pi}\int_{-\infty}^{\infty}\int_{-\infty}^{\infty} STFT_h(\tau, \omega)STFT^*{}_\gamma(\tau, \omega)d\tau d\omega
\end{aligned}$$

Similarly, we can obtain

$$\int_{-\infty}^{\infty} |s(t)|^2 dt = \frac{1}{2\pi}\int_{-\infty}^{\infty}\int_{-\infty}^{\infty} |STFT_{\gamma}(\tau, \omega)|^2 d\tau d\omega \tag{3.17}$$

Hence,

$$\frac{1}{2\pi}\int_{-\infty}^{\infty}\int_{-\infty}^{\infty} STFT^*_{\gamma}(\tau, \omega)[STFT_h(\tau, \omega) - STFT_{\gamma}(\tau, \omega)]d\tau d\omega = 0 \tag{3.18}$$

which implies that $STFT_{\gamma}(\tau,\omega)$ is orthogonal to $STFT_h(\tau,\omega)$-$STFT_{\gamma}(\tau,\omega)$. Then the length of $STFT_h(\tau,\omega)$ can be written as

$$\begin{aligned} &\frac{1}{2\pi}\int_{-\infty}^{\infty}\int_{-\infty}^{\infty} |STFT_h(\tau, \omega)|^2 d\tau d\omega \\ &= \frac{1}{2\pi}\int_{-\infty}^{\infty}\int_{-\infty}^{\infty} |STFT_{\gamma}(\tau, \omega)|^2 d\tau d\omega + \frac{1}{2\pi}\int_{-\infty}^{\infty}\int_{-\infty}^{\infty} |STFT_h(\tau, \omega) - STFT_{\gamma}(\tau, \omega)|^2 d\tau d\omega \end{aligned} \tag{3.19}$$

Obviously, the minimum length of $STFT_h(\tau,\omega)$ occurs only when the analysis function $h(t)$ is identical to the synthesis function $\gamma(t)$.

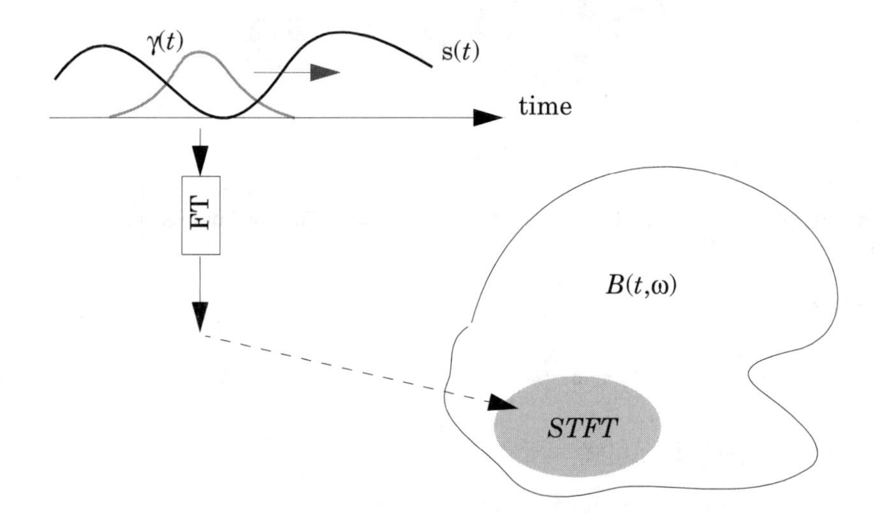

Figure 3-3 *STFT(t,ω)* is a subset of the entire set of 2D functions. An arbitrary two-dimensional time-frequency function may not be a valid *STFT(t,ω)*. The range of the mapping by performing the short-time Fourier transform is not the entire set of square integrable functions of two variables, but only a subset of it.

It is worth noting that $STFT_{\gamma}(\tau,\omega)$ is highly redundant. In fact, the signal $s(t)$ can be completely reconstructed merely from the sampled version of the short-time Fourier transform, $STFT(mT,n\Omega)$, where T and Ω denote the time and frequency sampling steps, respectively. In other words, we can use the sampled STFT to completely characterize the signal $s(t)$. By doing so, we can save considerable computation as well as memory. The trade-off is that the set of analysis functions and the set of synthesis functions will no longer be the same in general. For an arbitrary given analysis function, there might be no corresponding synthesis function that exists. The design of the desired analysis and synthesis functions will be the central topic of subsequent sections.

The short-time Fourier transform can also be viewed as a mapping between the time domain to the joint time-frequency domain, as illustrated in Figure 3-3. For any time domain function $s(t)$ and window function $\gamma(t)$, such a mapping always exists. But this is not always the case for the inverse transform. In other words, given a function $\gamma(t)$ and an arbitrary two-dimensional function $B(t,\omega)$, there may be no such signal $s(t)$ whose STFT is equal to $B(t,\omega)$. In this case, we say that $B(t,\omega)$ is not a valid short-time Fourier transform.

The simplest example is

$$B(t, \omega) = \begin{cases} 1 & for\ |t| < t_0\ |\omega| < \omega_0 \\ 0 & otherwise \end{cases} \tag{3.20}$$

Because no signal can be finitely supported in both the time and frequency domains, $B(t,\omega)$ in (3.20) is obviously not a valid time-frequency representation. As shown in Figure 3-3, the range of the mapping by performing the short-time Fourier transform is not the entire set of square integrable functions of two variables, but only a subset of it.

A valid short-time Fourier transform has to be such that its inverse Fourier transform is separable, that is

$$\frac{1}{2\pi}\int_{-\infty}^{\infty} B(t, \omega)e^{j\mu\omega}d\omega = s(\mu)\gamma(\mu - t) \qquad \forall\ t, \mu \tag{3.21}$$

For digital signal processing applications, it is necessary to extend the STFT framework to discrete-time signals. For practical implementation, each Fourier transform in the STFT has to be replaced by the discrete Fourier transform; the resulting STFT is discrete in both time and frequency and thus is suitable for digital implementation, i.e.,

$$STFT[m, n] = \sum_{k=0}^{L-1} s[k]\gamma[k-m]W_L^{-nk} = \sum_{k=0}^{L-1} s[k]\gamma_{m,n}[k] \tag{3.22}$$

where

$$STFT[m, n] \equiv STFT(t, \omega)\big|_{t = m\Delta t,\ \omega = 2\pi n/(L\Delta t)} \tag{3.23}$$

and

$$\gamma_{m,n}[k] \equiv \gamma[k-m]W_L^{nk} \tag{3.24}$$

where Δ_t denotes the time sampling interval. $\gamma[k] \equiv \gamma(k\Delta t)$ is the L-point window function. We call (3.22) the discrete STFT to distinguish it from the discrete-time STFT, which is continuous in frequency. It is rather easy to verify that the discrete STFT is periodic in frequency, that is,

$$STFT[m, n] = STFT[m, n + lL] \tag{3.25}$$

for $l = 0, \pm1, \pm2, \pm3\ldots$. Like the continuous-time STFT, an arbitrary 2D discrete function in general is not a valid discrete short-time Fourier transform.[2]

3.2 Gabor Expansion

Instead of representing a signal either as a function of time or as a function of frequency separately, in 1946 *Gabor*[3] suggested representing a signal in two dimensions, with time and frequency as coordinates [113]. Gabor named such two-dimensional representations the "information diagrams," as areas in them are proportional to the number of independent data that they can convey. Gabor pointed out that there are certain "elementary signals" that occupy the smallest possible area in the information diagram. Each elementary signal can be considered as conveying exactly one datum, or one "quantum of information." Any signal can be expanded in terms of these by a process that includes time analysis and frequency analysis as extreme cases. For a signal $s(t)$, the Gabor expansion is defined as

$$s(t) = \sum_{m=-\infty}^{\infty} \sum_{n=-\infty}^{\infty} c_{m,n} h_{m,n}(t) = \sum_{m=-\infty}^{\infty} \sum_{n=-\infty}^{\infty} c_{m,n} h(t - mT) e^{jn\Omega t} \tag{3.26}$$

where T and Ω denote the time and frequency sampling steps. The coefficients $c_{m,n}$ are traditionally called Gabor coefficients. Figure 3-4 illustrates the Gabor sampling grid.

In Gabor's original paper, he selected the Gaussian function as the elementary function, i.e.,

$$h(t) = g(t) = \sqrt[4]{\frac{\alpha}{\pi}} e^{-\alpha t^2/2} \tag{3.27}$$

because the Gaussian function is optimally concentrated in the joint time-frequency domain in terms of the *uncertainty principle*; that is,

$$\Delta_t \Delta_\omega = \frac{1}{\sqrt{\alpha}} \frac{\sqrt{\alpha}}{2} = \frac{1}{2} \tag{3.28}$$

2. One exception is when $B[m,n]$ is in the space formed by a basis. In this case, any pulse in the discrete time-frequency domain always has a corresponding time waveform – a weighted dual function.

3. Dennis Gabor was born on June 5, 1900, in Budapest, Hungary. His talent for memorization — an asset in any academic field — appeared at the age of twelve when he earned a prize from his father for learning by heart, in German, a 430-line poem. Gabor finished his doctorate in electrical engineering in 1927. His work in communication theory and holography started at the end of World War II. It was during that time that he wrote the famous Gabor expansion paper. In 1949 he joined the Imperial College of Science and Technology at London University and in the late sixties became a staff scientist at CBS Laboratories in the United States. While his formal education had been largely in the applied engineering fields, he had not neglected to study the basic physical and mathematical tools that would facilitate his life's work, which was mostly motivated by a desire to create or perfect a particular device invariably secured on a sound mathematical footing. His genius as an inventor lay in an innate ability to focus on a final goal, regardless of the difficulties. His endeavor paid off. In 1971 the Royal Swedish Academy of Sciences presented Dennis Gabor with the Nobel Prize for his discovery of the principles underlying the science of holography.

which is the lower bound of the uncertainty inequality.

Although Gabor restricted himself to an elementary signal that has a Gaussian shape, his signal expansion, in fact, holds for rather arbitrarily shaped signals. For almost any signal $h(t)$, its time-shifted and harmonically modulated version can be used as the *Gabor elementary function*. The necessary condition of the existence of the Gabor expansion is that the sampling cell $T\Omega$ must be small enough to satisfy

$$T\Omega \le 2\pi \tag{3.29}$$

Intuitively, if the sampling cell $T\Omega$ in Figure 3-4 is too large, we may not have enough information to completely recover the original signal. On the other hand, if the sampling cell $T\Omega$ is too small, the representation will be redundant. Traditionally, it is called *critical sampling* when $T\Omega = 2\pi$ and *oversampling* when $T\Omega < 2\pi$. A family of this kind of functions, $\{h_{m,n}\}_{m,n \in Z}$ in $l^2(Z)$, is often referred to as a *Weyl-Heisenberg* family. Although Gabor was not known to have investigated the existence of the formula (3.26), the sampling cell that he selected, $T\Omega = 2\pi$, happened to be the most compact representation.

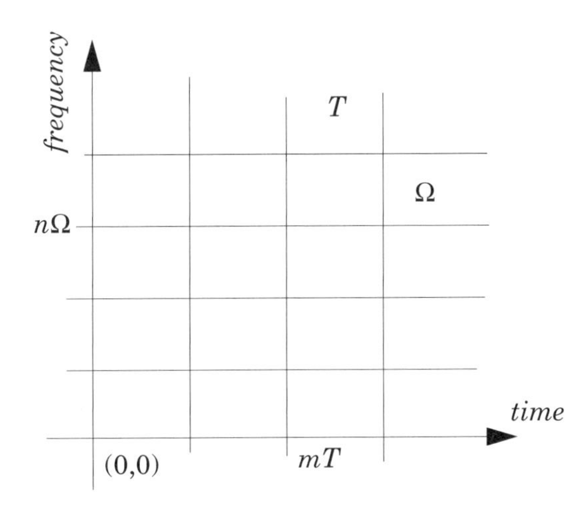

Figure 3-4 Gabor sampling lattice

Gabor was also not known to have published any practical algorithms of computing the Gabor coefficients.[4] Despite the earlier treatment by *Auslander* et al [65], Gabor's work was not widely known until 1980 when *Bastiaans* related the Gabor expansion and the short-time Fourier transform ([71], [73], and [74]). *Bastiaans* introduced the sampled short-time Fourier transform to compute the Gabor coefficients and successfully derived a closed form of the Gaussian function's dual function (or auxiliary functions, as initially named by *Bastiaans*).

4. In his notable paper, Gabor proposed an iterative approach to compute the coefficients $c_{m,n}$, which, however, in general has been found not to converge [114].

As mentioned in the preceding section, the continuous-time inverse STFT is a highly redundant expansion. In applications, for a compact representation, we usually use the sampled STFT. However, the imprudent choice of an analysis function $\gamma(t)$ and sampling steps, T and Ω, may lead to the sampled STFT being non-invertible. With the help of the Gabor expansion, we now can easily solve the problem of the inverse of the sampled STFT, even though it was apparently not Gabor's original motivation.

Based upon the frame theorem introduced in Chapter 2, if there is a set of dual functions $\{\gamma_{m,n}(t)\}_{m,n\in Z}$ corresponding to the set of the Gabor elementary functions $\{h_{m,n}(t)\}_{m,n\in Z}$, then the Gabor coefficients $c_{m,n}$ can be computed by a regular inner product operation, i.e.,

$$c_{m,n} = \int_{-\infty}^{\infty} s(t)\gamma^*{}_{m,n}(t)dt = \int_{-\infty}^{\infty} s(t)\gamma^*(t-mT)e^{-jn\Omega t}dt = STFT[mT, n\Omega] \tag{3.30}$$

which is a sampled STFT and also known as the *Gabor transform.* The Gabor coefficient $c_{m,n}$ computed by (3.30) is time- and frequency-shift invariant when the time-shift is a multiple of the time sampling step T and the frequency-shift is a multiple of the frequency sampling step Ω.

The dual functions $\gamma(t)$ in (3.30) and $h(t)$ in (3.26) are exchangeable. The Gabor expansion can be written in either way as

$$s(t) = \sum_{m,n} \langle s, \gamma_{m,n}\rangle h_{m,n}(t) = \sum_{m,n} \langle s, h_{m,n}\rangle \gamma_{m,n}(t) \tag{3.31}$$

Which one, $\gamma(t)$ or $h(t)$, is used for the analysis function to compute the Gabor coefficients depends on the application at hand. If we are mainly interested in the Gabor coefficients, then we may use $h_{m,n}(t)$ to calculate $c_{m,n}$ because it is selected first and thereby it is easier to make meet our requirements. In this case, once $h(t)$ is properly selected, the Gabor coefficients $c_{m,n}$ will well describe the signal's local time and frequency behaviors.

Unlike harmonically related complex sinusoidal functions used in the Fourier series, which are orthonormal, the set of the Gabor elementary functions with a desired time-frequency resolution usually does not form an orthogonal basis (which, in may applications, is even not a tight frame).[5] Consequently, for a given set of functions $\{h_{m,n}(t)\}_{m,n\in Z}$ the solution of the corresponding dual function $\{\gamma_{m,n}(t)\}_{m,n\in Z}$ won't be straightforward.

Substituting (3.30) into the right side of (3.26) yields

$$s(t) = \int_{-\infty}^{\infty} s(t') \sum_{m=-\infty}^{\infty} \sum_{n=-\infty}^{\infty} \gamma^*{}_{m,n}(t')h_{m,n}(t)dt' \tag{3.32}$$

which implies that the dual function exists if and only if the double summation is a delta function, that is,

5. The result on poor time-frequency localization of the Weyl-Heisenberg basis was first established for the Weyl-Heisenberg basis in $L^2(R)$ and is known as the *Balian-Low* theorem ([67] and [336]). An analogous problem with Weyl-Heisenberg bases in $l^2(Z)$ was pointed out by *Vetterli* [409].

$$\sum_{m=-\infty}^{\infty} \sum_{n=-\infty}^{\infty} \gamma^*_{m,n}(t')h_{m,n}(t) = \delta(t-t') \tag{3.33}$$

By the Poisson-sum formula, (3.33) can be reduced to a single integration [225], i.e.,

$$\frac{T_0\Omega_0}{2\pi}\int_{-\infty}^{\infty} h(t)\gamma^{0*}_{m,n}(t)\,dt = \delta(m)\delta(n) \tag{3.34}$$

where

$$\gamma^0_{m,n} = \gamma(t-mT_0)e^{jn\Omega_0} \tag{3.35}$$

where $T_0 = 2\pi/\Omega$ and $\Omega_0 = 2\pi/T$. Note that except for the critical sampling ($T\Omega = 2\pi$), $\gamma_{m,n}(t) \neq \gamma^0_{m,n}(t)$. In some literature, (3.34) is named the *Wexler-Raz identity,* which plays an important role in computing the dual functions.

It is obvious that given a function $h(t)$, the analytical solution of (3.34), $\gamma(t)$, is not straightforward. Except for a few specific functions, such as the Gaussian and exponential (two- or one-sided) at critical sampling, where the dual functions can be explicitly computed ([71], [103], and [112]), analytical solutions of $\gamma(t)$ are not generally available. Nevertheless, no known dual functions $\gamma(t)$ have a good time-frequency resolution. The difficulty associated with the design of a desirable dual function greatly motivates us to seek a numerical solution – the discrete Gabor expansion.

3.3 Periodic Discrete Gabor Expansion[6]

In addition to the solution of dual functions as discussed in the previous section, the development of the discrete Gabor expansion is also motivated by applications. Signals encountered in most applications today are discrete in time and the tool to perform signal processing is the digital computer. Hence, it is necessary and beneficial to extend Gabor's framework into the case of discrete-time and discrete-frequency.

The procedure for digitizing the continuous-time Gabor expansion (3.26) essentially is a standard sampling process. So, we leave it as an exercise for the reader. Note that sampling the time variables leads to periodicity in the frequency domain. Conversely, digitizing the frequency variable results in periodicity in the time domain. Because we digitize both time and frequency indices, strictly speaking, the discrete version of the Gabor expansion is only applicable to periodic discrete-time signals.[7]

6. Because we can easily extend a finite sequence to a periodic function, all results developed from periodic functions are automatically applicable for finite samples.
7. All other discrete forms, including the one that we introduce in the next section, are special cases of the periodic discrete Gabor expansion.

For a discrete-time signal $s[k]$ with period L, the discrete Gabor expansion is defined by

$$\tilde{s}[k] = \sum_{m=0}^{M-1} \sum_{n=0}^{N-1} \tilde{c}_{m,n} \tilde{h}[k - m\Delta M] W_L^{n\Delta Nk} \tag{3.36}$$

where Gabor coefficients are computed by the sampled short-time Fourier transform, i.e.,

$$\tilde{c}_{m,n} = \sum_{k=0}^{L-1} \tilde{s}[k] \tilde{\gamma}^*[k - m\Delta M] W_L^{-n\Delta Nk} \tag{3.37}$$

Note that the signal $s[k]$, the synthesis function $h[k]$, and the analysis function $\gamma[k]$ are all periodic and have the same period L. We name (3.36) the *periodic discrete Gabor expansion* to distinguish it from the *discrete Gabor expansion* in which neither the analyzed signal nor the window function need to be periodic. Figure 3-5 depicts the procedure for computing the periodic discrete Gabor coefficients.

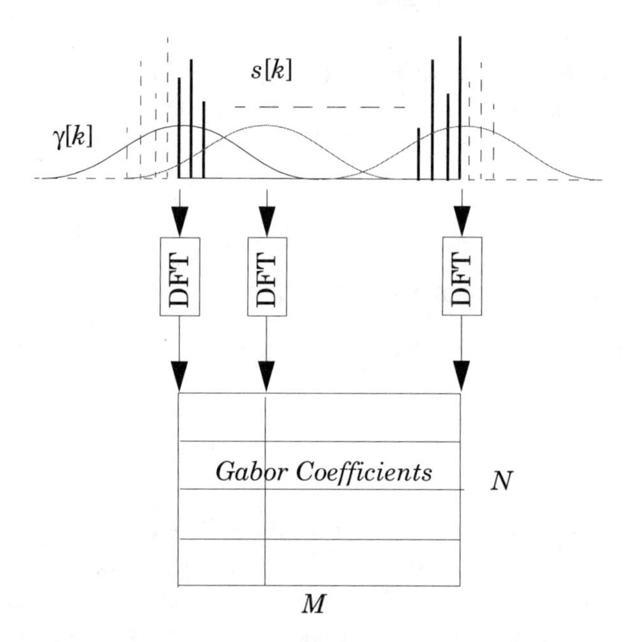

Figure 3-5 Periodic discrete Gabor transformation

The terms ΔM and ΔN in (3.36) denote discrete time and frequency sampling steps, respectively. The Gabor sampling lattice of discrete signals is plotted in Figure 3-6. The oversampling rate is defined by

$$a = \frac{L}{\Delta M \Delta N} \tag{3.38}$$

It is called critical sampling when $a = 1$ and oversampling when $a > 1$. For stable reconstruction, the sampling rate must be greater or equal to one. In *Wexler* and *Raz*'s original paper [225], it was required that $\Delta MM = \Delta NN = L$.[8] In this case, M and N are equal to the number of sampling points in time and frequency domains, respectively. The product MN is equal to the total number of the Gabor coefficients. Rewriting (3.38), we obtain

$$a = \frac{MN}{L} \tag{3.39}$$

which implies that the Gabor sampling rate is nothing more than the ratio between the total number of Gabor coefficients and the number of distinct samples. Critical sampling means that the number of Gabor coefficients is equal to the number of distinct samples. Oversampling means that the number of Gabor coefficients is greater than the number of samples. In this case, the resulting Gabor expansion is redundant.

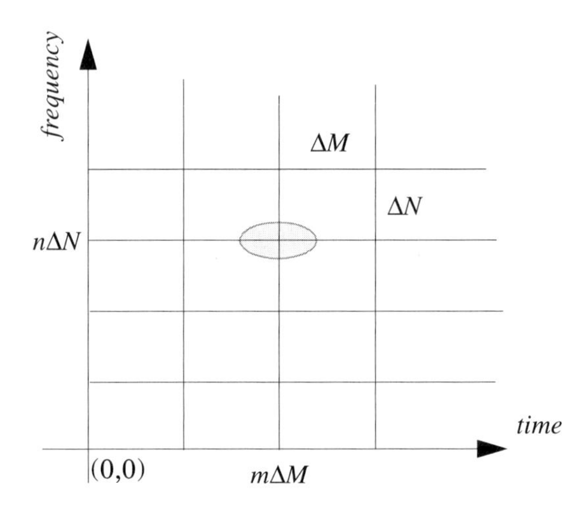

Figure 3-6 Discrete Gabor sampling lattice

If we restrict $\Delta NN = L$, (3.36) and (3.37) can be rewritten as

$$\tilde{s}[k] = \sum_{m=0}^{M-1} \sum_{n=0}^{N-1} \tilde{c}_{m,n} \tilde{h}[k - m\Delta M] W_N^{nk} \tag{3.40}$$

8. *Li* and *Qian* show that the requirement $\Delta NN = L$ is not necessary [167]. ΔN, in fact, can be any integer. Consequently, we could have more freedom in choosing the oversampling rate. On the other hand, the implementation may become complicated and less efficient. For example, we may no longer be able to use the efficient FFT algorithm to compute the Gabor expansion as well as the sampled short-time Fourier transform.

and

$$\tilde{c}_{m,n} = \sum_{k=0}^{L-1} \tilde{s}[k]\tilde{\gamma}^*[k - m\Delta M]W_N^{-nk} \tag{3.41}$$

where N can be considered as the number of frequency bins or number of frequency channels. Eq. (3.41) implies that the Gabor coefficients are periodic in n, i.e.,

$$\tilde{c}_{m,n} = \tilde{c}_{m,n+lN} \qquad \forall\, l \in Z \tag{3.42}$$

Substituting $L = \Delta NN$ into (3.38) yields

$$a = \frac{N}{\Delta M} \geq 1 \tag{3.43}$$

which says that for stable reconstruction, the time sampling step ΔM has to be less than or equal to the number of frequency bins or frequency channels N. Equations (3.38), (3.40), and (3.43) provide three different ways to describe the Gabor sampling rate.

As a matter of fact, the discrete Gabor transform (3.40) and Gabor expansion (3.41) can also be viewed as an N-channel digital filter bank system ([12] and [48]). While N is equal to the number of channels, ΔM characterizes the decimation/interpolation factor. For perfect reconstruction, the decimation/interpolation factor must be less than or equal to the number of channels. The reader can find a more detailed discussion of filter banks in Chapter 6.

A fundamental question for the Gabor transform and Gabor expansion is how to compute $\tilde{\gamma}[k]$ for a given $\tilde{h}[k]$ and sampling steps, ΔM and ΔN. Substituting (3.41) into the right side of (3.40) yields

$$\tilde{s}[k] = \sum_{k=0}^{L-1} \tilde{s}[k'] \sum_{m=0}^{M-1}\sum_{n=0}^{N-1} \tilde{\gamma}^*[k' - m\Delta M]\tilde{h}[k - m\Delta M]W_N^{n(k-k')} \tag{3.44}$$

Obviously, the periodic discrete Gabor expansion exists if and only if the double summation is equal to the delta function, i.e.,

$$\sum_{m=0}^{M-1}\sum_{n=0}^{N-1} \tilde{\gamma}^*[k' - m\Delta M]\tilde{h}[k - m\Delta M]W_N^{n(k-k')} = \delta[k - k'] \tag{3.45}$$

Unfortunately, the double summation form is not very pleasant to work with. Eq. (3.45) does not provide a clue to solving for $\tilde{\gamma}[k]$.

Expanding the left side of (3.45), we obtain

$$\sum_{m=0}^{M-1} \tilde{\gamma}^*[k' - m\Delta M]\tilde{h}[k - m\Delta M]N \sum_{q=0}^{\Delta N-1} \delta[k - k' - qN] \tag{3.46}$$

$$= N \sum_{q=0}^{\Delta N-1} \delta[k - k' - qN] \sum_{m=0}^{M-1} \tilde{\gamma}^*[k' - m\Delta M]\tilde{h}[k' + qN - m\Delta M]$$

Applying the discrete Poisson-sum formula to the second summation,

$$\sum_{m=0}^{M-1} \tilde{\gamma}^*[k - m\Delta M]\tilde{h}[k + qN - m\Delta M] \tag{3.47}$$

$$= \frac{1}{\Delta M} \sum_{k=0}^{\Delta M - 1} \left\{ \sum_{i=0}^{L-1} \tilde{h}[k + qN] W_{\Delta M}^{-pk} \tilde{\gamma}^*[k] \right\} W_{\Delta M}^{nk}$$

which implies that (3.45) holds if and only if

$$\sum_{k=0}^{L-1} \tilde{h}[k + qN] W_{\Delta M}^{-pk} \tilde{\gamma}^*[k] = \frac{\Delta M}{N} \delta[p]\delta[q] \qquad 0 \le p < \Delta M \qquad 0 \le q < \Delta N \tag{3.48}$$

which is usually considered as the discrete version of the *Wexler-Raz identity.* Note that (3.48) can be formulated in matrix form as

$$H\vec{\gamma}^* = \vec{\mu} \tag{3.49}$$

where H is a $\Delta M \Delta N$-by-L matrix, whose entries are defined as

$$h_{p\Delta M + q,\, k} \equiv \tilde{h}[k + qN] W_{\Delta M}^{-pk} \tag{3.50}$$

$\vec{\mu}$ is the $\Delta M \Delta N$ dimensional vector given by

$$\vec{\mu} = \left(\frac{\Delta M}{N}, 0, 0, \ldots\right)^T \tag{3.51}$$

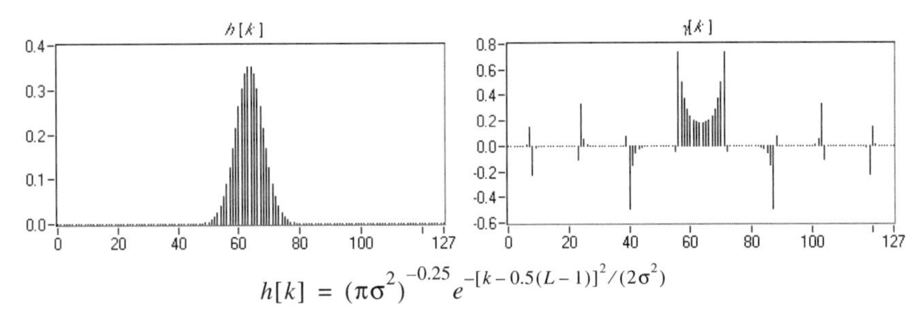

$$h[k] = (\pi\sigma^2)^{-0.25} e^{-[k - 0.5(L-1)]^2/(2\sigma^2)}$$

Figure 3-7 Discrete-time Gabor elementary function $h[k]$ and corresponding biorthogonal dual function $\gamma[k]$ at critical sampling. $L = 128$, $N = \Delta M = 16$, and $\sigma^2 = 128/2\pi$. Although the Gabor elementary function $h[k]$ is optimally concentrated in the joint time-frequency domain, the corresponding biorthogonal dual function $\gamma[k]$ is neither localized in time nor in frequency.

Eq. (3.49) shows that the dual function $\gamma[k]$ is nothing more than the solution of a linear system. The necessary and sufficient condition for the existence of the solution of (3.49) is that $\vec{\mu}$ is in the range of H. At the critical sampling case, $\Delta M \Delta N = L$, H is an L-by-L square matrix.

The Gabor elementary functions are linearly independent and form a basis. Consequently, the dual function $\gamma[k]$ is unique and biorthogonal to $h[k]$.

Figure 3-7 illustrates a discrete Gaussian window function $h[k]$ with critical sampling and corresponding dual function $\gamma[k]$. Note that although $h[k]$ (on the left) is concentrated in the joint time-frequency domain, the corresponding biorthogonal dual function $\gamma[k]$ (on the right) is concentrated neither in time nor in frequency.

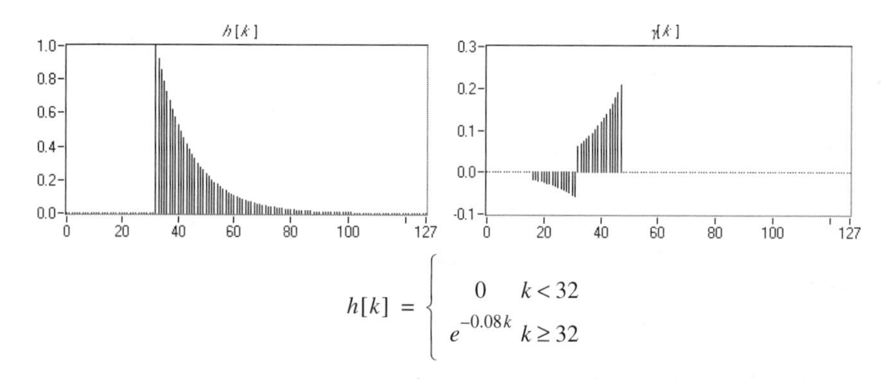

$$h[k] = \begin{cases} 0 & k < 32 \\ e^{-0.08k} & k \geq 32 \end{cases}$$

Figure 3-8 One-sided exponential function $h[k]$ and corresponding dual function $\gamma[k]$ at critical sampling. $N = \Delta M = 16$.

Figures 3-8 and 3-9 illustrate the one- and two-sided exponential sequences and their corresponding biorthogonal dual functions with critical sampling. Based on the algorithm described by (3.49), (3.50), and (3.51), we virtually can compute the dual function $\gamma[k]$ for an arbitrarily given Gabor elementary function $h[k]$. The limitations are that the signal $s[k]$, the function $h[k]$, and its dual function $\gamma[k]$ must be periodic and have the same length. For oversampling, $\Delta M \Delta N < L$, (3.49) is an underdetermined system and thereby the solution in general is not unique.

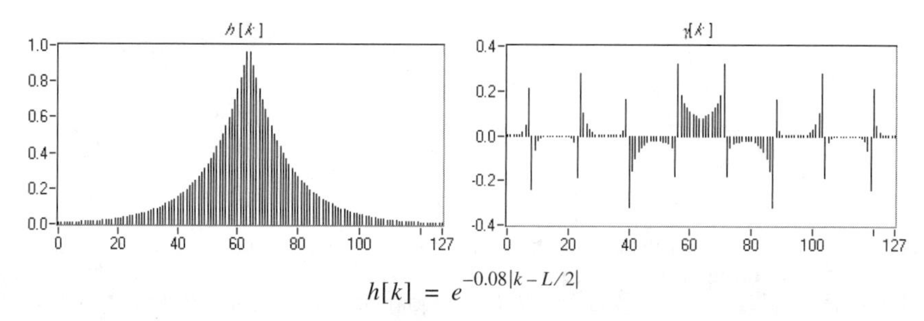

$$h[k] = e^{-0.08|k - L/2|}$$

Figure 3-9 Two-sided exponential function $h[k]$ and corresponding dual function $\gamma[k]$ at critical sampling. $N = \Delta M = 16$.

It is worth emphasizing that, unlike harmonically related complex sinusoidal functions used in the Fourier series, which are orthonormal, the set of desired Gabor elementary functions (with short time duration and narrow frequency bandwidth) does not constitute an orthogonal basis. In this case, the dual function $\gamma[k]$ will not be equal to the Gabor elementary function $h[k]$. The direct consequence is that although we could easily have the Gabor elementary functions optimally concentrated in the time-frequencies domain, the dual function $\gamma[k]$ may not be localized at all (as shown in Figure 3-7). Consequently, the Gabor coefficients $c_{m,n}$, the inner product of the signal and the dual functions, do not necessarily reflect the signal's behavior in the vicinity of $[mT-\Delta_t/2, mT+\Delta_t/2]\times[n\Omega-\Delta_\omega/2, n\Omega+\Delta_\omega/2]$. Whether or not the Gabor coefficients $c_{m,n}$ describe the signal's local behavior depends on the property of the dual function $\gamma[k]$. If $\gamma[k]$ is badly concentrated in the joint time-frequency domain, the Gabor coefficients $c_{m,n}$ will fail to describe the signal's local behavior. The two fundamental issues for a good time-frequency analysis scheme are:

- How to select the function $h[k]$
- How to compute the dual function $\gamma[k]$ if it is not unique

Over the years, many techniques have been successfully developed to implement the Gabor expansion, such as the Zak transform based method. Applying the Zak transform, similar to the convolution theorem (2.56), we can convert (3.77) and (3.78) into forms of products. By doing so, we may save considerable computation. However, such improvement is only significant when the signal and window lengths are closer. If the window length is much shorter than the signal, the computational improvement will be diminished due to the huge amount of zero-padding. Readers who are interested in the Zak transform based implementation can consult *Bastiaans* [75] and [76].

The main advantage of the method presented in this book is that for the given analysis or synthesis function, its dual function has the same length. This may be desired in hardware implementation. Moreover, as we shall introduce in Section 3.5, there is a very efficient method to compute a dual function.

3.4 Orthogonal-Like Gabor Expansion

As shown in the preceding section, given $h[k]$ and the sampling steps, the solution of $\gamma[k]$ in general is not unique. Then, the question is how to select the $\gamma[k]$ that best meets our goal. Recall that the Gabor coefficients $c_{m,n}$ are the sampled short-time Fourier transform with the window function $\gamma[k]$. This means that the window function $\gamma[k]$ has to be localized in the joint time-frequency domain. Otherwise, the Gabor coefficients $c_{m,n}$, the inner product of $s[k]$ and $\gamma[k]$, would not characterize the signal's local behavior.

Moreover, the behaviors of $\gamma[k]$ and $h[k]$, such as time/frequency centers and time/frequency resolution, have to be close. Suppose that $h[k]$ and its Fourier transform are centered at $k = 0$ and $\theta = 0$. If $\gamma[k]$ and $h[k]$ are significantly different — for instance, they have completely different time or frequency centers — then $c_{m,n}$ will not reflect the signal behavior in the vicinity of $(mT,n\Omega)$.

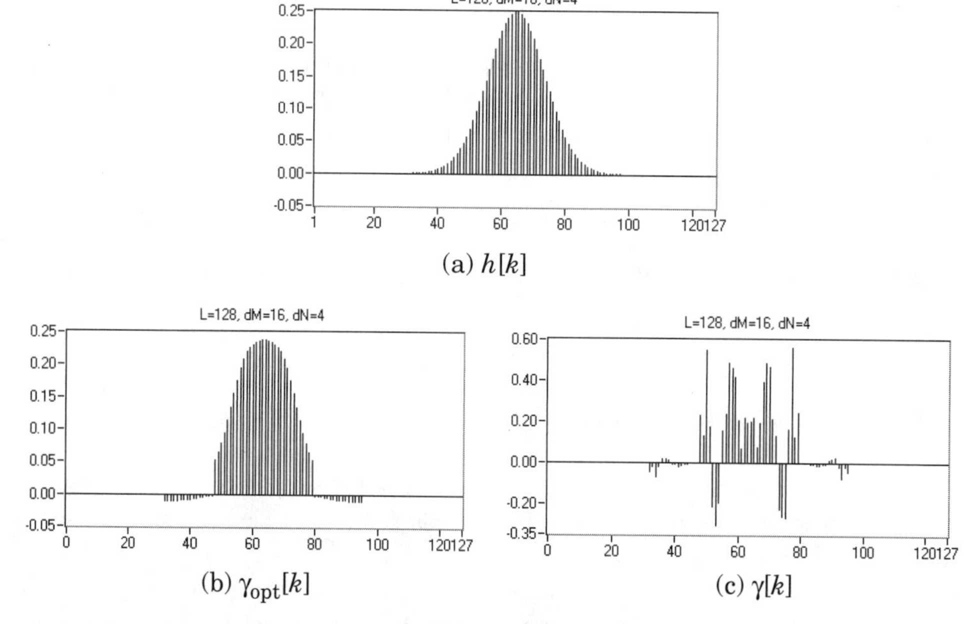

(a) $h[k]$

(b) $\gamma_{\mathrm{opt}}[k]$ (c) $\gamma[k]$

Figure 3-10 Parts (b) and (c) are dual functions of $h[k]$ in (a). They both yield perfect reconstruction. $\gamma_{\mathrm{opt}}[k]$ in (b) is optimally close to the Gaussian-type function $h[k]$ in (a).[9]

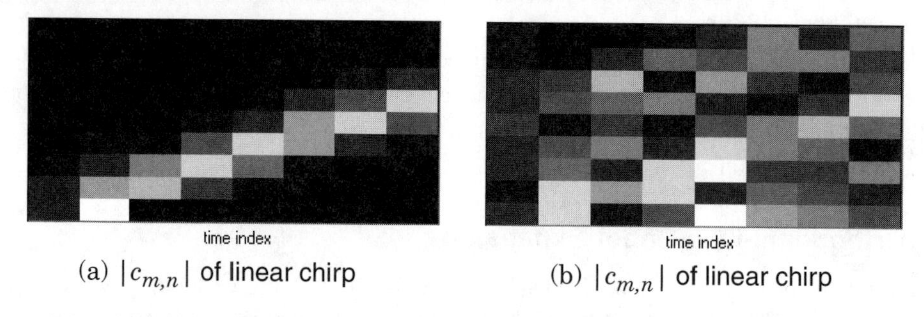

(a) $|c_{m,n}|$ of linear chirp (b) $|c_{m,n}|$ of linear chirp

Figure 3-11 (a) is computed by $\gamma_{\mathrm{opt}}[k]$, which well represents the linear chirp signal. (b) is computed by the $\gamma[k]$ plotted in Figure 3-10 (c), which does not provide the desired information of the linear chirp signal in the joint time-frequency domain, though both Gabor coefficients will lead to perfect reconstruction by using the same synthesis function $h[k]$.

Figure 3-10 (b) and (c) depict two different dual functions $\gamma[k]$ that both correspond to the same Gabor elementary function $h[k]$ in Figure 3-10 (a). The shape of $\gamma[k]$ in (b) is optimally close to that of $h[k]$, whereas the shape of $\gamma[k]$ in (c) significantly differs from that of $h[k]$.

9. The algorithm for computing different dual functions for the given function $h[k]$ is introduced in Appendix.

Although both $\gamma[k]$ will lead to perfect reconstruction with the same $h[k]$, the resulting Gabor coefficients are substantially different.

Figure 3-11 illustrates the magnitude of the Gabor coefficients computed by two different $\gamma[k]$ for the linear chirp signal. Figure 3-11(a) is computed by the optimal $\gamma[k]$. Because $h[k]$ in Figure 3-10(a) is concentrated in the joint time-frequency domain, the optimal $\gamma[k]$ is also concentrated. Consequently, the resulting Gabor coefficients well represent the monotone linear chirp signal. Figure 3-11(b) is computed by $\gamma[k]$ plotted in Figure 3-10(c), whose shape significantly differs from the Gabor elementary function $h[k]$ in Figure 3-10(a). Due to the bad shape of the analysis function, the resulting Gabor coefficients plotted in Figure 3-11(b) do not properly describe the time-varying nature of the linear chirp.

Because the Gabor elementary function $h[k]$ is the Gaussian function that is optimally concentrated in the joint time-frequency domain, the natural selection of $\gamma[k]$ is that whose shape is closest to $h[k]$, in the sense of the least mean square error (LMSE), i.e.,

$$\xi = \min_{H\vec{\gamma} = \vec{\mu}} \sum_{k=0}^{L-1} \left| \frac{\gamma[k]}{\|\vec{\gamma}\|} - h[k] \right|^2 \tag{3.52}$$

where

$$\|\vec{\gamma}\| = \sqrt{\sum_{k=0}^{L-1} |\gamma[k]|^2} \tag{3.53}$$

$h[k]$ is a normalized function; that is, $\|h[k]\|^2 = 1$. When ξ in (3.52) is small, $\gamma[k] \approx \alpha h[k]$ where a denotes a real-valued constant. In this case, the Gabor expansion can be written as

$$c_{m,n} \approx \alpha \sum_{k=0}^{L-1} s[k] h^*[k - m\Delta M] W_N^{-nk} \tag{3.54}$$

and

$$s[k] = \sum_{m=0}^{M-1} \sum_{n=0}^{N-1} c_{m,n} h[k - m\Delta M] W_N^{nk} \tag{3.55}$$

Obviously, as long as $h[k]$ is localized in the joint time-frequency domain, the Gabor coefficients $c_{m,n}$ will well depict the signal's local time-frequency properties. Although the set of $\{h_{m,n}[k]\}$ is not orthogonal and redundant, the coefficients $c_{m,n}$ are still good approximations of the orthogonal projections of the signal on $\{h_{m,n}[k]\}$. Therefore, (3.55) is called an *orthogonal-like Gabor expansion.*[10] In what follows we will derive the solution of (3.52).

10. Strictly speaking, it should be named *near tight frame*.

Expanding Eq. (3.52), one obtains [170]

$$\xi = \min_{H\vec{\gamma} = \vec{\mu}} \sum_{k=0}^{L-1} \left\{ \frac{\gamma^2[k]}{\|\vec{\gamma}\|} - 2\frac{h[k]\gamma[k]}{\|\vec{\gamma}\|} + h^2[k] \right\} \tag{3.56}$$

$$= \min_{H\vec{\gamma} = \vec{\mu}} \left\{ 2 - \frac{2}{\|\vec{\gamma}\|} \sum_{k=0}^{L-1} h[k]\gamma[k] \right\}$$

Note that the summation in (3.56) is a special case of (3.48) with $p = q = 0$. Hence, (3.56) can be rewritten as

$$\xi = \min_{H\vec{\gamma} = \vec{\mu}} \left(1 - \frac{1}{\|\vec{\gamma}\|} \frac{\Delta M}{N} \right) \tag{3.57}$$

which implies that the solution is the minimum energy of $\gamma[k]$. According to matrix analysis theory, $\gamma[k]$ exists as long as the vector $\vec{\mu}$ is in the range of matrix H. If matrix H is of full-row rank, the minimum energy of $\gamma[k]$ is equal to the pseudo inverse of H, i.e.,

$$\vec{\gamma*}_{opt} = H^T (HH^T)^{-1} \vec{\mu} \tag{3.58}$$

To faithfully characterize a signal's local properties in the joint time-frequency domain, it is critical to make $\gamma_{opt}[k]$ and $h[k]$ as close as possible. It has been discovered that the minimum difference between $\gamma_{opt}[k]$ and $h[k]$, ξ in (3.52), is related to the oversampling rate as well as the shape of the sampling grid. Generally speaking, the difference between $h[k]$ and $\gamma_{opt}[k]$ decreases as the oversampling rate increases. If the Gaussian function is used — for instance,

$$h[k] = \sqrt[4]{\frac{\alpha}{\pi}} e^{-\alpha k^2/2} \tag{3.59}$$

the minimum difference between $\gamma_{opt}[k]$ and $h[k]$ is observed when

$$\alpha_{opt} = \frac{2\pi}{\Delta MN} = \frac{2\pi}{L} \frac{\Delta N}{\Delta M} \tag{3.60}$$

which is proportional to the ratio between the frequency sampling step ΔN and the time sampling step ΔM. Moreover, it is also observed that the difference between $\gamma_{opt}[k]$ and $h[k]$ is related to the shape of the sampling lattice. For the same variance α_{opt}, ξ in (3.52) for the kite sampling lattice is smaller than that for the rectangle sampling lattice [161]. This is because the joint time-frequency distribution of the Gaussian function (3.59) consists of concentric ellipses (see Figures 3-2 and 7-1) that better fit the kite than the rectangle.

Figure 3-12 depicts the Gaussian function and corresponding dual functions at different sampling schemes. The discrete Gaussian function has the form

$$h[k] = \sqrt[4]{\frac{\alpha}{\pi}} e^{-\alpha(k - 0.5(L-1))^2/2} \qquad \alpha = \frac{2\pi}{N\Delta M} \tag{3.61}$$

In order to utilize the FFT (fast Fourier transform), the number of frequency bins (or channels) N is usually required to be equal to a power of two. For instance, $N = 2^k$, $k = 1, 2, 3,...$ The over-sampling rate is determined by $a = N/\Delta M$, where ΔM denotes the time-sampling step. The length of $h[k]$ and $\gamma_{opt}[k]$, L, has to be divided by both N and ΔM. That is, L/N and $L/\Delta M$ have to be integers.

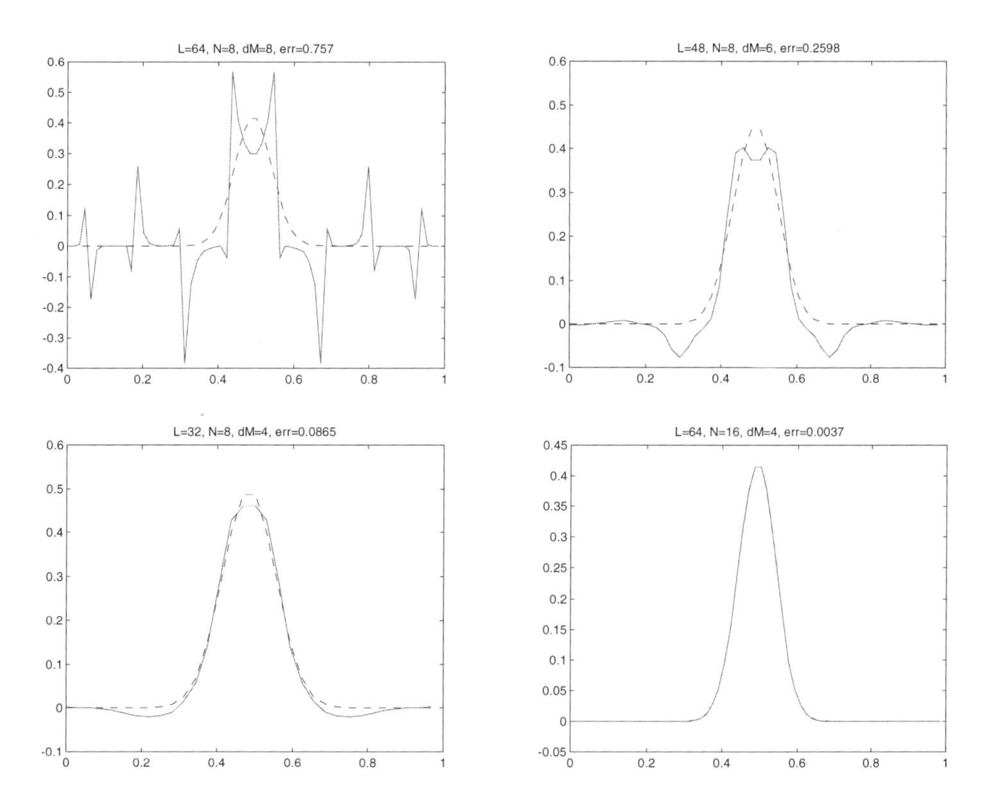

Figure 3-12 $h[k]$ (dashed line) vs. $\gamma_{opt}[k]$ (solid line). As the oversampling ratio N/dM increases, the dual functions get more and more close. When the oversampling is equal to four, the difference between $h[k]$ and $\gamma_{opt}[k]$ is almost negligible.

The discrete Gabor expansion and the concept of orthogonal-like Gabor expansion were first proposed by *Qian* et al ([194] and [195]). In addition to seeking $\gamma[k]$ that is optimally close to $h[k]$, the algorithm introduced in this section could be easily modified to have $\gamma[k]$ optimally close to an arbitrary function, i.e.,

$$\xi = \min_{H\vec{\gamma} = \vec{\mu}} \sum_{k=0}^{L-1} \left| \frac{\gamma[k]}{\|\vec{\gamma}\|} - d[k] \right|^2 \tag{3.62}$$

where $d[k]$ is a normalized target function. Such generalized optimization is very useful, in particular, for those applications where the analysis and synthesis functions are desired to have different properties. For instance, we could use the Gabor expansion for filter bank design. In many filter bank applications, the analysis and synthesis functions usually have different requirements. In this case, while one filter can be predetermined by $h[k]$, the other could be solved by (3.62). The general solution of (3.62) is discussed in Appendix.

3.5 A Fast Algorithm for Computing Dual Functions

Since the discrete Gabor expansion was developed, the design of dual functions, as introduced in Sections 3.3 and 3.4, is no longer a big problem. By using Eq. (3.49), essentially we could obtain a dual function $\gamma[k]$ for any given function $h[k]$ at the sampling steps, ΔM and ΔN, as long as the oversampling rate is larger than or equal to one. Based on (3.49), however, the dual function $\gamma[k]$ is the solution of the linear complex system, which is inefficient in terms of computation time and memory space. In this section, we will show that those complex matrices, in fact, could be converted into real-valued matrices and, thereby, save a considerable amount of computation [170].

Given the finite Gabor elementary function $h[k]$, the dual function $\gamma[k]$ for the periodic Gabor expansion can be computed through Eq. (3.48)

$$\sum_{k=0}^{L-1} \tilde{h}[k+qN] W_{\Delta M}^{-pk} \gamma^*[k] = \frac{\Delta M}{N} \delta[p]\delta[q] \tag{3.63}$$

where $0 \le p < \Delta M$ and $0 \le q < \Delta N$. Therefore, computing $\gamma[k]$ requires solving the linear complex system with an $\Delta M \Delta N$-by-L complex matrix H.

Let's take the inverse DFT with respect to (3.63), i.e.,

$$\sum_{k=0}^{L-1} \tilde{h}[k+qN]\gamma^*[k] \frac{1}{\Delta M} \sum_{p=0}^{\Delta M-1} W_{\Delta M}^{-p(k-k')} = \frac{\Delta M}{N} \delta[q] \frac{1}{\Delta M} \sum_{p=0}^{\Delta M-1} \delta[p] W_{\Delta M}^{-pk'} \tag{3.64}$$

Because

$$\frac{1}{\Delta M} \sum_{p=0}^{\Delta M-1} W_{\Delta M}^{-p(k-k')} = \delta(k-k'-l\Delta M) \tag{3.65}$$

Eq. (3.64) reduces to ΔM sub-systems,[11] i.e.,

$$\sum_{p=0}^{M-1} \tilde{h}[k+p\Delta M+qN]\gamma^*[k+p\Delta M] = \frac{1}{\Delta M}\delta[q] \tag{3.66}$$

11. Long before (3.66) was discovered, Portnoff had obtained a similar relationship [191]. However, unlike (3.66), which is finite summation, Portnoff's formula involves infinite summation. Hence, the solution of Portnoff's formula is much more involved than (3.66).

where $0 \le k < \Delta M$ and $0 \le q < \Delta N$. Eq. (3.66) suggests that $\gamma[k]$ can be computed by ΔM-separated real-valued linear systems, i.e.,

$$H_k \overrightarrow{\gamma^*}_k = \vec{\mu}_k \tag{3.67}$$

where H_k are ΔN-by-M matrices with the entries:

$$h_{q,p}(k) \equiv \tilde{h}[k + p\Delta M + qN] \tag{3.68}$$

and $\vec{\gamma}_k$ is an M-dimensional vector with the entries:

$$\gamma_p(k) = \gamma[k + p\Delta M] \tag{3.69}$$

and

$$\vec{\mu}_k = (\Delta M^{-1}, 0, 0\ldots)^T \tag{3.70}$$

with ΔN elements. While (3.49) is a $\Delta M \Delta N$-by-L linear *complex* system, the solution of (3.67) is a ΔM independent ΔN-by-M linear *real* system. Consequently, by using (3.67) we can save considerable computation. Moreover, because $\vec{\gamma}_k$ are independent to each other, the minimum norm of (3.49) can be computed by the minimum norms of each individual linear system in (3.67).

3.6 Discrete Gabor Expansion

For the periodic discrete Gabor expansion, the signal, analysis function, and synthesis function must have an identical length. This is rather inconvenient (even impractical) in many applications. When the number of data samples is large, to solve the dual function requires tremendous computation time and memory without mentioning numerical instability. It is desirable that the lengths of the analysis and synthesis functions are independent of the length of the samples so that we can use short windows to process arbitrarily long data. When this is achieved, the discrete Gabor expansion can then be used in many more signal processing applications where typical signals are long.

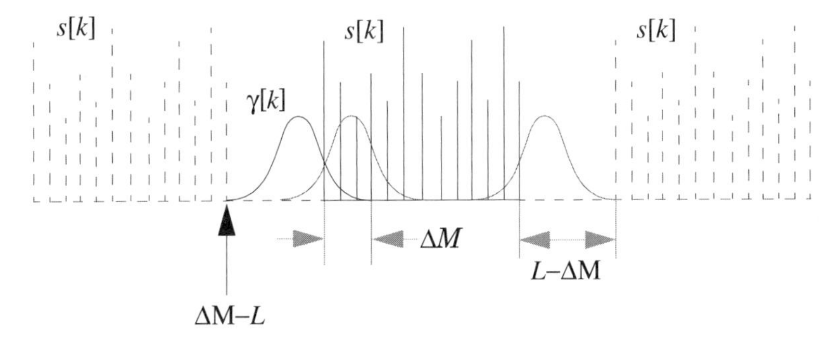

Figure 3-13 Due to zero padding, the Gabor transform can be computed without rolling over either the signal $s[k]$ or the analysis function $\gamma[k]$. The auxiliary periodic sequences actually release the periodic constraint.

Assume that the length of the signal $s[k]$ is L_s and the lengths of the Gabor elementary function $h[k]$ and the dual function $\gamma[k]$ are L. Let's build auxiliary periodic sequences as

$$\hat{s}[k] = \hat{s}[k+iL_0] = \begin{cases} 0 & -(L-\Delta M) \le k < 0 \\ s[k] & 0 \le k < L_s \end{cases} \tag{3.71}$$

$$\hat{h}[k] = \hat{h}[k+iL_0] = \begin{cases} h[k] & 0 \le k < L \\ 0 & L \le k < (L_0 - \Delta M) \end{cases} \tag{3.72}$$

$$\hat{\gamma}[k] = \hat{\gamma}[k+iL_0] = \begin{cases} \gamma[k] & 0 \le k < L \\ 0 & L \le k < (L_0 - \Delta M) \end{cases} \tag{3.73}$$

where all have the same period $L_0 = L_s + L - \Delta M$. The periodic discrete Gabor transform can then be plotted as in Figure 3-13.

Because of zero padding, we can compute the Gabor transform without rolling over either the signal $s[k]$ or the analysis function $\gamma[k]$. The auxiliary periodic sequences defined in (3.71), (3.72), and (3.73) actually release the periodic constraint. Substituting auxiliary periodic sequences into the form of the periodic discrete Gabor expansion (3.40) and (3.41) yields

$$\tilde{c}_{m,n} = \sum_{k=\Delta M - L}^{L_s - 1} \hat{s}[k]\hat{\gamma}^*[k - m\Delta M]W_N^{-nk} \tag{3.74}$$

and

$$\hat{s}[k] = \sum_{m=m_0}^{M-1} \sum_{n=0}^{N-1} \tilde{c}_{m,n}\hat{h}[k - m\Delta M]W_N^{nk} \tag{3.75}$$

where

$$m_0 = \frac{\Delta M - L}{\Delta M} \tag{3.76}$$

The oversampling rate is $a = N/\Delta M$. For perfect reconstruction, the time sampling step ΔM has to be less than or equal to the number of frequency channels N. The total number of time sampling points is the smallest integer that is larger than or equal to $L_0/\Delta M$. Because of zero padding, the oversampling rate is equal to the ratio between the number of Gabor coefficients and the number of zero padded samples $s[k]$.

With L remaining finite, letting $L_s \to \infty$ and thereby $L_0 \to \infty$, (3.74) and (3.75) directly lead to the Gabor expansion pair for discrete-time infinite sequences, i.e.,

$$\tilde{c}_{m,n} = \sum_{k=0}^{\infty} s[k]\gamma^*[k - m\Delta M]W_N^{-nk} \tag{3.77}$$

and

$$s[k] = \sum_{m=m_0}^{\infty} \sum_{n=0}^{N-1} \tilde{c}_{m,n} h[k - m\Delta M] W_N^{nk} . \tag{3.78}$$

The remaining question is how to compute the dual function $\gamma[k]$. Substituting (3.72) and (3.73) into (3.46) and writing the resulting formula in the form of matrices, we have

$$H\overrightarrow{\gamma_0}^* = \overrightarrow{\mu_0} \tag{3.79}$$

where $\vec{\gamma_0}$ is an L_0-by-1 vector and

$$\overrightarrow{\mu_0} = \left(\frac{\Delta M}{N}, 0, 0, \ldots\right)^T \tag{3.80}$$

H is a $\Delta M L_0/N$-by-L_0 matrix that can be written as

$$\begin{vmatrix} H_0 & H_1 & \ldots & H_{\Delta N-1} & 0 & \ldots & 0 & 0 \\ H_1 & \ldots & H_{\Delta N-1} & 0 & 0 & \ldots & 0 & H_0 \\ \ldots & H_{\Delta N-1} & 0 & 0 & \ldots & \ldots & H_0 & H_1 \\ H_{\Delta N-1} & 0 & 0 & 0 & 0 & \ldots & H_1 & \ldots \\ & \ldots & & \ldots & & \ldots & & \ldots \\ 0 & 0 & 0 & H_0 & H_1 & \ldots & H_{\Delta N-1} & 0 \\ 0 & 0 & H_0 & H_1 & \ldots & \ldots & 0 & 0 \\ 0 & H_0 & H_1 & \ldots & H_{\Delta N-1} & \ldots & 0 & 0 \end{vmatrix} \tag{3.81}$$

where H_i are ΔM-*by*-N block matrices whose entries $h_{p,k}(i)$ are

$$h_{p,k}(i) = h_{p,k}(q+l) \equiv \hat{h}[(q+l)N + k] W_{\Delta M}^{-pk} \tag{3.82}$$

Because $\Delta NN = L$, $h_{p,k}(i) = 0$ for $i = (p+l) \geq \Delta N$. That is, $H_i = 0$ for $i \geq \Delta N$. In order for the dual functions $\gamma[k]$ and $h[k]$ to have the same time support, we force the last $L_s - \Delta M$ elements of the vector $\vec{\gamma_0}$ to be zero; that is,

$$\vec{\gamma}_0 = \begin{vmatrix} \vec{\gamma} \\ \vec{0} \end{vmatrix} \qquad where\ \vec{0} = (0, 0, 0 \ldots 0)^T \tag{3.83}$$

$\vec{\gamma}$ and $\vec{0}$ are L-dimensional and $(L_s$-$\Delta M)$-dimensional vectors, respectively. Replacing $\vec{\gamma}_0$ in (3.79) by (3.83), (3.79) reduces to

$$\begin{vmatrix} H_0 & H_1 & \dots & H_{\Delta N-1} \\ H_1 & \dots & H_{\Delta N-1} & 0 \\ \dots & H_{\Delta N-1} & 0 & \dots \\ H_{\Delta N-1} & 0 & \dots & 0 \\ 0 & \dots & 0 & H_0 \\ \dots & & \dots & \\ 0 & H_0 & \dots & H_{\Delta N-2} \end{vmatrix} \bar{\gamma}^* = \overline{H}\bar{\gamma}^* = \bar{\mu} \tag{3.84}$$

If we define an auxiliary periodic sequence as

$$\bar{h}[k] = \bar{h}[k + l(2L - N)] \equiv \begin{cases} h[k] \; 0 \le k < L \\ 0 \quad L \le k < 2L - N \end{cases} \tag{3.85}$$

then the entries of $\overline{H}$ can be determined in the same manner as in the case of the periodic discrete Gabor expansion (3.50), i.e.,

$$\bar{h}_{p\Delta M + q,\, k} \equiv \bar{h}[k + qN] W_{\Delta M}^{-pk} \tag{3.86}$$

Consequently, (3.84) can be written as

$$\sum_{k=0}^{L-1} \bar{h}[k + qN] W_{\Delta M}^{-pk} \gamma^*[k] = \frac{\Delta M}{N} \delta[p]\delta[q] \tag{3.87}$$

where $0 \le p < \Delta M$ and $0 \le q < 2\Delta N - 1$. The significance of Eq. (3.87) is that it is independent of the signal length. It guarantees that the dual functions, $h[k]$ and $\gamma[k]$, have the same time support.

$\overline{H}$ in (3.85) is a K-by-L matrix, where

$$K = \Delta M(2\Delta N - 1) = 2L\frac{\Delta M}{N} - \Delta M = \frac{2L}{a} - \Delta M \tag{3.88}$$

where a denotes the oversampling rate. Therefore, (3.85) is an underdetermined system when

$$\frac{2L}{a} - \Delta M < L \tag{3.89}$$

That is, a is larger than $2L/(L+\Delta M)$.

It is interesting to note that the formula for computing the dual functions (3.87) for the discrete Gabor expansion and (3.48) for the periodic discrete Gabor expansion are identical except for the sequences $\bar{h}[k]$ and $\tilde{h}[k]$. While $\tilde{h}[k]$ is made up of $h[k]$ directly, $\bar{h}[k]$ is the zero padded window function $h[k]$ as shown by (3.85). Hence, the results obtained in Sections 3.4 and 3.5, with minor modifications, can be easily extended to the discrete Gabor expansion. For example, for the discrete Gabor expansion the size of the matrices H_k in (3.67) is $(2\Delta N-1)$-by-M and $0 \le q < 2\Delta N-1$. Consequently, the solution of the dual function for the discrete Gabor expansion is a ΔM independent $(2\Delta N-1)$-by-M linear *real* system.

For a given window function $h[i]$, with $0 \le i < L$, a general form of computing the dual function can be summarized as

$$H_k \overrightarrow{\gamma^*}_k = \vec{\mu}_k \qquad 0 \le k < \Delta M \tag{3.90}$$

where H_k are Δn-by-M matrices, $\Delta n = L_a/N$, and

$$L_a = \begin{cases} L & periodic \\ 2L - N & nonperiodic \end{cases} \tag{3.91}$$

The entries of H_k are defined as

$$h_k(q, p) \equiv \tilde{a}[k + p\Delta M + qN] \qquad 0 \le p < \frac{L}{\Delta M} \qquad 0 \le q < \frac{L_a}{N} \tag{3.92}$$

where $\tilde{a}[k]$ denotes a periodic auxiliary function. For the periodic discrete Gabor expansion,

$$\tilde{a}[i + nL_a] = h[i] \qquad n = 0, \pm 1, \pm 2 \ldots \qquad L_a = L \tag{3.93}$$

For a non-periodic discrete Gabor expansion,

$$\tilde{a}[i + nL_a] = \begin{cases} h[i] & 0 \le i < L \\ 0 & L \le i < L_a \end{cases} \qquad n = 0, \pm 1, \pm 2 \ldots \qquad L_a = 2L - N \tag{3.94}$$

$\vec{\gamma}_k$ in Eq. (3.90) are M-dimensional vectors with the entries:

$$\gamma_p(k) = \gamma[k + p\Delta M] \tag{3.95}$$

and $\vec{\mu}_k$ are Δn-dimensional vectors,

$$\vec{\mu}_k = (\Delta M^{-1}, 0, 0 \ldots)^T. \tag{3.96}$$

Hence, by a proper auxiliary function $\tilde{a}[k]$, we can use the uniform linear system Eq. (3.90) solve for the dual functions for both periodic or non-periodic discrete Gabor transforms. Because the solutions $\vec{\gamma}_k$ of Eq. (3.90) are independent of each other, the minimum distance between $\vec{\gamma}$ and $\vec{h}$ can be solved by the pseudo inverse of each individual linear system in (3.90).

Example 3.3 Compute the dual functions for a given normalized sequence

$$\vec{h} = \frac{1}{\sqrt{60}}(1, 2, 3, 4, 4, 3, 2, 1) \tag{3.97}$$

The number of frequency channels is $N = 4$ and the time sampling step is $\Delta M = 2$. Hence, the oversampling rate is 2. The dual functions can be solved by the $\Delta M = 2$ independent linear systems described by (3.67) to (3.69), where $k = 0, 1$ and $p = 0, 1, 2, 3$. For the periodic discrete Gabor expansion, $q = 0, 1$. We have

$$\frac{1}{\sqrt{60}} \begin{bmatrix} 1 & 3 & 4 & 2 \\ 4 & 2 & 1 & 3 \end{bmatrix} \begin{bmatrix} \gamma[0] \\ \gamma[2] \\ \gamma[4] \\ \gamma[6] \end{bmatrix} = \begin{bmatrix} \frac{1}{4} \\ 0 \end{bmatrix} \tag{3.98}$$

and

$$\frac{1}{\sqrt{60}}\begin{bmatrix} 2 & 4 & 3 & 1 \\ 3 & 1 & 2 & 4 \end{bmatrix}\begin{bmatrix} \gamma[1] \\ \gamma[3] \\ \gamma[5] \\ \gamma[7] \end{bmatrix} = \begin{bmatrix} \frac{1}{4} \\ 0 \end{bmatrix} \tag{3.99}$$

Then, the optimal dual function γ_{opt} is the LMS solution of (3.98) and (3.99). For the non-periodic discrete Gabor expansion, $q = 0, 1, 2$ and the auxiliary function is

$$\bar{h} = \frac{1}{\sqrt{60}}(1, 2, 3, 4, 4, 3, 2, 1, 0, 0, 0, 0) \tag{3.100}$$

Consequently, for $k = 0$,

$$\frac{1}{\sqrt{60}}\begin{bmatrix} 1 & 3 & 4 & 2 \\ 4 & 2 & 0 & 0 \\ 0 & 0 & 1 & 3 \end{bmatrix}\begin{bmatrix} \gamma[0] \\ \gamma[2] \\ \gamma[4] \\ \gamma[6] \end{bmatrix} = \begin{bmatrix} \frac{1}{4} \\ 0 \\ 0 \end{bmatrix} \tag{3.101}$$

and $k = 1$,

$$\frac{1}{\sqrt{60}}\begin{bmatrix} 2 & 4 & 3 & 1 \\ 3 & 1 & 0 & 0 \\ 0 & 0 & 2 & 4 \end{bmatrix}\begin{bmatrix} \gamma[1] \\ \gamma[3] \\ \gamma[5] \\ \gamma[7] \end{bmatrix} = \begin{bmatrix} \frac{1}{4} \\ 0 \\ 0 \end{bmatrix} \tag{3.102}$$

Then, the optimal dual function γ_{opt} is the LMS solution of (3.101) and (3.102).

Note that $\gamma[k]$ derived for the discrete Gabor expansion is a subset of $\vec{\gamma}_0$ in (3.83), which is a special solution of the periodic discrete Gabor expansion introduced in Section 3.3. Because we force the last L_s–ΔM elements of the vector $\vec{\gamma}_0$ to be zero, the existence of $\gamma[k]$ is much more restricted than that in periodic cases. Figure 3-14 illustrates dual functions for an identically given function. There is less freedom to select a dual function for the discrete Gabor expansion than for the periodic Gabor expansion. Hence, for the given length and sampling scheme the dual functions of the periodic Gabor expansion in Figure 3-14(a) are more similar than their counterparts in Figure 3-14(b). However, for the periodic Gabor expansion, the length of the signal has to be equal to the length of the windows, which is rather inconvenient if the number of samples is large.

In what follows, we shall further investigate the existence of the dual function at the critical sampling, $\Delta M = N$. In most cases, we only give results without derivations. The reader can find a rigorous mathematical treatment in [38].

For clarity of presentation, let's define the operation $\otimes$ by

$$e_0 \otimes e_1 \otimes e_2 \otimes \ldots e_m = \begin{cases} \textit{nonzero iff one term is not zero} \\ 0 \qquad \textit{otherwise} \end{cases} \tag{3.103}$$

If a set of numbers $\{e_0, e_1, e_2, ..., e_m\}$ satisfies the condition

$$e_0 \otimes e_1 \otimes e_2 \otimes \ldots \otimes e_m \neq 0 \tag{3.104}$$

then we call this set of numbers *exclusively non-zero.*

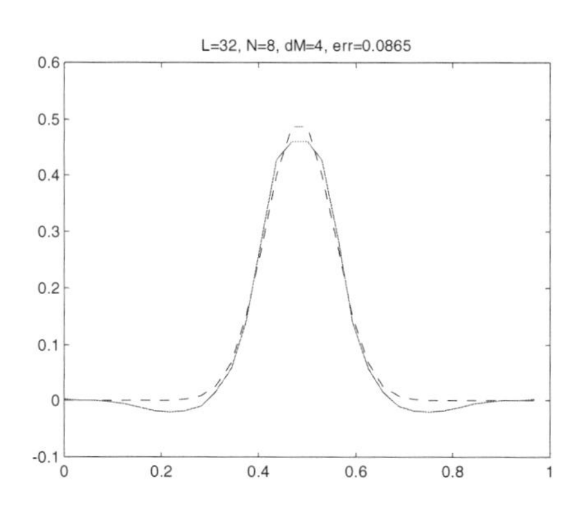

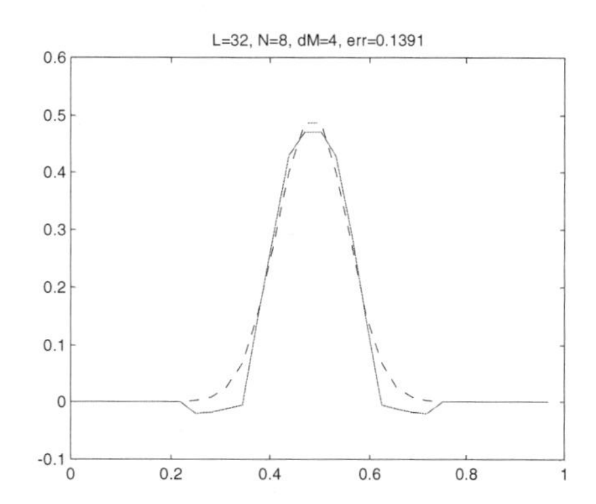

(a) For periodic Gabor expansion (b) For discrete Gabor expansion

Figure 3-14 There is less freedom to select a dual function for the discrete Gabor expansion than for the periodic Gabor expansion. Hence, for the given length and sampling scheme the dual functions of the periodic Gabor expansion (a) are more similar than their counterpart in (b). However, for the periodic Gabor expansion, the length of the signal has to be equal to the length of the windows, which is rather inconvenient if the number of samples is large.

Now, we state that for critical sampling, the biorthogonal dual function $\gamma[k]$ of the discrete Gabor expansion exists iff

$$h[k] \otimes h[N+k] \otimes h[2N+k] \otimes \ldots \otimes h[(\Delta N - 1)N + k] \neq 0 \tag{3.105}$$

where $0 \le k < N$ and $\Delta NN = L$. If $h[k]$ satisfies Eq. (3.105), then $\gamma[k]$ is uniquely determined by

$$\gamma[mN+k] = \begin{cases} \dfrac{1}{Nh[mN+k]} & h[mN+k] \neq 0 \\ 0 & otherwise \end{cases} \tag{3.106}$$

where $0 \le m < \Delta N$ and $0 \le k < N$.

Figure 3-15 depicts the locations of $h[mN+k]$. Eq. (3.105) implies that $h[mN+k]$ can only contain N non-zeros. Moreover, the non-zero point can only be one k for all different m. When $0 \le m < \Delta N = 1$ — that is, $N = L$ (the number of frequency channels is equal to the length of the function $h[k]$) — then the necessary and sufficient condition for the existence of the dual function is simply

$$h[k] \neq 0 \qquad \forall\, k \in [0, L) \tag{3.107}$$

which is illustrated in Figure 3-16. In fact, this is exactly the case for the non-overlapping windowed Fourier transform.

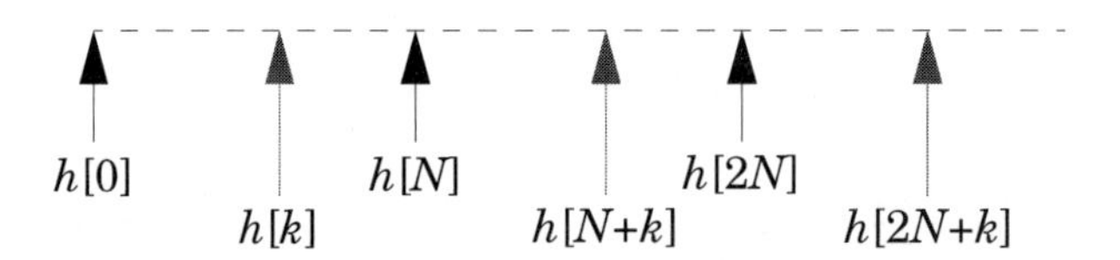

Figure 3-15 The locations of $h[mN+k]$ at the critical sampling, where $0 \le k < N$ and $0 \le m < \Delta N$

Figure 3-17 plots $h[k]$ and $\gamma[k]$ for $\Delta M = N = L/2 = 64$. In this case, the non-zero points are $h[32]$ to $h[63]$ and $h[64]$ to $h[64+31]$, which obviously satisfies the conditions described by (3.105) and (3.106).

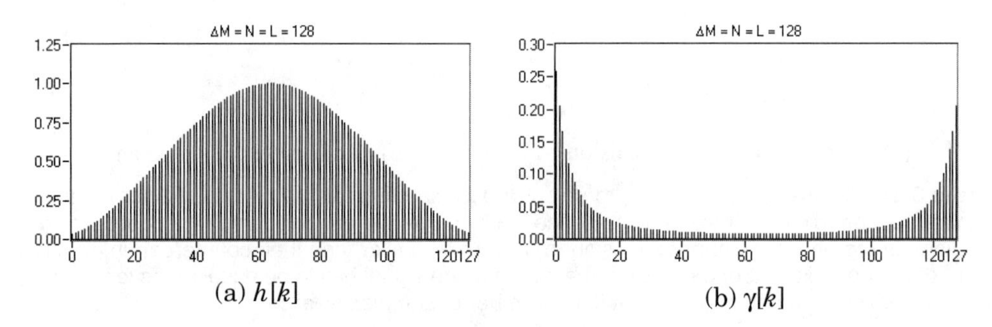

Figure 3-16 Biorthogonal dual sequences at the critical sampling ($\Delta M = N = L$)

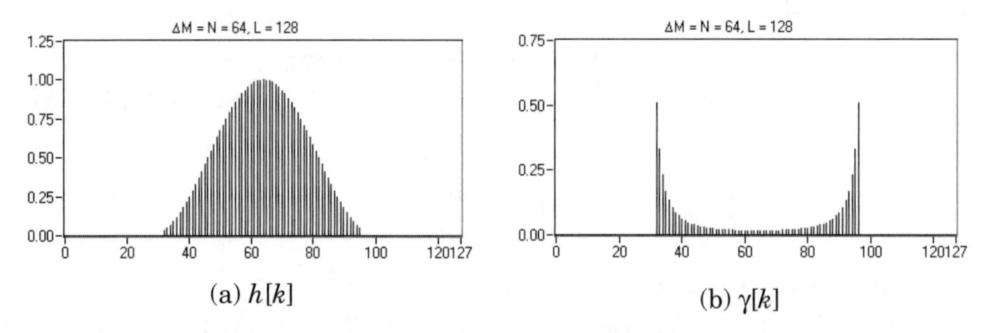

Figure 3-17 Biorthogonal dual sequences at the critical sampling ($\Delta M = N = L/2$)

CHAPTER 4

Linear Time-Variant Filters

For time-invariant filters, the filters' coefficients are independent of time. Hence, they can only be used for signals whose frequency contents do not evolve over time. One of the most important applications of time-frequency representations is for time-varying filtering. Because the frequency response of time-varying filters changes with time, these filters are usually more powerful than their time-invariant counterparts. Before going into any technical details, we want to show two simple examples regarding the advantages of using time-varying filters.

The first example is an application for time-varying harmonic analysis. Here, the term harmonic refers to frequencies that are integer (or fractional) multiples of a fundamental frequency. A simple example is the vibration observed from rotating machinery, such as the sound generated by a running engine. Although the causes of vibration are very different (some may be associated with bearings and others may be associated with the cooling fan), the vibration frequencies are all functions of a fundamental frequency — the engine rotation speed. For instance, the vibration frequency related to the bearing may be equal to the fundamental frequency multiplied by the number of balls inside the bearing housing. The vibration frequency related to the fan may be equal to the fundamental frequency multiplied by the number of blades. By applying the Fourier transformation to the time waveform of the vibration signal, we will obtain a group of harmonics in the frequency domain. Because all these harmonics have explicitly physical interpretations, by analyzing the amplitudes and phases of different harmonics, engineers can often determine whether the engine is running normally.

Figure 4-1 illustrates a sound waveform recorded from an electric motor running at a constant speed. Intuitively, not only will the motor rotation generate the sound, but also all the other parts that vibrate due to the motor rotation will make noise. The sound waveform plotted at the bottom, in fact, is a combination of all kinds of vibrations caused by the rotation of the motor.

Moreover, the vibration frequencies are multiples of the fundamental frequency – the motor rotation speed. In addition to the sound waveform, the bottom plot also depicts the tachometer pulses. Every two tachometer pulses indicate one revolution. The plot on the left shows the Fourier transform-based power spectrum. The plot in the middle is the magnitude of the corresponding Gabor coefficients. Since the electric motor contains four coils, we expect to observe harmonics with frequencies at four, eight, and twelve times the rotational speed. In addition, its cooling fan has seven blades. So, we will also expect to observe harmonics at seven, fourteen, and twenty-one times the rotational speed. As shown in Figure 4-1, when the rotational speed is constant, we can clearly see the fourth and seventh orders from both the Fourier transform-based power spectrum and the Gabor coefficients (in which they appear as horizontal white lines).

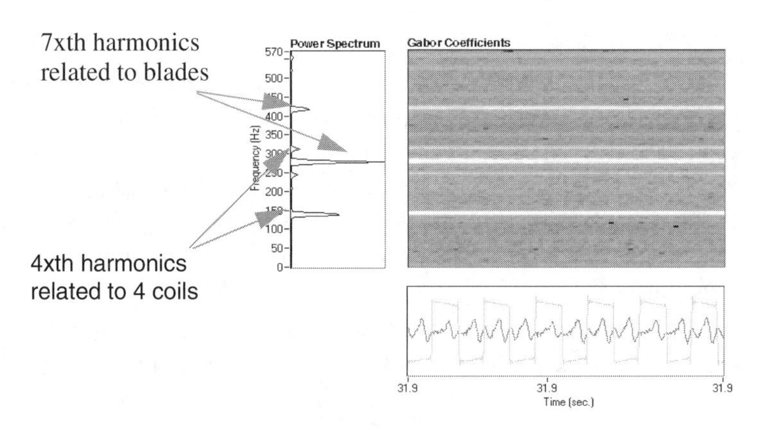

Figure 4-1 The physical characteristics of rotating machines can often be reflected in the harmonics. This figure shows the signal obtained from a four-phase electric motor with a seven-blade cooling fan, running at constant speed. While the 4xth harmonics correspond to vibrations related to the number of coils, the 7xth harmonics are related to vibrations associated with the number of blades of the cooling fan. At a constant speed, all these harmonics can be well identified from both the conventional power spectrum and the joint time-frequency plot.

However, the Fourier transform-based classical harmonic analysis only works when the fundamental frequency is constant (e.g., a motor running at a constant speed). It is not suitable for harmonics when the fundamental frequency evolves over time (e.g., motor speed changes). As discussed in Section 2.4, when the frequency changes, the frequency bandwidth becomes wide (see Figure 4-2). The frequency bandwidth is proportional to the rate at which the frequency changes (see Eq. (2.80)). The faster the change of frequency, the wider the corresponding frequency bandwidth. As the fundamental frequency bandwidth becomes wide, the bandwidth corresponding to the harmonics will become wide too. Finally, the harmonics will overlap in the frequency domain.

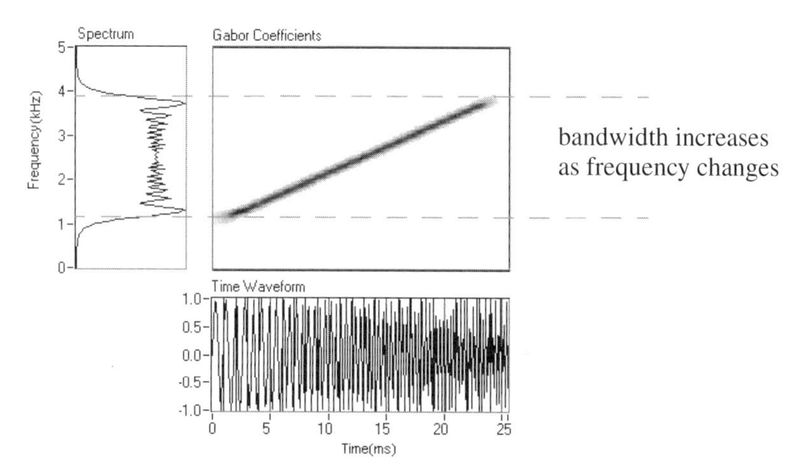

Figure 4-2 Frequency bandwidth increases as the frequency changes. The frequency bandwidth is proportional to the rate of frequency change. The faster the frequency change, the wider the corresponding frequency bandwidth.

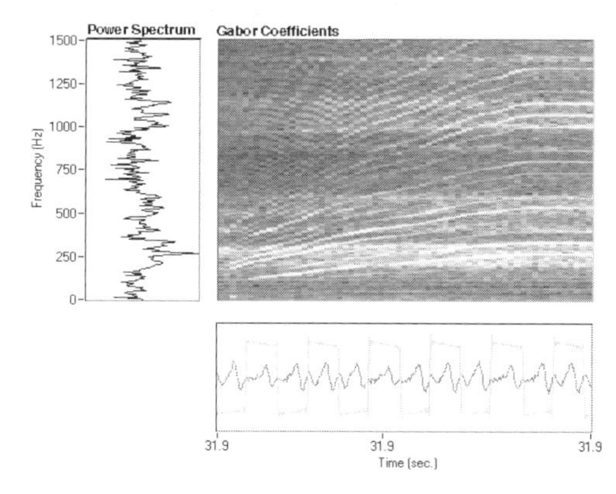

Figure 4-3 During the electrical motor run-up, the fundamental frequency bandwidth becomes wide and its harmonics overlap. Consequently, from the conventional power spectrum, we are no longer able to distinguish harmonics caused by different components. But this is not the case for the Gabor coefficients. Whether or not the electrical motor runs at constant speed, the signatures of different harmonics, in terms of the Gabor coefficients, are always obvious.

Figure 4-3 plots a sound time waveform recorded when the motor runs up. Because the harmonics overlap, the vibrations caused by different sources can no longer be distinguished from the conventional power spectrum. But this is not the case for the Gabor coefficients. Whether or not the motor runs at constant speed, the signatures of different harmonics, in terms of the Gabor coefficients, are always obvious. This observation suggests that we can use the Gabor transform to study such time-varying harmonics. For example, we can extract the Gabor coefficients that are associated with the desired time-varying harmonic. Then, we can take the inverse of the Gabor transform to obtain a corresponding time waveform. Such a method has been used in the automobile industry for engine run up/down testing.

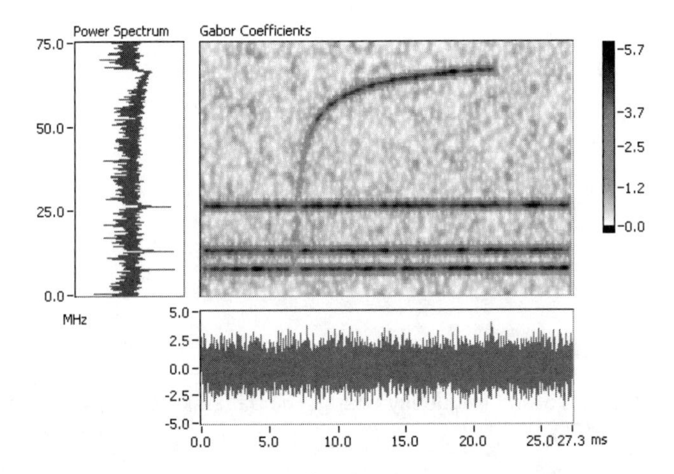

Figure 4-4 Simulated transionospheric pulse pair signal, SNR = -8 dB. Due to the low SNR, the ionized impulse signal cannot be recognized in either the time or the frequency domain. However, by employing time-frequency representation, we can readily distinguish it and establish a parametric model of the instantaneous frequency.

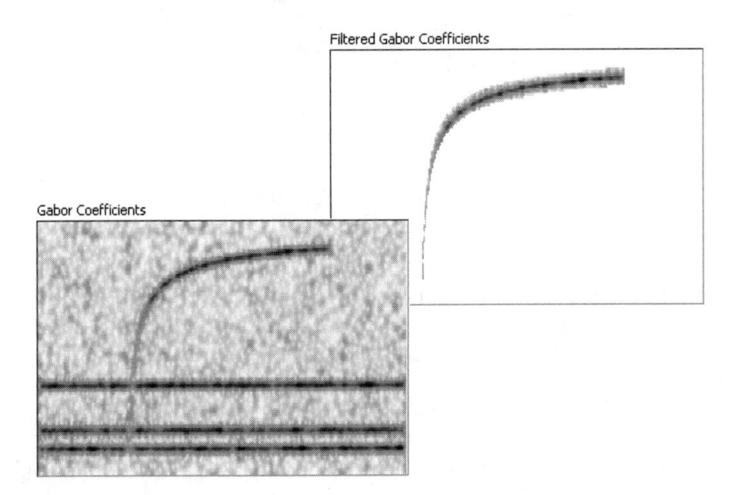

Figure 4-5 Masking the desired signal from the time-frequency representation

The second example shows how the time-varying filter improves the detection and estimation of the noise-corrupted signal. Generally speaking, random noise tends to spread evenly in the joint time-frequency domain, whereas the signal will concentrate in a relatively small range.

Although the overall SNR (signal-to-noise ratio) after the Gabor transform does not change, the local SNR in the joint time-frequency domain can be substantially improved. Consequently, we may be able to discover interesting features that are not obvious in either the time or frequency domain alone.

One application is detection and estimation of impulsive signals which have been dispersed by transmission through an ionized medium, such as impulsive signals received by RF satellite sensors. After passing through dispersive media, the impulse signal becomes a non-linear chirp signal. Figure 4-4 illustrates a simulated transionospheric pulse pair (TIPP). While the time waveform (the bottom plot) is severely corrupted by random noise, the power spectrum (the plot on the left) is mainly dominated by radio carrier signals that are basically unchanged over time. In this case, neither the time waveform nor the power spectrum indicates the existence of the impulse signal. However, when looking at the middle plot of the Gabor coefficients, we can immediately identify the presence of the chirp-type signal arching across the joint time-frequency domain. The curve in fact represents the TIPP's instataneous frequency, which can be approximated by

$$\varphi'(t) = f_{\infty} - \sqrt{\frac{K}{t}} \tag{4.1}$$

where $\varphi(t)$ denotes the signal's phase which asymptotically converges to f_{∞}, the minimum transmitted frequency[1] in the ionospheric path. The parameter K is the curvature constant for the non-linear chirp. By using (4.1), we can mask the desired Gabor coefficients from the background noise (see Figure 4-5) and compute the corresponding time waveform. The reader can find detailed descriptions about detection and estimation of the TIPP signal in [199]. In contrast to the conventional time-invariant filter, such a process can be considered as time-varying filtering.

A major challenge of the Gabor expansion-based time-variant filter is that the set of the Gabor elementary functions $\{h_{m,n}\}_{m,n \in Z}$ in general does not form a basis. To achieve good time and frequency localization, we often employ the oversampling scheme. Consequently, the range of the mapping through the Gabor transform is not the entire set of square integrable functions of two variables, but only a subset of it. For finite data samples, oversampling implies that the number of Gabor coefficients is more than the number of time domain data samples. Therefore, not every function in the joint time-frequency domain corresponds to a signal in the time domain. For an arbitrary 2D function, there may be no physically existing time domain signal whose Gabor coefficients are equal to the given 2D function.

A common approach for these kinds of problems is to invoke the least mean square error (LMSE) solution. That is, to find a time domain function whose Gabor coefficients are closest to, in the sense of the least mean square error, the given set of 2D functions. The LMSE is one of

1. The energy content of TIPPs has been measured from 25MHz to more than 100MHz. To reduce the sampling rate, the TIPP signal usually first passes through a mixer, which converts 25MHz and 100MHz into 75MHz and 0MHz, respectively. Consequently, the highest frequency f_{∞} in Figure 4-4 actually corresponds to the minimum frequency 25MHz.

the most well-known technologies. It has been extensively studied for many years, but it may not be the best solution for some applications. In particular, to solve the LMSE, we need to compute the pseudo inverse, which is demanding in terms of computation and memory. When the number of samples is large, it is almost impossible to apply the LMSE method if only conventional personal computers are available.

As an alternative, we will introduce an iterative algorithm. By repeatedly applying the Gabor transform, masking, and Gabor expansion, we will finally, under certain conditions, obtain a time waveform possessing a desired time-frequency support. We will see, by means of numerical simulations, the iterative algorithm not only yields a better signal-to-noise ratio but also is amenable for real-time implementation. The main issue regarding the iterative method is the convergence property. When does it converge? What does it converge to?

In this chapter, we only discuss the Gabor expansion-based time-varying filter. In fact, the time-variant filter can also be formulated in terms of other transform schemes, such as the wavelet transform, STFT spectrogram [118], scalogram, ambiguity function [213], and Wigner-Ville distribution ([82], [121], [123], [126], and [159]). Compared to bilinear time-frequency representations, the linear transform-based approaches are much simpler [231].

4.1 LMSE Method

Without loss of generality, we can always rearrange the 2D Gabor coefficients $c_{m,n}$ as an αL-by-*1* vector $\vec{c}$, where α and L denote the oversampling ratio and the data length, respectively. For example,

$$\vec{c} = \begin{bmatrix} c_{0,0} \\ \cdots \\ c_{0,N-1} \\ c_{1,0} \\ \cdots \\ c_{1,N-1} \\ \cdots \end{bmatrix} \tag{4.2}$$

By doing so, we can further rewrite the pair of the discrete Gabor transform (3.77) and discrete Gabor expansion (3.78) in matrix form, i.e.,

$$\vec{c} = G\vec{s} \tag{4.3}$$

where G represents an αL-by-L analysis matrix. Since generally $\alpha > 1$, G is over-determined. Then,

$$H^T\vec{c} = H^TG\vec{s} = \vec{s} \tag{4.4}$$

where H represents αL-by-L synthesis matrix. Eq. (4.4) implies that

$$H^TG = I_{L\times L} \tag{4.5}$$

Note that generally speaking,

$$GH^T \neq I_{\alpha L \times \alpha L} \tag{4.6}$$

unless $\alpha = 1$, that is, critical sampling. For critical sampling, however, the analysis and synthesis functions cannot both be localized in the joint time-frequency domain.

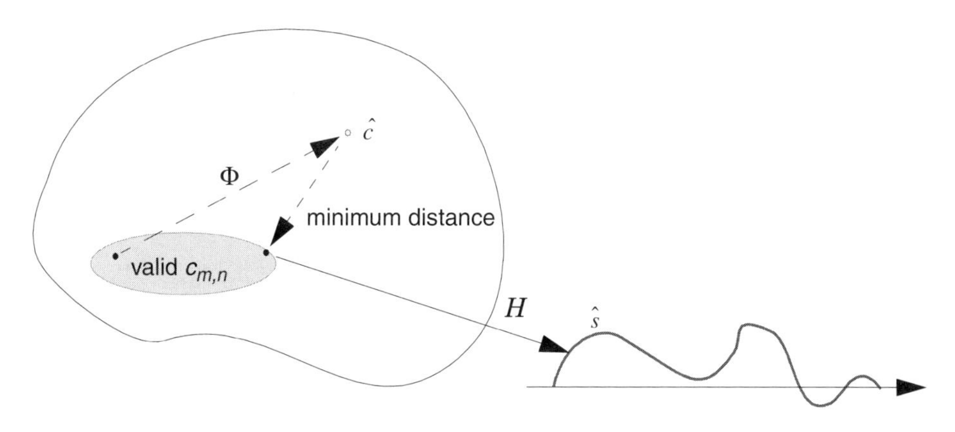

Figure 4-6 Map of LMSE filtering

Let's use $\hat{c}$ to represent the modified Gabor coefficient vector, i.e.,

$$\hat{c} = \Phi\tilde{c} = \Phi G\tilde{s}, \tag{4.7}$$

where Φ denotes an αL-by-αL mask matrix with entries,

$$\phi_{i,j} = \begin{cases} w_i & i = j \\ 0 & otherwise \end{cases} \tag{4.8}$$

where w_i denotes a weighting factor. Apply (4.4) to the modified Gabor coefficients $\hat{c}$ to obtain $\hat{s}$, i.e.,

$$\hat{s} = H^T\hat{c} \tag{4.9}$$

Because of (4.6),

$$G\hat{s} = GH^T\hat{c} \neq \hat{c} \tag{4.10}$$

This implies that the resulting time data samples $\hat{s}$ do not have the desired time-frequency properties characterized by $\hat{c}$. In fact, the modified Gabor coefficients $\hat{c}$ may not correspond to any physically existing time function. In this case, we can only find the time waveform that is the best in some sense. For example, we find the time waveform $\hat{s}$ whose Gabor coefficients are closest to the desired Gabor coefficients $\hat{c}$, in the sense of the LMSE,

$$\xi = \min_{\hat{s}} \|\hat{c} - G\hat{s}\|^2 \tag{4.11}$$

It is well-known that the solution of (4.11) is the pseudo inverse of the matrix G, i.e.,

$$\hat{s} = (G^T G)^{-1} G^T \hat{c} \tag{4.12}$$

Figure 4-6 illustrates the relationship between the estimated time waveform $\hat{s}$ and the desired Gabor coefficients $\hat{c}$.

Unfortunately, due to the computation involved in solving (4.12), the application of the LMSE solution is limited. For engine run up/down testing, *40* Khz sample rates for periods *20* to *30* seconds are considered modest. Hence, the total number of data samples is about *800,000* to *1,200,000,000*. If the oversampling ratio is four, then the dimension of the analysis matrix G is between *3,200,000*-by-*800,000* and *4,800,000*-by-*1,200,000*! The computation time involved in (4.12) with such a big matrix is unthinkable, if not impossible. Hence, the LMS solution (4.12), in fact, is only suitable for a small number of data samples.

Finally, it is worth noting that when $H = G$ (G forms a tight frame) Eq. (4.12) becomes

$$\hat{s} = (G^T G)^{-1} G^T \hat{c} = (H^T G)^{-1} H^T \hat{c} = H^T \hat{c} \tag{4.13}$$

which is identical to (4.9). In other words, when the analysis and synthesis functions are the same, the time waveform computed by the regular synthesis operation is equivalent to the LMSE solution computed by (4.12). But, the computation for the synthesis operation is only a fraction of that for computing the pseudo inverse.

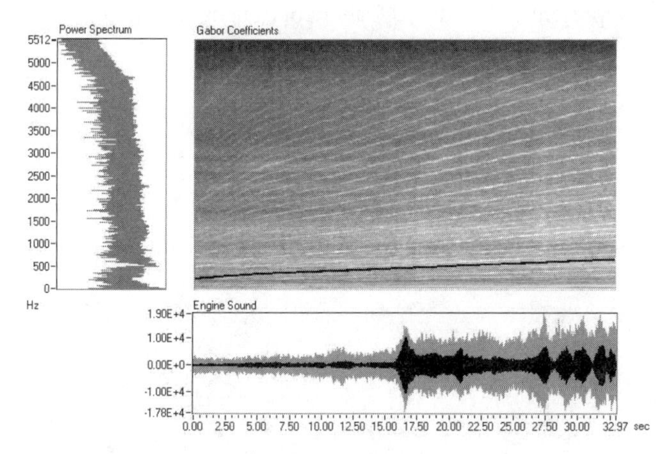

Figure 4-7 Gabor expansion-based time-varying filter. The bottom plot illustrates the electrical motor run up sound (light color) vs. the extracted seventh harmonic (dark color) time waveforms. The dark line in the Gabor coefficients plot marks the Gabor coefficients corresponding to the seventh harmonic.

For the Gabor transform, the tight frame usually implies considerable oversampling (that is, huge redundancy). Due to slow computation speed and huge memory consumption, the exact tight frame is not practical in most applications. Usually, we pursue the orthogonal-like representation introduced in Section 3.4. It has been found that for commonly used window functions, such as the Gaussian and Hanning windows, the difference between the analysis and synthesis windows is negligible when the oversampling rate is four.

Figure 4-7 depicts the results of the Gabor expansion-based time-varying filtering and shows how the seventh order is extracted. In this example, the analysis function is a 2048-point Hanning window. The window length L is equal to the number of frequency bins N. The oversampling rate is four (or 75% overlap). Since the analysis and synthesis windows are almost identical, the regular synthesis operation (4.13) yields the LMSE solution. That is, the Euclidean distance between the masked Gabor coefficients and that of the extracted time waveform, in terms of the mean square error, is minimum.

In this example, the mask function Φ in (4.7) is binary. Its entries, w_i in (4.8), are either zero or one. However, this is not necessary for (4.13). When the analysis and synthesis functions have the same form, the fact that the regular synthesis operation leads to the LMSE solution is held for any mask function Φ.

4.2 Iterative Method

Although the LMSE method is the most popular solution, it may not be the best criterion for some applications. In many applications, engineers and scientists may be more interested in SNR rather than LMSE. Moreover, the LMSE method usually requires tremendous computation unless the set of Gabor elementary functions forms a tight frame. As an alternative, in this section, we will introduce an iterative approach.

The so-called iterative method was first investigated by *Xia* and *Qian* [231] and can be described as follows. First, map the noisy signal into the joint time-frequency domain through the Gabor transform G, i.e.,

$$\tilde{c} = G\tilde{s} \tag{4.14}$$

If we do not alter the Gabor coefficients $\tilde{c}$, then we can recover the original signal by the Gabor expansion matrix H; that is,

$$H^T G\tilde{s} = \tilde{s} \tag{4.15}$$

If we apply a mask function Φ to filter out some noise, the modified Gabor coefficients become

$$\hat{c} = \Phi\tilde{c} \tag{4.16}$$

Note that the entries of the mask function Φ are limited to either zero or one. Hence, the process described by (4.16) is incredibly simple. That is, all coefficients inside the desired area are kept unchanged and the rest are set to zero.

Compute the Gabor expansion by

$$\hat{s}_1 = H^T\hat{c} = H^T\Phi\tilde{c} = H^T\Phi G\tilde{s} \tag{4.17}$$

Because of (4.6),

$$\tilde{c}_1 = G\hat{s}_1 = GH^T\Phi\tilde{c} = GH^T\hat{c} \neq \hat{c} \tag{4.18}$$

This means that the Gabor coefficients of $\hat{s}_1$ are not equal to the desired Gabor coefficients $\hat{c}$.

Repeat the process described by (4.17) and (4.18).

$$\hat{c}_1 = \Phi\check{c}_1 = \Phi G H^T \hat{c} \tag{4.19}$$

$$\hat{s}_2 = H^T \hat{c}_1 = (H^T \Phi G)^2 \check{s} \tag{4.20}$$

$$\check{c}_2 = G\hat{s}_2 = (G H^T \Phi)^2 \check{c} \tag{4.21}$$

Continue this process and after i iterations, we will have

$$\hat{s}_i = (H^T \Phi G)^i \check{s} \tag{4.22}$$

$$\check{c}_i = (G H^T \Phi)^i \check{c} \tag{4.23}$$

$$\hat{c}_i = (\Phi G H^T)^i \hat{c} \tag{4.24}$$

It can be shown [231] that if and only if

$$\sum_{i=0}^{L/N-1} \gamma^*[iN+k]h[iN+k+m\Delta M] = \sum_{i=0}^{L/N-1} h^*[iN+k]\gamma[iN+k+m\Delta M] \tag{4.25}$$

then

- The Gabor coefficients $\check{c}_i$ and the time function $\check{s}_i$ converge.
- As the number of iterations i goes to infinity, $\hat{c}_i = \Phi\check{c}_i$, which implies that the support of the Gabor coefficients $\check{c}_i$ after i iterations is indeed inside the masked area.

The proof employs an alternating projection principle [33], which is beyond the scope of this book. The reader who is interested in this topic can consult [33] and [231]. There are two trivial cases for condition (4.25):

1. Critical sampling, $N = \Delta M$. Note that in this case, the analysis and synthesis window functions cannot both be localized in the joint time-frequency domain. Hence, it has very limited application.
2. $\gamma[k] = h[k]$; for instance, $\{\gamma_i\}_{i \in Z}$ forms a tight frame. It is interesting to note that in this case, the result after the first iteration is equivalent to the LMSE solution (4.13). That is, $\check{c}_1$ in Eq. (4.18), the Gabor coefficients of the time functions after one iteration, has a minimum distance to $\hat{c}$ in Eq. (4.16), the desired Gabor coefficients. Note that this is true irrespective of whether or not the mask function Φ is binary.

As discussed earlier, the tight frame usually demands considerable oversampling (that is, huge redundancy). It is not practical. Usually, we pursue the orthogonal-like representation introduced in Section 3.4.

With the above result, we might ask whether $\tilde{c}_i = \Phi\tilde{c}_i$ violates the known fact that any signal cannot be of compact support in both time and frequency domains simultaneously. The answer is no. As a matter of fact, the concept of compact support is proved (see Section 2.5) in the time and frequency domains separately rather than jointly. Whether such a concept applies for the joint time-frequency domain so far is still an open question. While it seems true for the continuous time-frequency transforms, such as the continuous STFT and Wigner-Ville distribution, we do find counter examples for the signals that are measured in the discrete time-frequency grids, such as the Gabor transform.

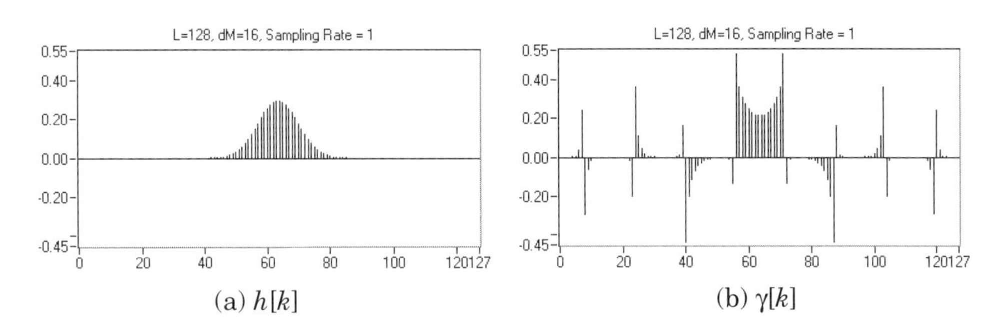

(a) $h[k]$ (b) $\gamma[k]$

Figure 4-8 *h*[*k*] and *γ*[*k*] are biorthogonal

For example, $h[k]$ and $\gamma[k]$, illustrated in Figure 4-8, are biorthogonal to each other, i.e.,

$$\langle \gamma_{m,n}[k], h[k] \rangle = \delta[m]\delta[n] \tag{4.26}$$

Assume that the signal $s[k] = h[k]$, then the corresponding Gabor coefficients are

$$c_{m,n} = \langle s[k], \gamma_{m,n}[k] \rangle = \langle h[k], \gamma_{m,n}[k] \rangle = \delta[m]\delta[n] \tag{4.27}$$

which indicates that the Gabor coefficients are a pulse at (0,0), the origin of the joint time-frequency domain.

Another trivial example is that the analysis function $\gamma(t)$ is a rectangular pulse, such as

$$\gamma(t) = \begin{cases} 1 & 0 \le t < T \\ 0 & \text{otherwise} \end{cases} \tag{4.28}$$

Assume that the signal $s(t) = \gamma(t)$, then the Gabor coefficients are

$$c_{m,n} = \langle s, \gamma_{m,n} \rangle = \langle \gamma, \gamma_{m,n} \rangle = \int \gamma(t)\gamma^*(t - mT)e^{-jn\Omega t}dt = \delta(m)\delta(n) \tag{4.29}$$

which indicates that, except for $c_{0,0}$, all other Gabor coefficients are equal to zero.

Figure 4-9 plots the masked Gabor coefficients for the example illustrated in Figure 4-4. In this example, the set of Gabor elementary functions forms an orthogonal-like transform (a near tight frame). Due to the nature of the tight frame (2.6) in this case, the overall SNR in the joint time-frequency domain is the same as that in the time domain, but the local SNR is substantially improved. This is because the Gabor transform spreads random noise over the entire time-frequency domain and, meanwhile, contains signal information in some localized areas. Hence, we can easily identify those Gabor coefficients that are related to the signal, as shown in Figure 4-9. Figure 4-10 depicts the reconstructed signal with five iterations. After five iterations, we find that the mean square error between the Gabor coefficients $\check{c}_i$ and the masked Gabor coefficients $\hat{c}_i = \Phi\check{c}_i$ becomes negligible.

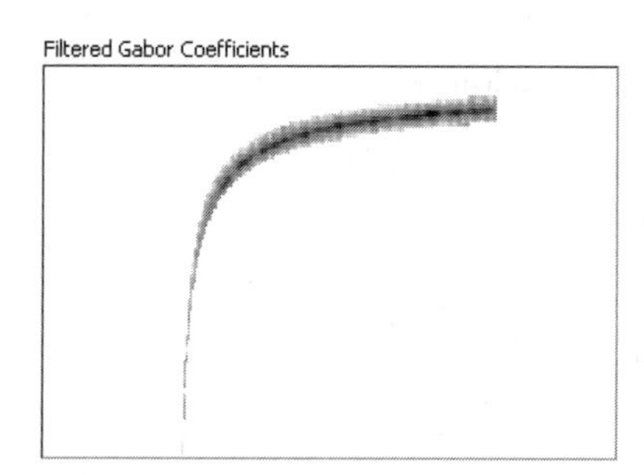

Figure 4-9 Filtered Gabor coefficients of ionized impulse signal in Figure 4-4

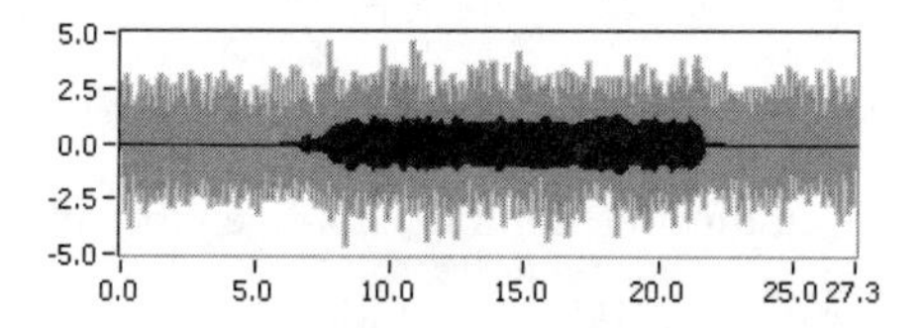

Figure 4-10 Comparison of estimated and noise corrupted signals. In this example, the SNR of the estimated signal is equal to 10.33 dB, an improvement of more than 18 dB.

For the LMSE method, the Gabor coefficients of the estimated signal have minimum distance to the desired Gabor coefficients, but in general they do not completely fall into the desired time-frequency region determined by the mask function Φ. On the other hand, the solution given by the iteration method ensures that the time-frequency support of $\hat{s}$ indeed is inside of the desired area.

4.3 Selection of Window Functions

In Sections 4.1 and 4.2 we discussed two types of the Gabor expansion-based time-varying filtering. In this section we will investigate, through numerical simulations, the significance of window functions on the performance of the time-varying filter.

For the sake of simplicity in our study we use a known noise-corrupted linear chirp signal, as plotted in Figure 4-11. For this signal, the SNR is close to 0 dB. Figure 4-12 illustrates the Gabor analysis function. Due to the low SNR, it is difficult to identify the signal from either the time or frequency domain alone. But with the Gabor transform we can clearly see the signature of the linear chirp signal in the joint time-frequency domain.

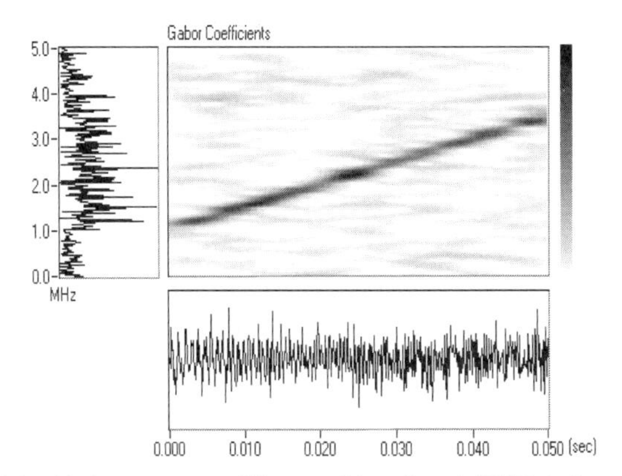

Figure 4-11 Noise-corrupted linear chirp signal (SNR is less than 0 dB)

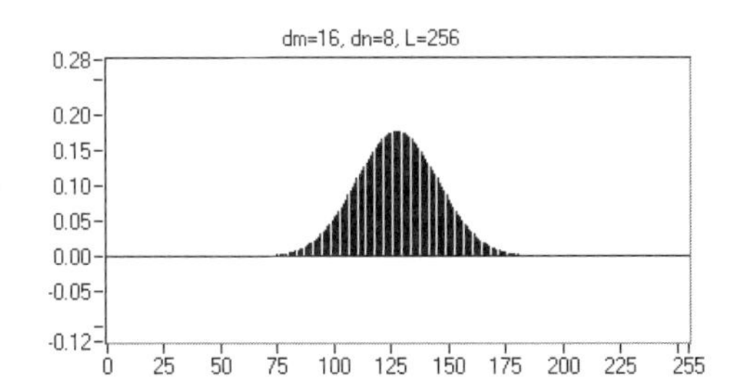

Figure 4-12 Gabor analysis function

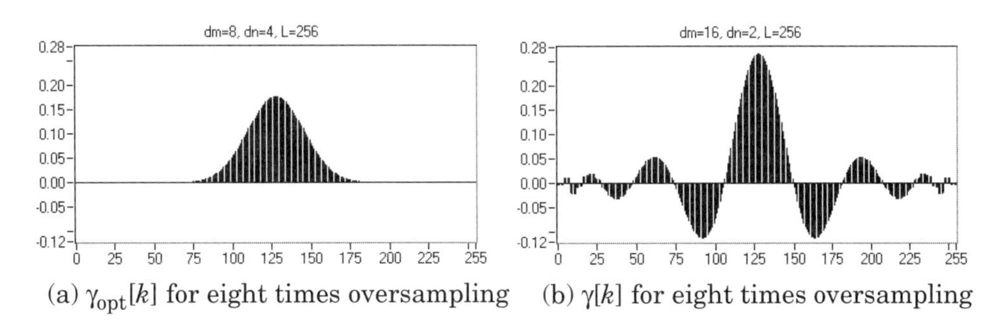

(a) $\gamma_{\text{opt}}[k]$ for eight times oversampling (b) $\gamma[k]$ for eight times oversampling

Figure 4-13 Dual functions for the Gabor analysis function in Figure 4-12. While the function in (a) forms an orthogonal-like Gabor transform, the function in (b) does not.

Figure 4-13 depicts two dual functions for the Gabor analysis function in Figure 4-12. Both of them use eight times oversampling. While the function in Figure 4-13(a) forms an orthogonal-like Gabor transform, the function in Figure 4-13(b) does not. Although both of them will lead to perfect reconstruction, the function corresponding to the orthogonal-like Gabor transform in Figure 4-13(a) is much more similar to the Gabor analysis function in Figure 4-12 than its counterpart in Figure 4-13(b). Figure 4-14 depicts the relationship between the number of iterations and the SNR. Obviously, the SNR with the orthogonal-like Gabor transform (represented by the symbol "o") is higher than that with the nonorthogonal-like Gabor transform (represented by the symbol "+").

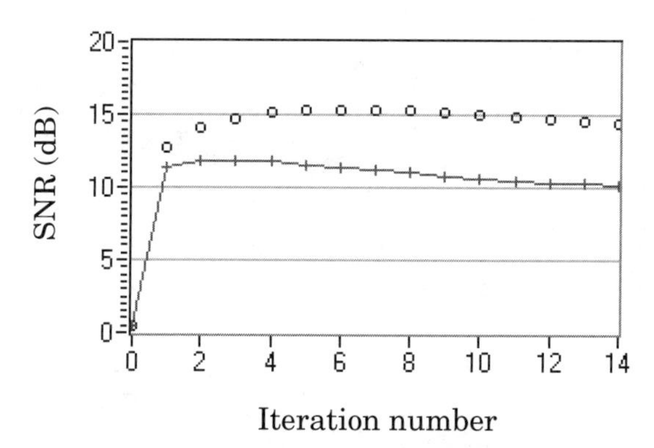

Figure 4-14 SNR vs. number of iterations. "o" corresponds to $\gamma_{opt}[k]$ in Figure 4-13 (a), whereas "+" corresponds to $\gamma[k]$ in Figure 4-13 (b). Although both $\gamma[k]$ are for eight times oversampling, the orthogonal-like dual function $\gamma_{opt}[k]$ yields much better results.

Figure 4-14 indicates that after five iterations, the SNR can be improved up to 15 dB. It is also interesting to note that the result after the first iteration is equivalent to the LMSE. Although the LMSE solution gives a time waveform whose Gabor coefficients are closest, in the mean square error sense, to the desired Gabor coefficients, the corresponding SNR is about 3 dB below that after five iterations. The peak SNR in Figure 4-14 is 15.5 dB, that is, approximately 25% higher than that achieved by LMSE. The LMSE method is the most popular estimation criterion, but in many applications people may be more interested in the high SNR rather than the small Euclidean distance.

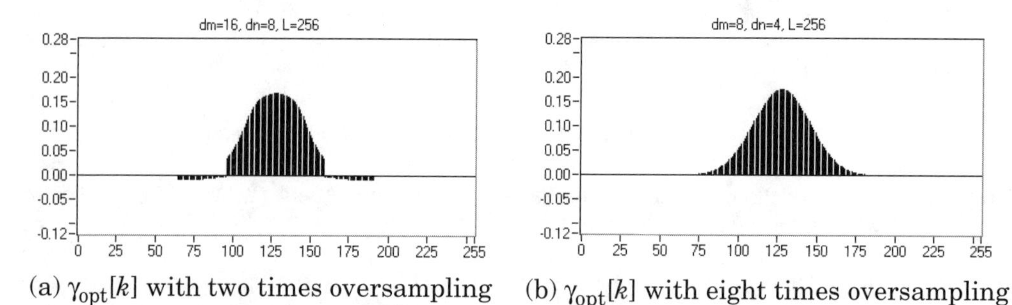

(a) $\gamma_{opt}[k]$ with two times oversampling (b) $\gamma_{opt}[k]$ with eight times oversampling

Figure 4-15 Dual functions for the Gabor analysis function in Figure 4-12. Both of them yield orthogonal-like Gabor transforms. While the function in (a) is doubly oversampled, the function in (b) is eight times oversampled.

Figure 4-15 depicts two other dual functions for the Gabor analysis function in Figure 4-12. While the function in Figure 4-15(a) is doubly oversampled, the function in Figure 4-15(b) is eight times oversampled. Although both of them form an orthogonal-like Gabor transform, the function with eight times oversampling in Figure 4-15(b) is much more similar to the Gabor analysis function in Figure 4-12 than its counterpart in Figure 4-15(a) with two times oversampling. Figure 4-16 depicts the relationship between the number of iterations and the SNR. Obviously, the SNR with eight times oversampling (represented by the symbol "o") is higher than that with two times oversampling (represented by the symbol "+").

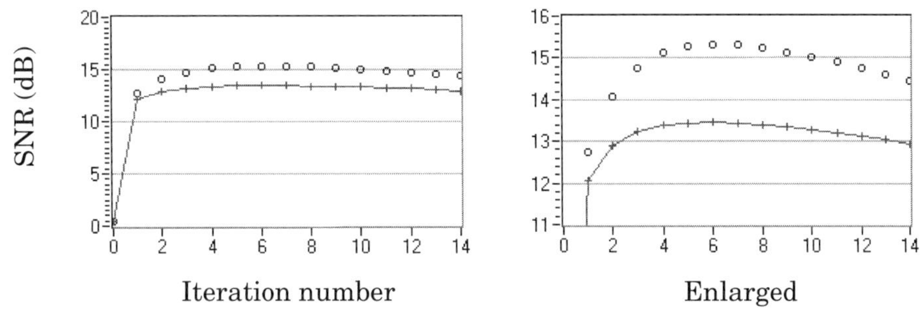

Figure 4-16 SNR vs. number of iterations. The plot on the right is the enlarged portion of the plot on the left plot; "+" = double oversampling, and "o" = eight times oversampling.

In short, the outcome of the iterative time-variant filter is related to the closeness between the analysis and synthesis functions. The closer these two functions (tight frame), the better the performance. An exact tight frame usually requires a huge amount of oversampling (huge redundancy). For computation considerations, we would like the sampling rate to be as low as possible. There is a trade-off between the computational burden and the performance of the time-variant filter. For a given oversampling rate, the orthogonal-like Gabor expansion introduced in Section 3.4 produces the $\gamma_{opt}[k]$ that is most similar to $h[k]$.

CHAPTER 5

Fundamentals of the Wavelet Transform

A common technique for studying certain properties of a signal is to compare the given signal with a set of carefully selected functions. For example, to explore a signal's periodic property, traditionally we compare the time function $s(t)$ with a set of harmonically related complex sinusoidal functions $exp\{j2\pi nt/T\}$, where T denotes the period. Because each individual complex sinusoidal function $exp\{j2\pi nt/T\}$ corresponds to a particular frequency $2\pi n/T$, the Fourier transform, the measure of similarity between $s(t)$ and $exp\{j2\pi nt/T\}$, characterizes the signal's behavior at frequency $2\pi n/T$. By varying the parameter n, we can get all different frequency tick marks and thereby obtain a complete signal frequency spectrum.

As introduced in Example 2.11, in addition to complex sinusoidal functions we can also build frequency tick marks by scaling the time index t of a given function $\psi(t)$. This is because scaling a signal in the time domain results in inverse scaling in the frequency domain. By changing the scale factor, we can obtain a set of frequency tick marks to cover the entire frequency domain. When such a set of dilated (scaled) and translated (time-shifted) functions $\psi(a^{-1}(t-b))$ are employed to measure the given signal, the resulting representation is named as the *time-scale representation* and the weight is computed by the *wavelet transform*. The function $\psi(t)$ is traditionally called the *mother wavelet*.

Although the original idea can be traced back to the *Haar* transform first introduced at the beginning of the last century [305], wavelets were not popular until the early eighties when researchers from geophysics, theoretical physics, and mathematics developed a mathematical foundation ([297] and [343]). Since then, this topic has been treated in considerable detail by numerous researchers in both the mathematics and engineering literature. In particular, *Mallat*

1. This chapter is co-authored with Xiang-Gen Xia, University of Delaware, Newark, Delaware, USA.

([338] and [339]) and *Meyer* [350] discovered a close relationship between wavelets and the structure of mutiresolution analysis. Their work, *multiresolution analysis*, not only led to a simple way of calculating the mother wavelet, but also established a connection between continuous-time wavelets and digital filter banks. Following *Mallat* and *Meyer*'s work, *Daubechies* further developed a systematic technique for generating finite-duration orthonormal wavelets with finite impulse response (FIR) filter banks ([266], [267], and [13]).

It is hard to exaggerate the significance of the contribution made by Mallat, Meyer, Daubechies, and many other researchers to the development of wavelet analysis. As *Barbara Burke Hubbard* wrote [22], "Meyer's original orthogonal wavelets had emerged almost miraculously from his computations... With Meyer's infinite orthogonal wavelets, calculating a single wavelet coefficient was a lot of work... It is much less evident how to construct (wavelets); until Mallat and Meyer developed their multiresolution theory..." Amazingly, with the help of the multiresolution theory, "to compute the wavelet transform of a signal we need neither scaling functions nor wavelets: just very simple digital filters."

These results have triggered tremendous interest in the signal processing as well as mathematics communities. However, unlike most traditional expansion systems (such as the Fourier series), the basis functions of wavelet analysis are not solutions of differential equations. Therefore, wavelets appear highly unlikely to have the revolutionary impact upon pure mathematics and physics that Fourier analysis has had. On the other hand, due to its efficient representation of highly non-stationary signals and other interesting properties, the wavelet transform has been found very useful for many signal processing applications, such as wavelet-based denoising and image/video compression. The reader can find an excellent sketch of the philosophy of wavelet analysis and the history of its development, in plain English, from Hubbard's book [22].

To assist engineers and scientists in applying the wavelet analysis for their applications, Chapter 5 and Chapter 6 provide a brief introduction to the fundamentals of wavelet analysis. While Chapter 5 focuses on the basic concepts of wavelet analysis, Chapter 6 is dedicated to the numerical implementation of wavelet transforms. The material presented in these two chapters is fundamental for understanding wavelet analysis. There is no intention to investigate the mathematical details of wavelet analysis. This is also not meant to be a handbook for the design of mother wavelet functions. The goal of these chapters is merely to serve as an introduction for those engineers and scientists who want to use this exciting technology in their applications and for those students who have never before been taught this topic. The reader can find a comprehensive treatment of wavelet analysis and wavelet design in Burrus et al. [6], Daubechies [267], Mallat [27], Strang and Nguyen [47], and Vetterli and Kovacevic [49].

Section 5.1 starts with a discussion of the basic concepts of wavelet analysis and continuous-time wavelet transforms. Although both the short-time Fourier transform and the wavelet transform can be used for time-frequency analysis, their interpretations and applications are rather different. While time-frequency transforms, such as the short-time Fourier transform, are suited for signals with narrow instantaneous frequency bandwidth (e.g., the chirp type of signal as shown in Figure 4-3 and Figure 4-4), the time-scale transform, such as wavelets, are suited for signals with sudden peaks or discontinuities (e.g., the sound generated by engine knocks, as shown in Figure 5-18 and Figure 5-19).

For the Gabor expansion, virtually any function can be used as a window function, whereas for the wavelet transform the valid mother wavelet has to satisfy the so-called admissibility condition. It was not evident how to construct an orthogonal mother wavelet with discrete sampling grids until the multiresolution theory was developed. To facilitate the reader in understanding multiresolution analysis, in Section 5.2 our presentation begins with the simplest multiresolution analysis — piecewise constant approximation, and the oldest wavelet — the Haar wavelet. Based on multiresolution theory, in Section 5.3 we derive a systematic approach to design orthogonal mother wavelets. However, due to the infinite product operation involved in computing the continuous-time mother wavelet, analytical forms of continuous-time mother wavelets essentially are only possible for very few simple cases, such as for the Haar and sinc wavelets. Therefore, there is a need for developing the discrete-time wavelet transform for implementation on digital computers. In Section 5.4 we discuss the relationship between the wavelet transform and digital filter banks. It turns out that the discrete wavelet transform can be directly computed by digital filter banks; there is no need for computing the mother wavelet at all!

At the end of this chapter, in Section 5.5, we briefly address the issue of the applications of wavelet analysis. After more than ten years of excitement, many researchers seem to now agree that wavelets cannot solve all of the world's problems, but they can be powerful tools for certain types of applications. In general, the wavelet transform is most suitable for signals with short time duration, such as engine knocks, and applications in which the inverse wavelet transform is required, such as image/video compression. As *Donoho* points out [276], wavelet systems have some inherent generic advantages and are near optimal for a wide class of problems.

5.1 Continuous Wavelet Transform

In Chapter 2, we learned two approaches for building frequency tick marks to measure the frequency content of a signal. While one approach is to use *harmonically related complex sinusoidal functions*, such as those employed in the Fourier transform and the Gabor transform, the other is achieved by scaling the time variable t of a given function $\psi(t)$ (see Section 2.3). As introduced in Example 2.11, if the center frequency (or the mean frequency) of the function $\psi(t)$ is ω_0, then the center frequencies of its time-scaled (or dilated) version $\psi(t/a)$ will become a scaled ω_0; that is, $\langle\omega\rangle = \omega_0/a$. When such dilated and translated functions $\psi(a^{-1}(t-b))$ are used as tick marks to measure signals, the resulting inner products are named the *wavelet transform* (WT), or the *continuous-time wavelet transform* (CWT) if the signal under consideration is a function of continuous-time, i.e.,

$$CWT(a, b) = \frac{1}{\sqrt{|a|}}\int_{-\infty}^{\infty} s(t)\psi^*\left(\frac{t-b}{a}\right)dt \qquad a \neq 0 \tag{5.1}$$

Because all tick marks are simply dilated[2] and translated versions of a single prototype function

$\psi(t)$, traditionally $\psi(t)$ in (5.1) is called the *mother wavelet.* The parameter a represents the scale index, determining the center frequency of the function $\psi(a^{-1}(t-b))$. The parameter b indicates the time shifting (or translation). The wavelet transform (5.1) takes $s(t)$, a member of the set of square integrable functions of one real variable t in $L^2(R)$, and transforms it to $CWT(a,b)$, a member of the set of functions of two real variables (a,b).

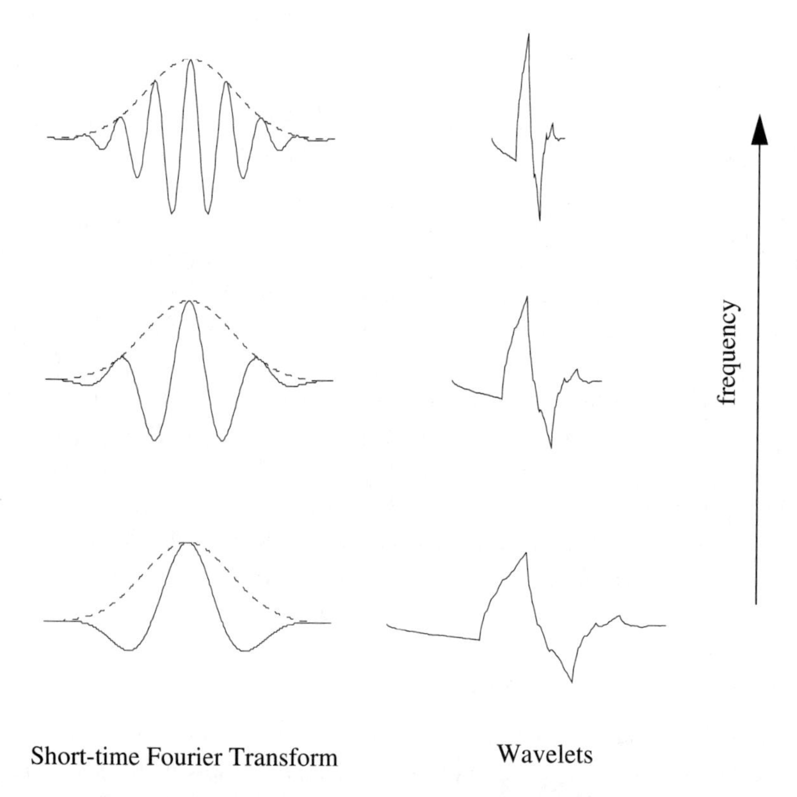

Figure 5-1 In the short-time Fourier transform (on the left), the size of the window is fixed and the number of oscillations varies. A small window is "blind" to low frequencies, which are too large for the window. But if one uses a large window, information about a brief change will be lost in the information present during the entire interval corresponding to the window. On the other hand, the wavelet (on the right) keeps the number of oscillations constant and varies the width of the window. For the wavelet transform, the higher the frequency, the better the time resolution (the poorer the frequency resolution), and vice versa. Hence, wavelet analysis is sensitive to rapid changes in the signal, such as the sound generated by engine knocks.

Suppose that $\psi(t)$ is centered at $t = 0$ and its Fourier transform $\Psi(\omega)$ is concentrated at $\omega = \omega_0$. Then, the time and frequency centers of its dilated and translated version $\psi(a^{-1}(t-b))$ are b

2. Mathematicians use the word "dilation" to refer to both expansion and compression.

and ω_0/a, respectively. Hence, the quantity $CWT(a,b)$, the inner product of $s(t)$ and $\psi(a^{-1}(t-b))$, can be considered as a quantitative measure of the signal's behavior in the vicinity of $(b, \omega_0/a)$. For example,

$$CWT(a,b)\Big|_{a = \frac{\omega_0}{\omega},\ b = t} = TF\left(t, \frac{\omega_0}{\omega}\right) \tag{5.2}$$

which suggests that the wavelet transform is also a time-frequency representation. The square of the wavelet transform is commonly called the *scalogram*, i.e.,

$$SCAL(a,b) = |CWT(a,b)|^2 \tag{5.3}$$

A natural question at this point must be what is the difference between the short-time Fourier transform and the wavelet transform? One major difference is that the short-time Fourier transform has constant time and frequency resolution but the wavelet transform does not.

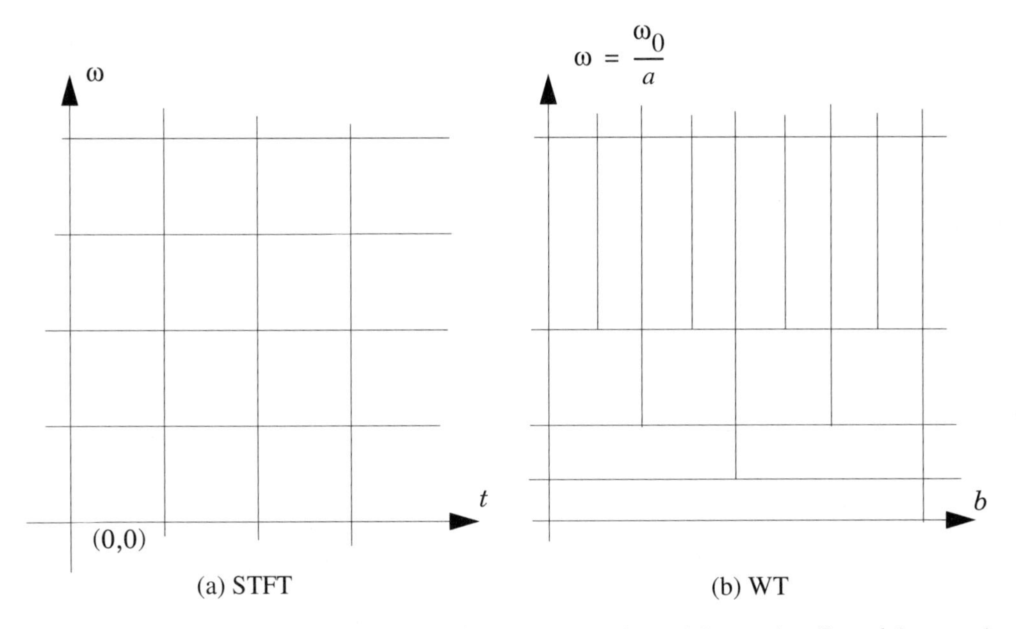

Figure 5-2 While the tiling of the short-time Fourier transform is linear, the tiling of the wavelet transform is logarithmic. Although the shapes of the sampling grids of the wavelet transform vary with the scaling factor, the area of each tile of the grids is the same.

Figure 5-1 illustrates the tick marks used in the STFT and the wavelet transform. In the STFT, the "ruler" $\{\gamma(\tau - t)e^{j\omega\tau}\}_{\tau,\,\omega \in R}$ used to measure the signal's time-frequency property is made up of a time-shifted and frequency-modulated single prototype function $\gamma(t)$. Therefore, all elementary functions have the same envelope, as shown in Figure 5-1. Once $\gamma(t)$ is chosen, both the time and frequency resolutions of the elementary functions $\gamma(\tau-t)e^{j\omega\tau}$ are fixed. On the other hand, the "ruler" $\{\psi(a^{-1}(t-b))\}_{a,\,b \in R,\,a \neq 0}$ used in the wavelet transform is obtained by dilating and translating a mother wavelet $\psi(t)$. Consequently, the time and frequency resolutions of the elementary function $\psi(a^{-1}(t-b))$ vary with the scaling factor a. If the standard deviations of the mother wavelet $\psi(t)$ are Δ_t in time domain and Δ_ω in frequency domain, then the corresponding time and frequency deviations of $\psi(a^{-1}(t-b))$ are $a\Delta_t$ and Δ_ω/a (see Example 2.11). In other words, the smaller the scaling factor a, the better the time resolution $a\Delta_t$ (the poorer the frequency resolution Δ_ω/a), and vice versa. However, the products of the corresponding time and frequency resolutions remain constant, i.e.,

$$\alpha\Delta_t\frac{\Delta_\omega}{\alpha} = \Delta_t\Delta_\omega \tag{5.4}$$

which implies that the wavelets also obey the uncertainty principle introduced in Section 2.5.

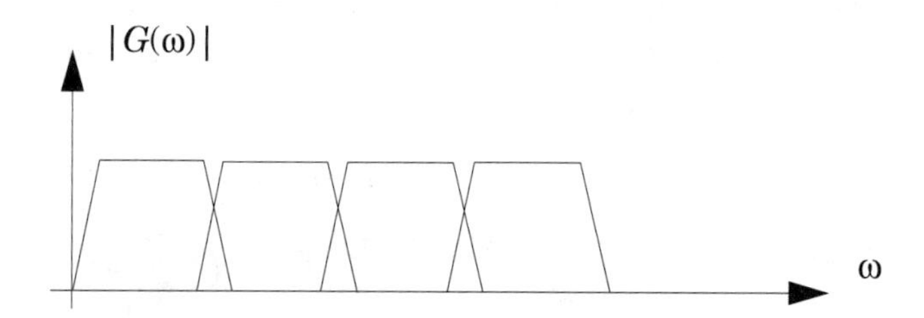

Figure 5-3 For the short-time Fourier transform, the window function $\gamma(t)$ has uniform frequency bandwidth. $G(\omega)$ denotes the Fourier transform of $\gamma(t)$.

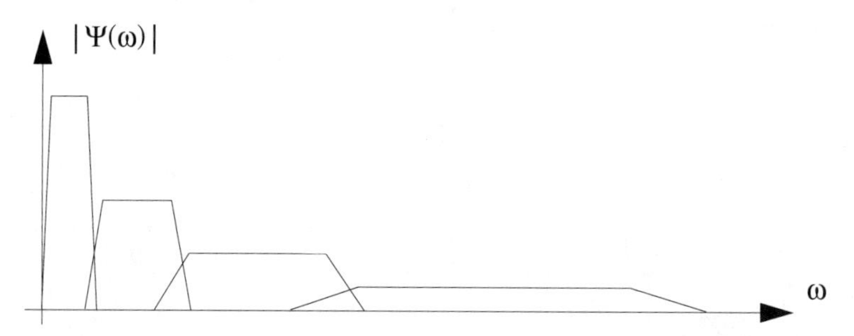

Figure 5-4 The frequency bandwidth of wavelets $\psi(a^{-1}t)$ increases as frequencies increase, but the ratio between the frequency bandwidth and the center frequency is constant. $\Psi(\omega)$ denotes the Fourier transform of $\psi(t)$.

Figure 5-2 illustrates the tiling of the short-time Fourier transform and the wavelet transform. While the tiling of the short-time Fourier transform is linear, the tiling of the wavelet transform is logarithmic. Although the shapes of the sampling grids of the wavelet transform vary with the frequency (or scaling factor), the area of each of the sampling grids is the same.

Finally, unlike the short-time Fourier transform, for the wavelet transform the ratio between the frequency bandwidth $2\Delta_\omega/a$ (twice the standard deviation in the frequency domain) and the mean frequency ω_0/a is constant, i.e.,

$$Q = \frac{2\Delta_\omega / a}{\omega_0 / a} = \frac{2\Delta_\omega}{\omega_0} \qquad (5.5)$$

which is independent of the scale factor a. Figure 5-3 and Figure 5-4 illustrate the frequency responses of the analysis function $\gamma(t)$ employed in the short-time Fourier transform and $\psi(t)$ used in the wavelet transform, respectively. Due to the relationship (5.5), the wavelet transform can also be considered to be a constant Q analysis.

It is also interesting to note that the elementary functions used in the short-time Fourier transform are complex sinusoidal functions. Hence, the short-time Fourier transform is always complex. On the other hand, the wavelet transform can be real-valued if both the signal $s(t)$ and the mother wavelet $\psi(t)$ are real. This makes the wavelet transform more attractive than the STFT in many applications.

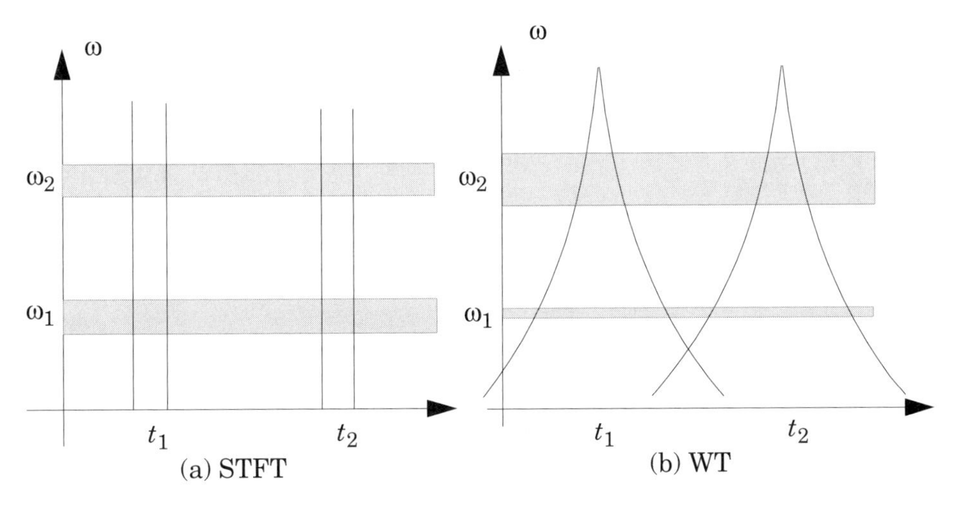

Figure 5-5 While the time and frequency resolutions of the short-time Fourier transform are uniform throughout the entire time-frequency domain, they vary for the wavelet transform. Frequency (time) resolution gets better at low (high) frequencies and becomes worse at high (low) frequencies.

In order to get a better feeling about the differences between the short-time Fourier transform and the wavelet transform, let's examine a simple example. Suppose that we have a signal that contains two time domain impulses and two frequency domain impulses, such as

$$s(t) = \delta(t - t_1) + \delta(t - t_2) + e^{j\omega_1 t} + e^{j\omega_2 t} \qquad (5.6)$$

Then its frequency representation will be

$$S(\omega) = e^{j\omega t_1} + e^{j\omega t_2} + 2\pi\delta(\omega - \omega_1) + 2\pi\delta(\omega - \omega_2) \qquad (5.7)$$

Figure 5-5 plots the resulting STFT and WT for the signal in (5.6). While the time and frequency resolution of the STFT is uniform in the entire time-frequency domain, it varies in the WT. At high frequencies, we have good time resolution and bad frequency resolution. At low frequencies, we have good frequency resolution and bad time resolution. However, the ratio of the bandwidth to the center frequency is a constant. Figures 5-6 and 5-7 illustrate the square of the STFT and the continuous wavelet transform, respectively, of a frequency hopper signal.

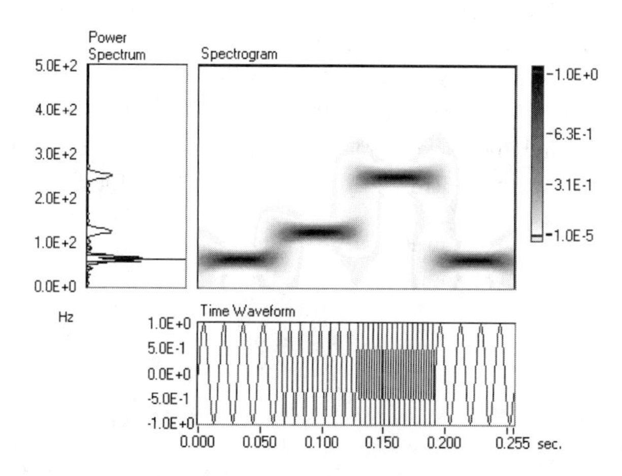

Figure 5-6 Square of the STFT of a frequency hopper signal

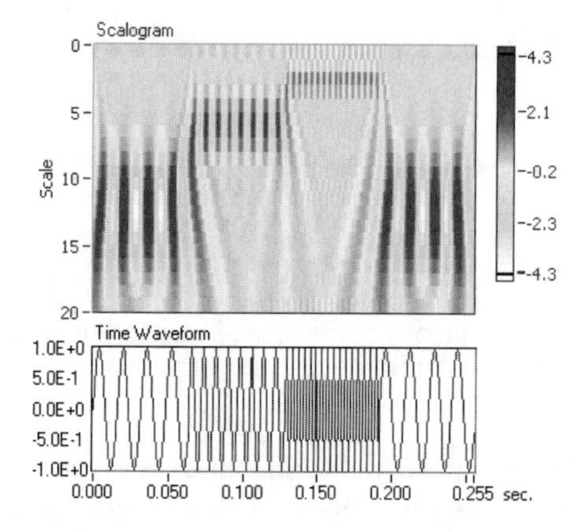

Figure 5-7 Continuous wavelet transform (with Morlet wavelet) of a frequency hopper signal. Compared with the STFT spectrogram, the scalogram has better frequency resolution (but poor time resolution) in the low frequency band and better time resolution (but poor frequency resolution) in the high frequency band.

It is worthwhile to note that if we only want to analyze the signal and do not want to recover the original signal based upon the transformations, then the mother wavelet $\psi(t)$ in (5.1) could be any function we like. However, when perfect reconstruction is needed, the selection of the mother wavelet $\psi(t)$ is restricted. In general, the function $\psi(t)$ has to be such that

$$C_{\Psi} = \int_{-\infty}^{\infty} \frac{|\Psi(\omega)|^2}{|\omega|} d\omega < \infty \tag{5.8}$$

where $\Psi(\omega)$ is the Fourier transform of the mother wavelet $\psi(t)$. Traditionally, Eq. (5.8) is the *admissibility condition*. It is easy to see that the admissibility condition implies $\Psi(0) = 0$. That is, the mother wavelet $\psi(t)$ has to be bandpass in the frequency domain. The integral of the mother wavelet $\psi(t)$ over the entire time domain has to be zero. Once $\psi(t)$ meets the admissibility condition, we can use *CWT(a,b)* as a weighting function to synthesize the original signal $s(t)$ from the dilated and translated versions of the mother wavelet $\psi(t)$, i.e.,

$$s(t) = \frac{1}{C_{\Psi}} \int_{-\infty}^{\infty} \int_{-\infty}^{\infty} \frac{1}{a^2} CWT(a, b) \psi\left(\frac{t-b}{a}\right) dadb \tag{5.9}$$

As far as the continuous-time wavelet transform is concerned, practically any function can call itself a wavelet, as long as it has a zero integral.

Just like the continuous-time short-time Fourier transform discussed in Chapter 3, the representation of the signal in terms of continuous-time wavelets is redundant. The original signal can be completely reconstructed by a sampled version of *CWT(a,b)*. Traditionally, we sample *CWT(a,b)* in a *dyadic grid*, i.e.,

$$a = 2^{-m} \qquad and \qquad b = n2^{-m} \tag{5.10}$$

Substituting (5.10) into (5.1), we have

$$d_{m,n} = CWT(2^{-m}, n2^{-m}) = \int_{-\infty}^{\infty} s(t)\psi^*_{m,n}(t)dt \tag{5.11}$$

where $\psi_{m,n}(t)$ is the dilated and translated mother wavelet $\psi(t)$ defined by

$$\psi_{m,n}(t) = 2^{m/2}\psi(2^m t - n) \tag{5.12}$$

Note that while the continuous-time wavelet transform *CWT(a,b)* in (5.1) is shift-invariant (or translation-invariant in mathematics), its sampled version $d_{m,n}$ in (5.10) is shift-variant. The quantity $d_{m,n}$ is subject to exactly where on the signal one starts the processing. For certain applications, such as for the pattern recognition, the property of being shift-variant is highly undesirable. "*Shifting over a little changes the coefficients completely, making pattern analysis hazardous*" [22].

As we have introduced in Section 2.1, if the set of $\{\psi_{m,n}\}_{m,n \in Z}$ forms a frame, then the original signal $s(t)$ can be recovered from the sampled wavelet transform $d_{m,n}$ by

$$s(t) = \sum_{m=-\infty}^{\infty} \sum_{n=-\infty}^{\infty} d_{m,n} \hat{\psi}_{m,n}(t) \tag{5.13}$$

where $\{\hat{\psi}_{m,n}\}_{m,n \in Z}$ denotes a dual frame of $\{\psi_{m,n}\}_{m,n \in Z}$. A central issue of the wavelet transform is how to build dual frames $\{\psi_{m,n}\}_{m,n \in Z}$ and $\{\hat{\psi}_{m,n}\}_{m,n \in Z}$ with desired properties. For the sake of simplicity, we will limit our discussion to the orthonormal frame. In this case, (5.13) becomes

$$s(t) = \sum_{m=-\infty}^{\infty} \sum_{n=-\infty}^{\infty} d_{m,n} \psi_{m,n}(t) \tag{5.14}$$

Eqs. (5.11) and (5.14) form a pair of the wavelet transform and its inverse, the *wavelet series.*

In this section we have introduced the fundamentals of the wavelet transform and the continuous-time wavelet transform. The question remaining now is how do we build a set of orthogonal wavelets $\{\psi_{m,n}\}_{m,n \in Z}$ so that we can apply (5.11) and (5.14) to wavelet analysis. The earlier wavelets, such as Meyer's orthogonal wavelets, had emerged almost miraculously from the computations of some mathematicians. With Meyer's infinite orthogonal wavelets, calculating a single wavelet coefficient was a lot of work. It was not evident how to construct the discrete wavelet transform, particularly the orthogonal wavelets, until Mallat [339] and Meyer [350] developed their multiresolution theory. It is their multiresolution analysis that links the continuous- and discrete-time wavelet transforms and provides a systematic way to design orthogonal wavelets with desired properties.

5.2 Piecewise Approximation

In this section and the next, we will introduce the basics of multiresolution theory. Our presentation will start with the simplest multiresolution analysis — piecewise approximation.

The basic concept of piecewise approximation is to apply piecewise constant functions (bar graphs) to approximate the continuous-time waveforms. Figure 5-8 (*b*) and (*f*) show how a given function $s(t)$ can be approximated at two different scales. Obviously, the approximation depicted in Figure 5-8 (*b*) can be described by a translated piecewise constant function $\phi(t-n)$, where

$$\phi(t) = \begin{cases} 1 & 0 \le t < 1 \\ 0 & otherwise \end{cases} \tag{5.15}$$

Figure 5-8 (*a*) depicts the waveform of $\phi(t)$. The approximation can be written as

$$s(t) \approx \sum_{n} c_{0,n} \phi(t-n) = \sum_{n} c_{0,n} \phi(2^0 t - n) \tag{5.16}$$

where $c_{0,n}$ denote the weights of the function $\phi(t-n)$. Similarly, the approximation plotted in Figure 5-8 (*f*) can be described by translated piecewise constant function $\phi(2t)$ in Figure 5-8 (*e*), that is

$$s(t) \approx \sqrt{2} \sum_{n} c_{1,n} \phi(2t-n) = \sqrt{2} \sum_{n} c_{1,n} \phi(2^1 t - n) \tag{5.17}$$

where $\phi(2t)$ is the compressed version of $\phi(t)$. Because the interval of $\phi(2t)$ is half that of $\phi(t)$, the piecewise approximation illustrated in Figure 5-8 (*f*) has smaller error (or better resolution) than that in Figure 5-8 (*b*).

Intuitively, Eqs. (5.16) and (5.17) can be generalized to

$$s(t) \approx 2^{\frac{m}{2}} \sum_{n} c_{m,n} \phi(2^m t - n) \tag{5.18}$$

As intervals of piecewise constant functions $2^{m/2}\phi(2^m t)$ get smaller and smaller, the resolution becomes better and better. When m goes to infinity, the approximation converges to $s(t)$. The factor $2^{m/2}$ ensures that the function $2^{m/2}\phi(2^m t - n)$ has unit energy. When m goes to minus infinity, the basis function $2^{m/2}\phi(2^m t\text{-}n)$ converges to zero.

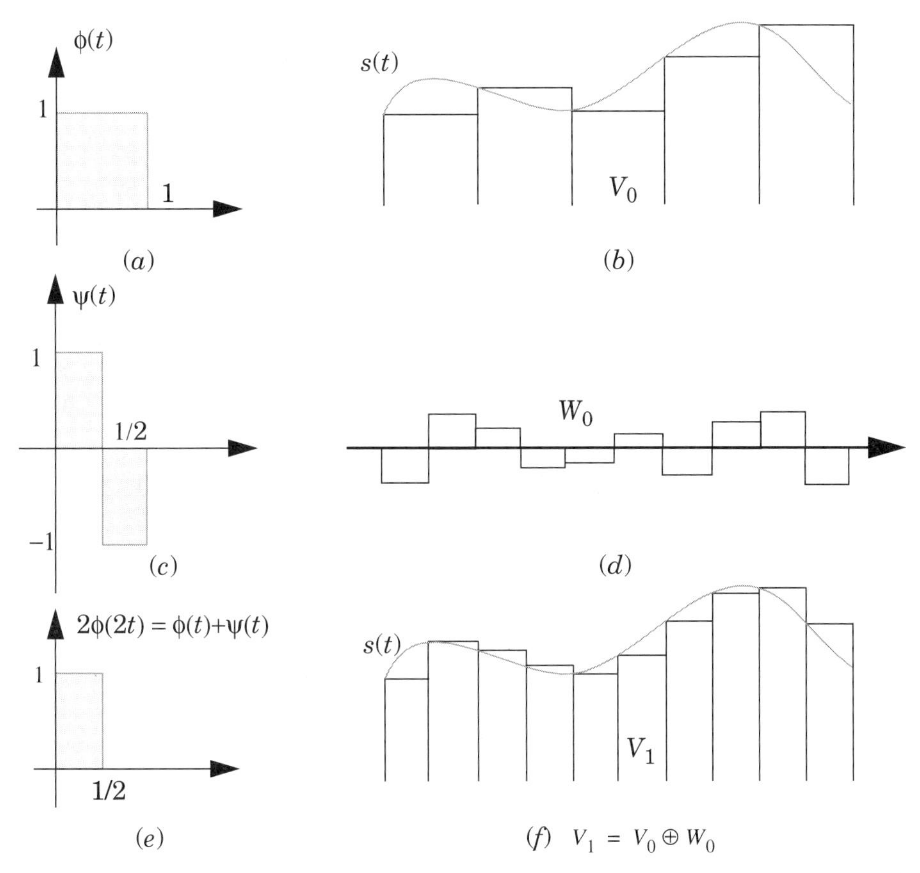

Figure 5-8 Piecewise approximation: (a) elementary function of V_0, (b) V_0, (c) elementary function of W_0 ($\psi(t)$ is orthogonal to $\phi(t)$), (d) W_0, (e) elementary function of V_1, and (f) V_1 can be achieved by the coarse approximation V_0 plus the detail W_0.

Figure 5-8 also shows that the function $\phi(t)$ in Figure 5-8 (*a*) can be written in terms of $\phi(2t)$ in Figure 5-8 (*e*), i.e.,

$$\phi(t) = \phi(2t) + \phi(2t - 1) \tag{5.19}$$

It says that the low resolution basis function $\phi(t)$ can be completely determined by the high resolution basis $\phi(2t)$. If V_m denotes a space determined by sets of translated functions $2^{m/2}\phi(2^m t - n)$, then the relationship (5.19) can be remembered as $V_0 \subset V_1$.

It is also interesting to note that the difference between Eqs. (5.16) and (5.17) (see Figure 5-8 (*d*)) can be represented by a set of translated functions $\psi(t\text{-}n)$, where

$$\psi(t) = \begin{cases} 1 & 0 \le t < \frac{1}{2} \\ -1 & \frac{1}{2} \le t < 1 \\ 0 & otherwise \end{cases} \tag{5.20}$$

In contrast to (5.19), we can also write the high resolution function $\phi(2t)$ in terms of the low resolution function $\phi(t\text{-}n)$ (see Figure 5-9), i.e.,

$$\phi(2t) = \frac{\phi(t) + \psi(t)}{2} \tag{5.21}$$

and

$$\phi(2t-1) = \frac{\phi(t) - \psi(t)}{2} \tag{5.22}$$

While the set of functions $\phi(t\text{-}n)$ preserves coarse information about the signal $s(t)$, the set of functions $\psi(t\text{-}n)$ provides detailed information about the signal $s(t)$. By using (5.21) and (5.22), we can rewrite (5.17) as

$$s(t) \approx \sqrt{2}\sum_n c_{1,n}\phi(2t-n) = \sum_n \frac{c_{1,2n} + c_{1,2n+1}}{2}\phi(t-n) + \frac{c_{1,2n} - c_{1,2n+1}}{2}\psi(t-n) \tag{5.23}$$

If W_0 denotes a space that consists of the set of functions $\psi(t\text{-}n)$, then the operation described by (5.23) can be remembered by

$$V_1 = V_0 \oplus W_0 \tag{5.24}$$

where V_0 denotes the space determined by a set of translated functions $\phi(t\text{-}n)$ and V_1 denotes the space determined by a set of translated functions $\sqrt{2}\phi(2t-n)$. The subspace W_0 is said to be an orthogonal complementary space of V_0 in V_1. $V_0 \subset V_1$ and $W_0 \subset V_1$. If we continue carrying over the decomposition (5.24), we will have

$$V_1 = V_0 \oplus W_0 = V_{-1} \oplus W_{-1} \oplus W_0 = V_m \oplus W_m \ldots W_{-2} \oplus W_{-1} \oplus W_0 \tag{5.25}$$

The space W_m consists of the set of dilated and translated functions $\{2^{m/2}\psi(2^m t - n)\}_{m,\, n \in Z}$.

As mentioned earlier, $2^{m/2}\phi(2^m t)$ converges to zero as m goes to minus infinity. In other words, V_m will become a space with only zero when m approaches minus infinity. Consequently, (5.25) can be rewritten as

$$V_1 = V_{-\infty} \ldots \oplus W_{-2} \oplus W_{-1} \oplus W_0 \tag{5.26}$$

which can be further generalized to

$$V_m = \ldots \oplus W_{m-3} \oplus W_{m-2} \oplus W_{m-1} \tag{5.27}$$

On the other hand, as the interval of the function $2^{m/2}\phi(2^m t)$ is reduced (m is increased), the accuracy of the approximation will be improved. Finally,

$$V_\infty = \lim_{m \to \infty} V_m = \ldots \oplus W_{-2} \oplus W_{-1} \oplus W_0 \oplus W_1 \oplus \ldots \tag{5.28}$$

which implies that V_∞ covers the entire signal space.

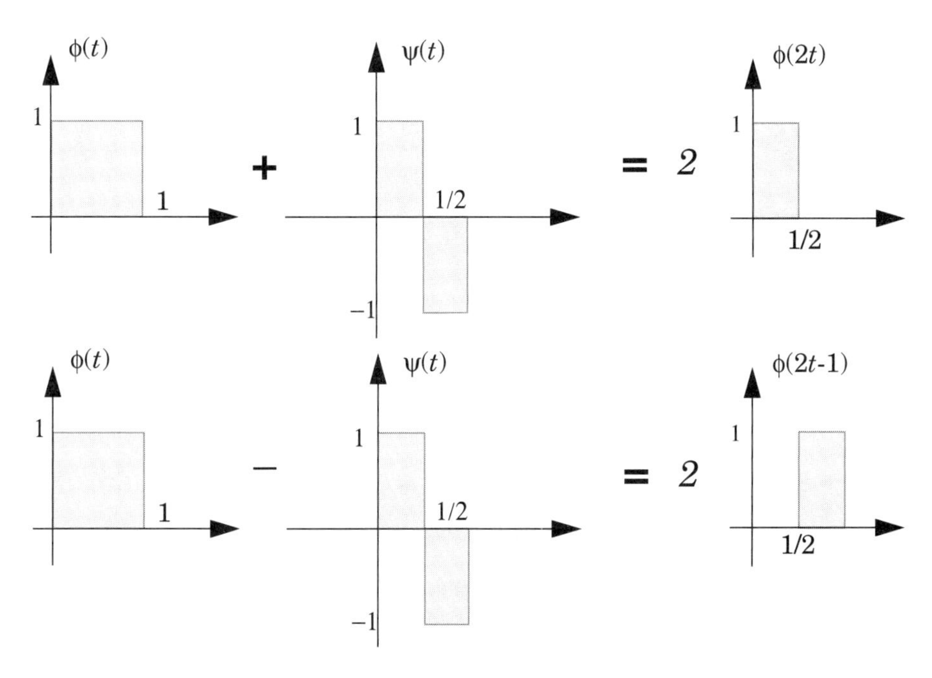

Figure 5-9 The set $\{\phi(2t-n)\}$ can be completely represented by the set $\{\phi(t-n)\}$ plus $\{\psi(t-n)\}$.

Eq. (5.28) shows that the piecewise approximation involves an approximation of signals in a sequence of nested linear vector spaces. In one direction, these successive sets $\{2^{m/2}\psi(2^m t - n)\}_{m,\, n \in Z}$ approximate the signal with greater and greater precision, approaching the original. In the other direction, the sets $\{2^{m/2}\phi(2^m t - n)\}_{m,\, n \in Z}$ approach zero as m goes to minus infinity, containing less and less information. Such an approximation technique has traditionally been given the name *multiresolution analysis*. The piecewise approximation is the simplest case of multiresolution analysis. What are the general conditions for the function to be used for multiresolution analysis? This will be the central topic of the rest of the chapter.

Before ending this section, we want to further establish the relationship between the piecewise approximation and the wavelet transform. As demonstrated in Figure 5-8, Eq. (5.18), and Eq. (5.28), as the width of the bars gets smaller, the bar graph will become closer and closer to the continuous-time waveform, i.e.,

$$2^{m/2}\sum_n c_{m,n}\phi(2^m t - n) \to s(t) \qquad for \qquad m \to \infty \tag{5.29}$$

Applying (5.23) and (5.27), we can rewrite (5.29) as

$$s(t) = \sum_{m=-\infty}^{\infty}\sum_{n=-\infty}^{\infty} d_{m,n} 2^{m/2}\psi(2^m t - n) \tag{5.30}$$

which has exactly the same form as the wavelet series (5.14). The function $\psi(t)$ defined in (5.20) is actually the simplest wavelet — a *Haar wavelet* — introduced in the beginning of the twentieth century [305]. Figure 5-10 depicts a typical Haar wavelet at two different scales.

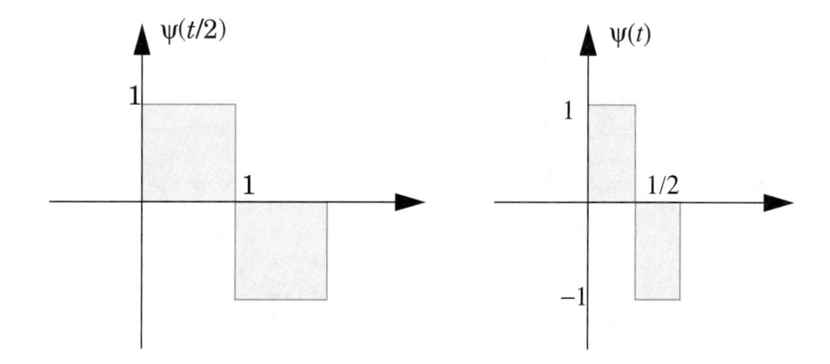

Figure 5-10 The set of Haar wavelets forms an orthogonal space. The Haar wavelet is not only orthogonal to its translated versions, but is also orthogonal to its dilated versions.

From the definition of the Haar wavelet $\psi(t)$ (5.20), it is easy to see that the translations of $\psi(t)$ in (5.20) are orthonormal, i.e.,

$$\int_{-\infty}^{\infty} \psi(t-n)\psi^*(t-n')dt = \delta(n-n') \tag{5.31}$$

Based on (5.31), the translations of dilated versions $2^{m/2}\psi(2^m t - n)$ for any fixed m are also orthonormal, because

$$\int_{-\infty}^{\infty} \psi_{m,n}(t)\psi^*_{m,n'}(t)dt = 2^m\int_{-\infty}^{\infty} \psi(2^m t - n)\psi^*(2^m t - n')dt \tag{5.32}$$

$$= \int_{-\infty}^{\infty} \psi(t-n)\psi^*(t-n')dt$$

$$= \delta(n-n') \qquad \forall\, m \in Z$$

Moreover, the space W_m is perpendicular to the space W_m for $m \neq m'$. Consequently,

$$\int_{-\infty}^{\infty} \psi_{m,n}(t)\psi_{m',n'}(t)dt = 2^{(m+m')/2}\int_{-\infty}^{\infty} \psi(2^m t - n)\psi^*(2^{m'}t - n')dt \tag{5.33}$$

$$= \delta(m - m')\delta(n - n')$$

which means that the set of Haar wavelets $\{2^{m/2}\psi(2^m t - n)\}_{m,\, n \in Z}$ forms an orthonormal basis. Hence, the corresponding wavelet series coefficients $d_{m,n}$ in (5.30) can be readily computed by the regular inner product operation,

$$d_{m,n} = 2^{m/2}\int_{-\infty}^{\infty} s(t)\psi^*(2^m t - n)dt \tag{5.34}$$

which is exactly the sampled version of *CWT(a,b)* in (5.11).

In this section we introduced, through piecewise approximation, the simplest orthogonal wavelet — the Haar wavelet. Due to the lack of continuity, however, the application of the Haar wavelet is rather limited. For example, it is not suitable for higher order polynomials. So, are there mother wavelets other than the Haar that can be used to express a function using Eqs. (5.30) and (5.34)? The answer is yes, and the solution is multiresolution analysis — a generalized piecewise approximation.

5.3 Multiresolution Analysis

As demonstrated in Section 5.2, piecewise approximation involves the approximation of signals in a sequence of nested linear vector spaces. In one direction, these successive sets approximate the signal with greater and greater precision, approaching the original. In the other direction, these sets approach zero, containing less and less information. Obviously, not every sequence of nested vector spaces possesses such properties. Representations, such as Eqs. (5.30) and (5.34), arise in the context of what is called multiresolution analysis (MRA), which essentially amounts to constructing a hierarchy of approximations to functions in various subspaces of a linear vector space. While piecewise approximation is a special case of MRA, the general definition of MRA is as follows:

A multiresolution analysis consists of the nested linear vector space $\ldots V_{m-1} \subset V_m \subset V_{m+1} \ldots$ such that:

(1) The function $\phi(t) \in V_0$ is orthogonal to its translates $\phi(t-n)$, $n \in Z$, i.e.,

$$\int_{-\infty}^{\infty} \phi(t)\phi^*(t-n)dt = \delta(n) \tag{5.35}$$

and for any $s(t)$ in V_0,

$$s(t) = \sum_n c_n \phi(t-n) \tag{5.36}$$

where

$$c_n = \int_{-\infty}^{\infty} s(t)\phi^*(t-n)dt \tag{5.37}$$

$\phi(t)$ is called a *scaling function.*

(2) $s(t) \in V_m$ if and only if $s(2t) \in V_{m+1}$ for $m \in Z$. This means that the signal at a given resolution contains all the information of the signal at coarser resolutions.

(3) The intersection of the sequence V_m has only the zero signal, i.e.,

$$\bigcap_m V_m = \{0\} \tag{5.38}$$

In other words, zero is the only object common to all the spaces V_m.

(4) Any signal in L^2 can be approximated by signals in the union of the spaces V_m, i.e.,

$$lim_{m \to \infty} V_m = L^2(R) \tag{5.39}$$

This implies that any signal in L^2 can be approximated with arbitrary precision.

It is easy to verify that the piecewise approximation introduced in Section 5.2 is an MRA. For the Haar wavelet, the scaling function $\phi(t)$ in Eq. (5.15) has the value 1 for $0 \le t < 1$, and 0 for all other values of t. In this case, condition (1) — the scaling function $\phi(t)$ is orthogonal to its translates — is very obvious. However, it will be hard to verify this condition for functions with a support greater than 1. When we shift such a function by 1, the two functions overlap, and we need delicate cancellations of positive and negative terms in order to avoid correlation. It is desirable to have a systematic way of creating the scaling function $\phi(t)$ and the corresponding mother wavelet $\psi(t)$.

Based on condition (2), if $\phi(t/2) \in V_{-1}$ then $\phi(t/2) \in V_0$. By combining condition (1), we can further define a filter $h_0[n]$ such that[3]

$$\boxed{\phi\left(\frac{t}{2}\right) = 2\sum_n h_0[n]\phi(t-n) \qquad n \in Z} \tag{5.40}$$

which is called the *dilation equation* or *refinement equation*. In the case of the piecewise approximation (see Eq. (5.19)), $h_0[0] = h_0[1] = 1/2$.

Taking the Fourier transform of both sides of (5.40) yields

$$\int_{-\infty}^{\infty} \phi\left(\frac{t}{2}\right) e^{-j\omega t} dt = 2\sum_n h_0[n] \int_{-\infty}^{\infty} \phi(t-n) e^{-j\omega t} dt \tag{5.41}$$

By replacing the variable of integration $t/2$ by t, we can rewrite (5.41) as

$$2\int_{-\infty}^{\infty} \phi(t) e^{-j2\omega t} dt = 2\sum_n h_0[n] \int_{-\infty}^{\infty} \phi(t) e^{-j\omega(t+n)} dt = 2\sum_n h_0[n] e^{-j\omega n} \int_{-\infty}^{\infty} \phi(t) e^{-j\omega t} dt \tag{5.42}$$

In other words,

$$\Phi(2\omega) = H_0(\omega)\Phi(\omega) \tag{5.43}$$

3. In this book, we always use the subscript 0 to denote a lowpass filter and 1 to denote a highpass filter.

or

$$\Phi(\omega) = H_0\left(\frac{\omega}{2}\right)\Phi\left(\frac{\omega}{2}\right) \tag{5.44}$$

where $\Phi(\omega)$ denotes the Fourier transform of $\phi(t)$. $H_0(\omega)$ is the discrete Fourier transform of $h_0[n]$, which is periodic in frequency. Moreover, $H_0(0) = 1$ as long as $\Phi(0) \neq 0$. This says that the frequency response of $H_0(\omega)$ at DC is unity (lowpass filter). If we continue to carry out such a decomposition, then

$$\Phi(\omega) = H_0\left(\frac{\omega}{2}\right)\Phi\left(\frac{\omega}{2}\right) = H_0\left(\frac{\omega}{2}\right)H_0\left(\frac{\omega}{4}\right)\Phi\left(\frac{\omega}{4}\right) = \prod_{k=1}^{\infty} H_0\left(\frac{\omega}{2^k}\right)\Phi(0) \tag{5.45}$$

Without loss of generality, let $\Phi(0) = 1$; that is,

$$\Phi(0) = \int_{-\infty}^{\infty} \phi(t)dt = 1 \tag{5.46}$$

In this case, $\phi(t)$ is a normalized scaling function. Substituting (5.46) into (5.45) yields

$$\boxed{\Phi(\omega) = \prod_{k=1}^{\infty} H_0\left(\frac{\omega}{2^k}\right)} \tag{5.47}$$

which shows that instead of the recursive method (5.40), the scaling function $\phi(t)$ can also be computed from a product of the lowpass filters $H_0(2^{-k}\omega)$ in the frequency domain. Then, what properties does the lowpass filter $H_0(\omega)$ need to have so that the scaling function $\phi(t)$ is orthogonal to its translates $\phi(t-n)$?

Because $\{\phi(t-n)\}_{n \in Z}$ is orthogonal in terms of

$$\int_{-\infty}^{\infty} \phi(t)\phi^*(t-n)dt = \delta(n) \qquad n \in Z \tag{5.48}$$

by Parseval's relationship (2.60), we have

$$\frac{1}{2\pi}\int_{-\infty}^{\infty} \Phi(\omega)\Phi^*(\omega)e^{-jn\omega}d\omega = \delta(n) \tag{5.49}$$

Taking the summation of both sides with respect to n yields

$$\frac{1}{2\pi}\sum_n \int_{-\infty}^{\infty} \Phi(\omega)\Phi^*(\omega)e^{-j\omega n}d\omega = 1 \tag{5.50}$$

That is,

$$\frac{1}{2\pi}\int_{-\infty}^{\infty} \Phi(\omega)\Phi^*(\omega)\sum_n e^{-j\omega n}d\omega = \int_{-\infty}^{\infty} \Phi(\omega)\Phi^*(\omega)\sum_n \delta(\omega - 2n\pi)d\omega = 1 \tag{5.51}$$

Therefore,

$$\sum_k |\Phi(\omega + 2k\pi)|^2 = 1 \tag{5.52}$$

which is referred to as the *Poisson summation formula* [7]. Substituting (5.44) into (5.52) yields

$$\sum_k |\Phi(\omega + 2k\pi)|^2 = \sum_k \left| H_0\left(\frac{\omega}{2} + k\pi\right) \Phi\left(\frac{\omega}{2} + k\pi\right) \right|^2 \tag{5.53}$$

$$= \sum_{n=-\infty}^{\infty} \left| H_0\left(\frac{\omega}{2} + 2n\pi\right) \Phi\left(\frac{\omega}{2} + 2n\pi\right) \right|^2 + \sum_{n=-\infty}^{\infty} \left| H_0\left(\frac{\omega}{2} + (2n+1)\pi\right) \Phi\left(\frac{\omega}{2} + (2n+1)\pi\right) \right|^2$$

$$= 1$$

where the variable k is partitioned into even and odd parts. Because $H_0(\omega)$ is periodic in frequency, that is, $H_0(\omega) = H_0(\omega+2\pi)$, the identity (5.53) reduces to

$$\left| H_0\left(\frac{\omega}{2}\right) \right|^2 \sum_{n=-\infty}^{\infty} \left| \Phi\left(\frac{\omega}{2} + 2n\pi\right) \right|^2 + \left| H_0\left(\frac{\omega}{2} + \pi\right) \right|^2 \sum_{n=-\infty}^{\infty} \left| \Phi\left(\frac{\omega}{2} + (2n+1)\pi\right) \right|^2 = 1 \tag{5.54}$$

By the relation (5.52), the identity (5.54) reduces to

$$|H_0(\omega)|^2 + |H_0(\omega+\pi)|^2 = H_0(\omega)H_0^*(\omega) + H_0(\omega+\pi)H_0^*(\omega+\pi) = 1 \tag{5.55}$$

Because $H_0(0) = 1$, as was shown earlier, (5.55) implies that $H_0(\pi) = 0$. Moreover, (5.55) also indicates that $|H_0(\omega+\pi)|$ is a reflection of $|H_0(\omega)|$ about the quadrature frequency (namely, $\omega = \pi/2$). So, $H_0(\omega)$ is referred to as a *quadrature mirror filter* (QMF). Figure 5-11 illustrates the relationship between $|H_0(\omega)|$ and $|H_0(\omega+\pi)|$. This condition is also known as the *halfband condition.* A filter with a frequency response equal to $|H_0(\omega)|^2$ is thus said to be a halfband filter.

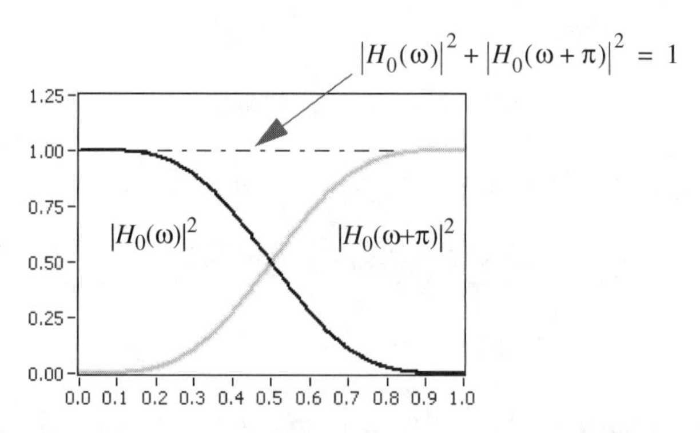

Figure 5-11 $|H_0(\omega+\pi)|$ is a reflection of $|H_0(\omega)|$ about the quadrature frequency (namely, $\omega = \pi/2$).

So far, we have proved that the orthonormal scaling function $\phi(t)$ can be generated by a lowpass quadrature mirror filter $H_0(\omega)$, for $H_0(0) = 1$ and $H_0(\pi) = 0$. To compute orthogonal mother wavelets, we need another function $H_1(\omega)$ such that

$$H_0(\omega)H_1{}^*(\omega) + H_0(\omega+\pi)H_1{}^*(\omega+\pi) = 0 \tag{5.56}$$

One solution of (5.56) is

$$H_1(\omega) = -e^{-j\omega}H_0{}^*(\omega+\pi) \tag{5.57}$$

Substituting $H_0(0) = 1$ and $H_0(\pi) = 0$ into (5.57) yields $H_1(0) = 0$ and $H_1(\pi) = 1$, respectively. This means that $H_1(\omega)$ in (5.57) is a highpass filter. Moreover, replacing $H_0(\omega)$ in (5.55) by $H_1(\omega)$, we have

$$|H_1(\omega)|^2 + |H_1(\omega+\pi)|^2 = 1 \tag{5.58}$$

which shows that $H_1(\omega)$ in (5.57) is also QMF.

Because translating the Fourier transform by π radians in the frequency domain is equivalent to multiplying the corresponding time domain sequence by $(-1)^k$, the corresponding time domain function of $H_1(\omega)$ can be easily computed by

$$h_1[k] = (-1)^k h_0[1-k] \tag{5.59}$$

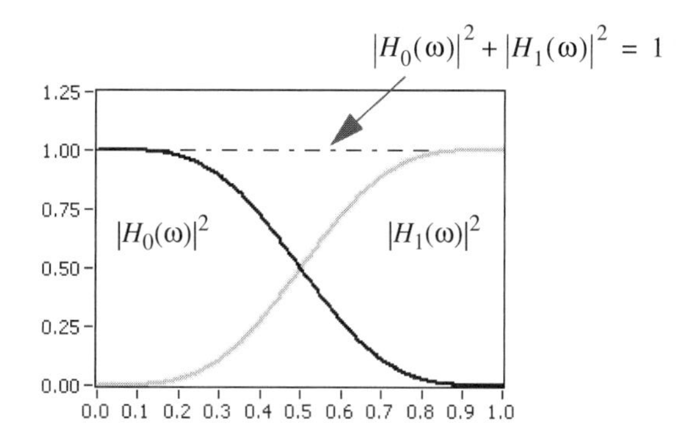

Figure 5-12 The transformation formed by the lowpass filter $H_0(\omega)$ and the highpass filter $H_1(\omega)$ is energy conserving.

Note that Eqs.(5.55), (5.56), and (5.58) can be written compactly in matrix form as

$$\begin{bmatrix} H_0(\omega) & H_0(\omega+\pi) \\ H_1(\omega) & H_1(\omega+\pi) \end{bmatrix} \begin{bmatrix} H_0{}^*(\omega) & H_1{}^*(\omega) \\ H_0{}^*(\omega+\pi) & H_1{}^*(\omega+\pi) \end{bmatrix} = \begin{bmatrix} 1 & 0 \\ 0 & 1 \end{bmatrix} \tag{5.60}$$

By exchanging the matrices on the left-hand side of (5.60) we have[4]

$$\begin{bmatrix} H_0^*(\omega) & H_1^*(\omega) \\ H_0^*(\omega+\pi) & H_1^*(\omega+\pi) \end{bmatrix} \begin{bmatrix} H_0(\omega) & H_0(\omega+\pi) \\ H_1(\omega) & H_1(\omega+\pi) \end{bmatrix} = \begin{bmatrix} 1 & 0 \\ 0 & 1 \end{bmatrix} \tag{5.61}$$

which is known as the *paraunitary condition* ([6], [48], and [49]). Eq. (5.61) implies that

$$H_0(\omega)H_0^*(\omega+\pi) + H_1(\omega)H_1^*(\omega+\pi) = 0 \tag{5.62}$$

and

$$H_0(\omega)H_0^*(\omega) + H_1(\omega)H_1^*(\omega) = 1 \tag{5.63}$$

Eq. (5.63) is often called the *power complementarity condition* ([48] and [49]). As shown in Figure 5-12, the transformation formed by the set of filters $H_0(\omega)$ and $H_1(\omega)$ is energy conserving.

Assume that $\psi(t)$ is a function whose Fourier transform $\Psi(\omega)$ satisfies

$$\boxed{\Psi(\omega) = H_1\left(\frac{\omega}{2}\right)\Phi\left(\frac{\omega}{2}\right) = H_1\left(\frac{\omega}{2}\right)\prod_{k=2}^{\infty} H_0\left(\frac{\omega}{2^k}\right)} \tag{5.64}$$

then the corresponding time relationship is

$$\boxed{\psi(t) = 2\sum_k h_1[k]\phi(2t-k)} \tag{5.65}$$

It is interesting to note that $\phi(t)$, $h_1[k]$, and $\psi(t)$ all are related to the lowpass filter $h_0[k]$. Once $h_0[k]$ is determined, we can readily obtain $\phi(t)$, $h_1[k]$, and $\psi(t)$ by Eqs.(5.47), (5.59), and (5.64), respectively.

One can prove ([13] and [339]) that under minor conditions on $H_0(\omega)$, $\psi(t-n)$ for all integers n form an orthonormal basis for the orthogonal complementary space W_0 of V_0 in V_1, i.e., $V_1 = V_0 \oplus W_0$.

Because $\{\psi(t-n)\}_{n \in Z}$ forms an orthonormal space W_0, $2^{m/2}\{\psi(2^m t - n)\}_{m, n \in Z}$ at a fixed m must also form an orthonormal basis for the orthogonal complementary space W_m of V_m in V_{m+1} (see Eq. (5.33)). By MRA conditions (2) to (4),

$$L^2 = \ldots \oplus W_{m-1} \oplus W_m \oplus W_{m+1} \otimes \ldots \tag{5.66}$$

Therefore, the dilated and translated functions $\psi_{m,n}(t)$ form an orthonormal basis for the signal space L^2. This means that for any signal $s(t)$, we have

$$s(t) = \sum_{m=-\infty}^{\infty} \sum_{n=-\infty}^{\infty} d_{m,n}\, \psi_{m,n}(t) \tag{5.67}$$

4. Generally speaking, matrix multiplication is not commutative unless the product is a multiple of the identity matrix, as in this case.

where

$$d_{m,n} = \int_{-\infty}^{\infty} s(t)\psi^*_{m,n}(t)dt \tag{5.68}$$

which is exactly what we want. Therefore, $\psi(t)$ is the mother wavelet. The constants $d_{m,n}$ are the *wavelet series coefficients.*

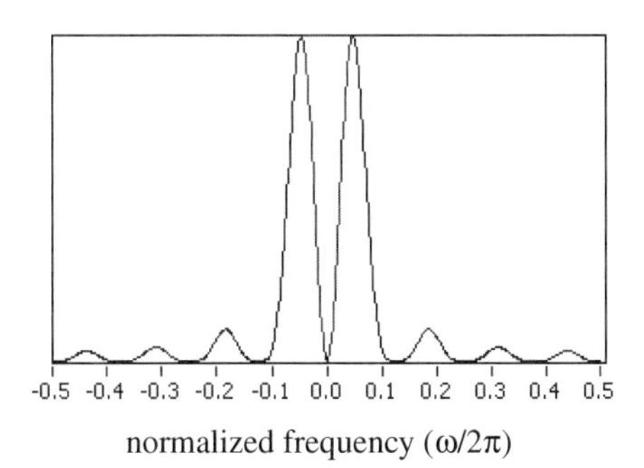

Figure 5-13 The Fourier spectrum of the Haar wavelet. In the frequency domain, the Haar wavelet exhibits strong ripples.

Example 5.1 Haar wavelet

$$H_0(\omega) = \frac{1}{2}(1 + e^{-j\omega}) \tag{5.69}$$

Obviously, the function $H(\omega)$ in (5.69) is lowpass with $H_0(0) = 1$ and $H_0(\pi) = 0$. Based on Eq. (5.45), the Fourier transform of the scaling function is

$$\Phi(\omega) = \prod_{k=1}^{\infty} H_0\left(\frac{\omega}{2^k}\right) = \prod_{k=1}^{\infty} \frac{1}{2}(1 + e^{-j\omega/2^k}) \tag{5.70}$$

Note that

$$\Phi(\omega) = \prod_{k=1}^{\infty} e^{-j\omega/2^{k+1}} \frac{1}{2}(e^{j\omega/2^{k+1}} + e^{-j\omega/2^{k+1}}) = \prod_{k=1}^{\infty} e^{-j\omega/2^{k+1}} \cos\left(\frac{\omega}{2^{k+1}}\right) \tag{5.71}$$

$$= \exp\left\{-j\omega \sum_{k=1}^{\infty} \frac{1}{2^{k+1}}\right\} \prod_{k=1}^{\infty} \cos\left(\frac{\omega}{2^{k+1}}\right) = e^{-j\omega/2} \prod_{k=1}^{\infty} \cos\left(\frac{\omega}{2^{k+1}}\right)$$

Because[5]

5. See formula 1.439, p. 38, *Table of Integrals, Series, and Products*, by I. S. Gradshteyn and I. M. Ryshik, Academic Press, New York, 1965.

$$\prod_{k=1}^{\infty} \cos\left(\frac{\omega}{2^{k+1}}\right) = \frac{\sin(\omega/2)}{\omega/2} \tag{5.72}$$

we have

$$\Phi(\omega) = e^{-j\omega/2} \frac{\sin(\omega/2)}{\omega/2} \tag{5.73}$$

From (5.57), the highpass filter corresponding to $H_0(\omega)$ in (5.69) is

$$H_1(\omega) = -e^{-j\omega} H_0^{\ *}(\omega + \pi) = \frac{1}{2}(1 - e^{-j\omega}) \tag{5.74}$$

Obviously, $H_0(\omega)$ and $H_1(\omega)$ constitute quadrature mirror filters (5.56). By (5.64) and (5.74), we can compute the Fourier transform of the wavelet, i.e.,

$$\Psi(\omega) = H_1\left(\frac{\omega}{2}\right)\Phi\left(\frac{\omega}{2}\right) = \frac{1}{2}(1 - e^{-j\omega/2})e^{-j\omega/4} \frac{\sin(\omega/4)}{\omega/4} \tag{5.75}$$
$$= \frac{1}{2}(e^{-j\omega/4} - e^{-j3\omega/4}) \frac{\sin(\omega/4)}{\omega/4}$$

The magnitude is

$$|\Psi(\omega)| = \sqrt{2\left(1 - \cos\frac{\omega}{2}\right)\left(\frac{2\sin(\omega/4)}{\omega}\right)^2} \tag{5.76}$$
$$= \sqrt{(2\sin(\omega/4))^2\left(\frac{2\sin(\omega/4)}{\omega}\right)^2} = \frac{(\sin\omega/4)^2}{|\omega/4|}$$

which is sketched in Figure 5-13. The inverse Fourier transform of $\Psi(\omega)$ is

$$\psi(t) = \begin{cases} 1 & 0 \le t < \frac{1}{2} \\ -1 & \frac{1}{2} \le t < 1 \\ 0 & otherwise \end{cases} \tag{5.77}$$

which is exactly the Haar wavelet. Figure 5-14 illustrates the Haar wavelet and its corresponding scaling function. Although the Haar wavelet is compactly supported in time, it has strong ripples in the frequency domain. Because of the discontinuity, the application of the Haar wavelet is rather limited.

By a similar procedure, the reader can compute a mother wavelet based on the ideal lowpass filter

$$H_0(\omega) = \begin{cases} 1 & \omega \le \left|\frac{\pi}{2}\right| \\ 0 & otherwise \end{cases} \tag{5.78}$$

The resulting wavelet is called a *sinc wavelet.*

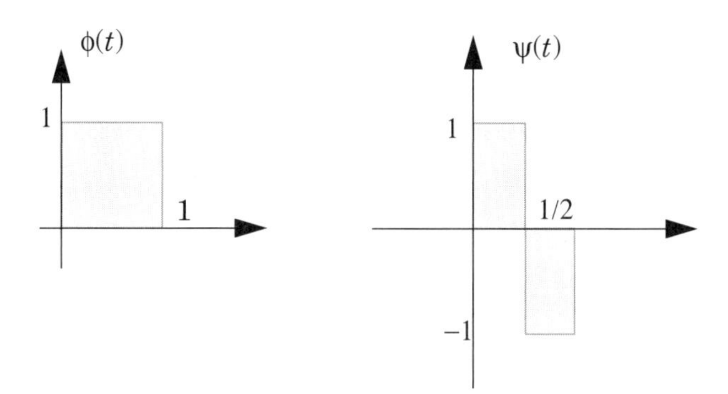

Figure 5-14 The Haar scaling function $\phi(t)$ and wavelet $\psi(t)$. The Haar wavelet is compactly supported in the time domain, but its application in signal processing is rather limited due to the lack of continuity.

It is worth noting that neither the Haar wavelet nor the sinc wavelet is an ideal wavelet for many applications. While the Haar wavelet is too jerky, the sinc wavelet does not have compact support in the time domain. Generally speaking, the ideal scaling function and mother wavelet should be smooth and compactly supported. It can be shown that both the smoothness and compact support of a function are related to its frequency decay.[6] The decay toward high frequencies corresponds to the smoothness in the time domain. The smoother the time function, the faster the decay. If the decay is exponential, the time function is infinitely differentiable. The decay toward low frequencies corresponds to the number of *vanishing* moments of the time function. For a wavelet $\psi(t)$, the N vanishing moments means that $\psi(t)$ has up to N zero moments $m_1(k) = 0$, for $0 \le k < N$, where

$$m_1[k] = \int_{-\infty}^{\infty} t^k \psi(t) dt \tag{5.79}$$

Traditionally, the kth scaling function moment is defined as

$$m_0[k] = \int_{-\infty}^{\infty} t^k \phi(t) dt \tag{5.80}$$

The wavelet vanishing moment determines what the wavelet "doesn't see." It is also weakly related to the number of oscillations. In general, the more vanishing moments a wavelet has, the more it oscillates.

In the time domain, the smoothness of the wavelet $\psi(t)$ is usually characterized by the order of *regularity*, that is, the number of its continuous derivatives. To have a regularity greater than n, a wavelet $\psi(t)$ must have at least $n + 1$ vanishing moments (a necessary condition).

6. The frequency properties in which we are most interested are the Fourier transforms of the scaling function and mother wavelet, but not the Fourier transforms of the lowpass filters. They are related, as shown in Eqs.(5.47) and (5.64), but are not the same.

In this section, we introduced the MRA concept. Based on the MRA, we derived a systematic way to build the scaling function $\phi(t)$. With the pair of quadrature mirror filters, $H_0(\omega)$ and $H_1(\omega)$, we further obtained the orthogonal mother wavelet $\psi(t)$. While $H_0(\omega)$ is lowpass with $H_0(\pi) = 0$ and $H_0(0) = 1$, $H_1(\omega)$ is highpass with $H_1(0) = 0$ and $H_1(\pi) = 1$. Example 5.1 demonstrated that the Haar wavelet can be generated from $H_0(\omega)$ given in (5.56). By using a similar approach, we can also obtain the sinc wavelet by $H_0(\omega)$ given in (5.78). Unfortunately, due to the infinite products involved in solving $\Phi(\omega)$ in (5.47), except for very simple $H_0(\omega)$ (such as in Eqs. (5.69) and (5.78)), analytical solutions of the scaling function $\phi(t)$ and mother wavelet $\psi(t)$ for an arbitrary lowpass function $H_0(\omega)$ in general do not exist. However, the MRA links the wavelet transform and digital filter banks together. Because of the MRA, as we will see shortly, to compute the wavelet transform of a signal we need neither scaling functions nor wavelets: just very simple digital filters. The multiresolution theory not only provides a systematic way to construct orthogonal wavelets, but also leads to a fast wavelet transform.

5.4 Wavelet Transformation and Digital Filter Banks

In the last three sections, all the signals that we have discussed have been continuous-time signals. As a matter of fact, the majority of signals that we deal with these days are a function of discrete time. Therefore, it is important to develop the discrete-time wavelet transform. However, unlike the development of the discrete Fourier transform, the discrete wavelet transform cannot be directly derived from its continuous-time counterpart. For example, simply replacing t by kT for wavelets $2^{m/2}\psi(2^m(t-n))$, where T denotes the time sampling interval, cannot maintain the time index $2^m(t-n)$ as an integer when $m < 0$. The transition from the continuous-time wavelet transform to the discrete wavelet transform is much more involved and needs to utilize the MRA. But the result turns out to be extremely simple. Unlike the continuous-time wavelet transform, to compute the wavelet transform of a signal we need neither scaling functions nor wavelets: just simple digital filter banks.

Based upon MRA, we can conclude that if the signal $s(t)$ is in V_m for finite m, then $s(t)$ should be completely determined by[7]

$$s(t) = \sum_{n=-\infty}^{\infty} c_{m,n}\phi_{m,n}(t) \tag{5.81}$$

Because $V_m = V_{m-1} \oplus W_{m-1}$, (5.81) can be rewritten as

$$s(t) = \sum_{n} c_{m_0,n}\phi_{m_0,n}(t) + \sum_{k=m_0}^{m-1}\sum_{n} d_{k,n}\psi_{k,n}(t) \qquad m > m_0 \tag{5.82}$$

Coefficients $d_{m,n}$ and $c_{m,n}$ are inner products between $s(t)$ and $\psi_{m,n}(t)$ and $\phi_{m,n}(t)$, respectively.

7. See Eq. (5.18) for the case of piecewise constant approximation.

While $d_{m,n}$ are *wavelet series coefficients*, the set of coefficients $c_{m,n}$, in fact, is the approximation of the signal $s(t)$ at scale m.

By Parseval's equality (2.60), we have

$$c_{m,n} = 2^{m/2}\int_{-\infty}^{\infty} s(t)\phi^*(2^m t - n)dt = \frac{1}{2\pi}2^{-m/2}\int_{-\infty}^{\infty} S(\omega)\Phi^*(2^{-m}\omega)e^{-j2^{-m}\omega n}d\omega \tag{5.83}$$

For a large scale m and $\Phi(0) = 1$ (normalized $\phi(t)$), (5.83) can be written as

$$c_{m,n} \approx \frac{1}{2\pi}2^{-m/2}\int_{-\infty}^{\infty} S(\omega)e^{-j2^{-m}n\omega}d\omega = 2^{-m/2}s(2^{-m}n) \tag{5.84}$$

This means that $c_{m,n}$ is approximately equal to the sample of $s(t)$ at $t = 2^{-m}n$ with a scale factor of $2^{-m/2}$. The higher the resolution is (that is, large m), the smaller the error. The reader can find related error analysis in [426] and [427].

Without loss of generality, let

$$c_{m,n} \equiv s[n] \equiv s(t)\Big|_{t = 2^{-m}n} \tag{5.85}$$

for large m. Because

$$c_{m-1,n} = \int_{-\infty}^{\infty} s(t)\phi^*_{m-1,n}(t)dt = 2^{\frac{m-1}{2}}\int_{-\infty}^{\infty} s(t)\phi^*\left(\frac{2^m t - 2n}{2}\right)dt \tag{5.86}$$

$$= 2^{\frac{m-1}{2}}\int_{-\infty}^{\infty} s(t)2\sum_i h_0[i]\phi^*(2^m t - (2n+i))dt$$

where we use the dilation equation (5.40). By exchanging the summation and integration, (5.86) becomes

$$c_{m-1,n} = \sqrt{2}\sum_i h_0[i]\int_{-\infty}^{\infty} s(t)\phi^*_{m,2n+i}(t)dt = \sqrt{2}\sum_i h_0[i]c_{m,2n+i} = \sqrt{2}\sum_i h_0[i-2n]c_{m,i} \tag{5.87}$$

which implies that once $c_{m,n}$ is known, we can recursively compute $c_{k,n}$, for $k < m$, by a lowpass filter $H_0(\omega)$. The operation (5.87) is depicted in Figure 5-15, where the block following the lowpass filter denotes downsampling by two.

$c_{m,n}$ → $H_0^*(\omega)$ → ↓2 → $c_{m-1,n}$

Figure 5-15 Low resolution coefficients $c_{m-1,n}$ can be recursively computed by passing the high resolution coefficients $c_{m,n}$ through lowpass filters $H_0(\omega)$.

Similarly, we can prove that

$$d_{m-1,n} = \sqrt{2}\sum_i h_1[i-2n]c_{m,i} \tag{5.88}$$

where $h_1[i]$ is defined in (5.59), which is a highpass filter. Note that $d_{m,n}$ are wavelet series coefficients. Eqs. (5.87) and (5.88) imply that $d_{m,n}$ can be obtained by using a filter bank, which is illustrated in Figure 5-16. The outputs of the highpass filters are the wavelet series coefficients $d_{m,n}$. It is interesting to note that for discrete-time samples, the wavelet transform can be accomplished by directly applying filter banks, without computing the mother wavelet function $\psi(t)$.

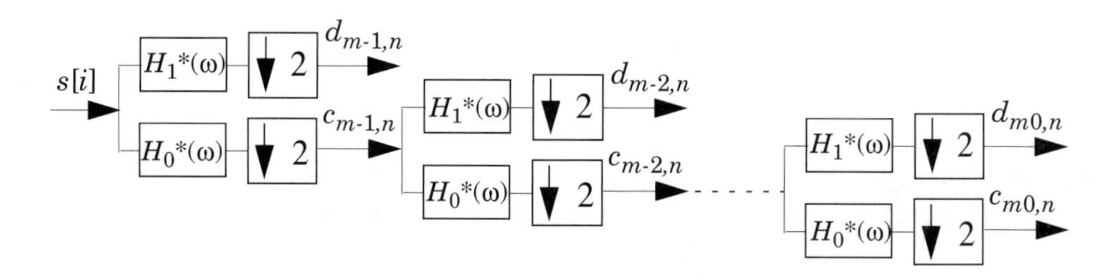

Figure 5-16 Implementation of the discrete wavelet transform through digital filter banks.

The relations (5.87) and (5.88) show that, given high-resolution coefficients, we can directly compute the low-resolution coefficients. Conversely, we can compute the high-resolution coefficients based upon the low-resolution coefficients, i.e.,

$$c_{m,n} = \sqrt{2}\left(\sum_i h_0[n-2i]c_{m-1,i} + \sum_i h_1[n-2i]d_{m-1,i}\right) \tag{5.89}$$

which implies that we can recover the original signal $s[i]$ by using a filter bank. The process in (5.89) is illustrated by Figure 5-17.

Proof

From (5.87) and (5.88), the right side of (5.89) can be written as

$$\begin{aligned}
&\sqrt{2}\left(\sum_i h_0[n-2i]c_{m-1,i} + \sum_i h_1[n-2i]d_{m-1,i}\right) \qquad (5.90)\\
&= 2\left(\sum_i h_0[n-2i]\sum_k h_0[k-2i]c_{m,k} + \sum_i h_1[n-2i]\sum_k h_1[k-2i]c_{m,k}\right)\\
&= 2\sum_k c_{m,k}\sum_i (h_0[n-2i]h_0[k-2i] + h_1[n-2i]h_1[k-2i])\\
&= 2\sum_k c_{m,k}\sum_i (h_0[n-2i]h_0[k-2i] + (-1)^{(n+k)}h_0[1-n+2i]h_0[1-k+2i])
\end{aligned}$$

where we use the relation described in (5.59). In order to have (5.89), we have to prove that

$$2\sum_i (h_0[n-2i]h_0[k-2i] + (-1)^{(n+k)}h_0[1-n+2i]h_0[1-k+2i]) = \delta(k-n)\,. \tag{5.91}$$

There are two cases: $n+k$ is odd and $n+k$ is even. When $n+k$ is odd, i.e., $n+k = 2p+1$, for an integer p, the left side of (5.91) reduces to

$$2\sum_i (h_0[n-2i]h_0[(2p+1)-n-2i] - h_0[1-n+2i]h_0[1-(2p+1)+n+2i]) \tag{5.92}$$
$$= 2\sum_i h_0[n-2i]h_0[(2p+1)-n-2i] - 2\sum_i h_0[n+2i-2p]h_0[2i-n+1]$$
$$= 2\sum_i h_0[2i+n]h_0[2i+2p+1-n] - 2\sum_i h_0[2i+n]h_0[2i+2p+1-n] = 0$$

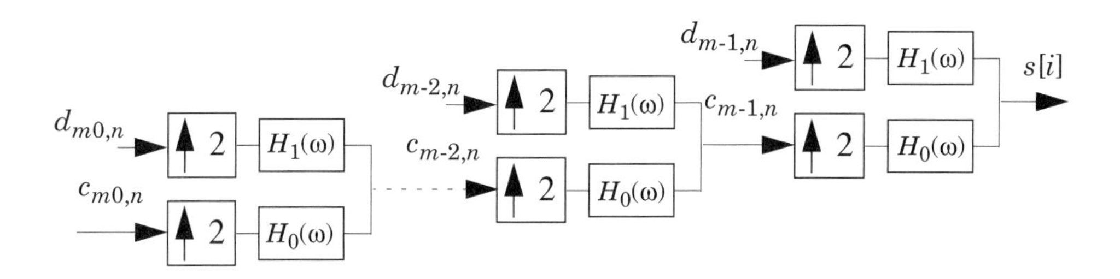

Figure 5-17 The original sample can be recovered by digital filter banks.

When $n+k$ is even, i.e., $n+k = 2p$, for an integer p, the left side of (5.69) reduces to

$$2\sum_i (h_0[n-2i]h_0[2p-n-2i] + h_0[1-n+2i]h_0[1-2p+n+2i]) \tag{5.93}$$
$$= 2\sum_i h_0[2i+n]h_0[2i+2p-n] + 2\sum_i h_0[2i+1-n]h_0[2(i-p)+1+n]$$
$$= 2\sum_i h_0[2i+n]h_0[2i+2p-n] + 2\sum_i h_0[2i+2p+1-n]h_0[2i+1+n]$$
$$= 2\sum_i h_0[2i+n]h_0[2i+2p-n] + 2\sum_i h_0[2i+1+2p-n]h_0[2i+1+n]$$

Because $2i$ is even and $2i+1$ is odd, we can group the two summations in (5.93) into

$$2\sum_i h_0[i+n]h_0[i+2p-n] = 2\sum_i h_0[i]h_0[i+2p-2n] = 2\sum_i h_0[i]h_0[i+2(p-n)]\,. \tag{5.94}$$

Because the lowpass filter $H_0(\omega)$ satisfies Eq. (5.55), i.e.,

$$H_0(\omega)H_0{}^*(\omega) + H_0(\omega+\pi)H_0{}^*(\omega+\pi) = 1 \tag{5.95}$$

taking the inverse Fourier transform of both sides yields

$$2\sum_i h_0[i]h_0[i-2n] = \delta(n) \tag{5.96}$$

for all integers n. Substituting (5.96) into (5.94) leads to

$$2\sum_i h_0[i]h_0[i+2(p-n)] = \delta(p-n)\delta(k-n) \tag{5.97}$$

Hence,

$$c_{m,n} = \sqrt{2}\left(\sum_i h_0[n-2i]c_{m-1,i} + \sum_i h_1[n-2i]d_{m-1,i}\right) \tag{5.98}$$

In this section, we derived the discrete wavelet transform and the wavelet series. As shown by Eqs. (5.88), (5.89), and (5.90), to compute the wavelet transform and wavelet series of discrete-time samples we need neither scaling functions nor wavelets: just regular digital filter banks!

5.5 Applications of the Wavelet Transform

"During the 1980s, many researchers saw promise in wavelets as an alternative to traditional Fourier analysis. Although this promise fueled an incredible amount of research, interest has waned somewhat in recent years. Did it turn out that wavelets weren't as useful as originally considered? The answer is no. As wavelet theory matured, more and more researchers seemed to recognize that wavelets could not solve all the world's problems, but they can be a powerful tool for certain types of applications."[8]

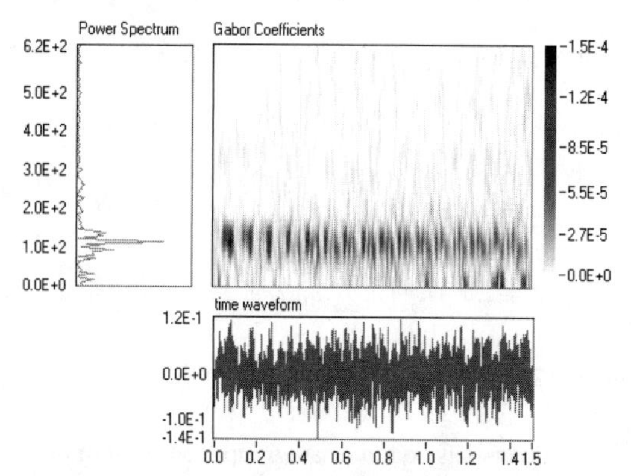

Figure 5-18 Because the energy of the sound produced by engine knocks is relatively small, the presence of engine knocks is completely overwhelmed in the averaged spectra computed by the Fourier as well as the Gabor transforms.

First of all, unlike most traditional expansion systems (such as the Fourier series), the basis functions of wavelet analysis are not solutions of differential equations. Hence, wavelets appear highly unlikely to have the revolutionary impact upon pure mathematics and physics that Fourier analysis has had. As far as signal processing is concerned, the wavelet analysis is most suitable for signals with a short time duration (wide frequency bandwidth) and applications where the inverse transform is required, such as image/video compression and denoising.

8. Jim Lewis, "Applying Wavelets for Denoising," *NI Week 2001*, National Instruments Corp., Austin, Texas, August 15 - 17, 2001.

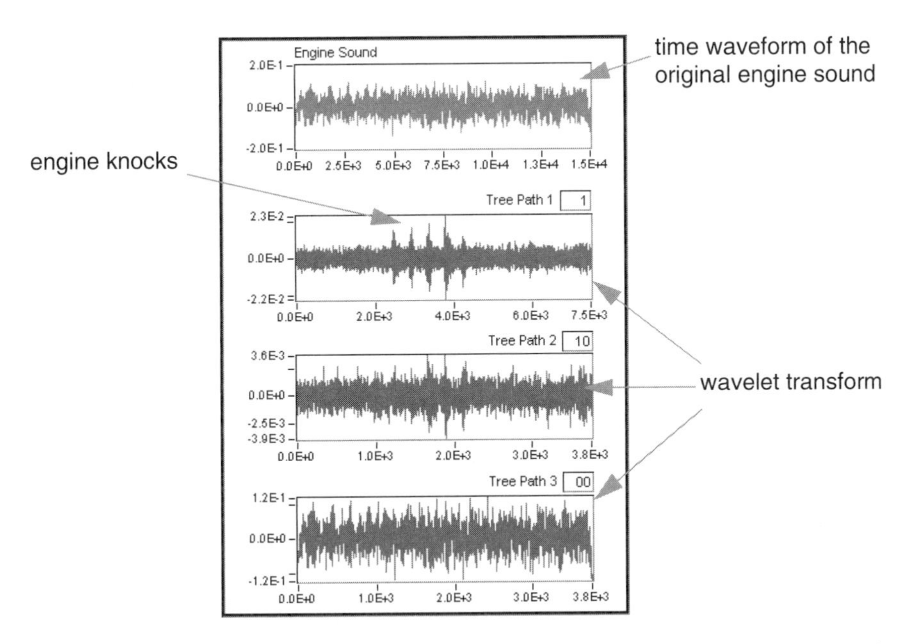

Figure 5-19 While the sound produced by engine knocks is completely concealed in background noise in a plot of the time waveform (top), the wavelet transform (the second from the top) clearly indicates the existence of engine defects.

Figure 5-18 illustrates the results of a time-frequency analysis of an engine sound. When listening to this record, we can clearly identify several knocking sounds caused by out of phase firing inside the engine. However, neither the frequency spectrum (on the left) nor the Gabor coefficients (in the middle) show the existence of such a hidden problem. This is because, to compute the Fourier transform, we have to include the signal before knocking takes place and also the signal after the knocking ends. Although the magnitude of the sound produced by an engine knocks could be rather large in a very short time period (as shown by the second plot of Figure 5-19), the energy of the sound, compared to the entire background noise, is negligible. Consequently, there will be no obvious signatures in the spectrum to show the presence of engine knocking. The Fourier transform smears the signal's local behavior globally.

With the Gabor coefficients (computed by the short-time Fourier transform), the window has to be long enough to cover a certain number of oscillations. Although the window employed in Figure 5-18 is very narrow, it is still too large to match the abrupt sound caused by engine knocking.

However, there is no such limitation for the wavelet transform. We can make the wavelets virtually as narrow as we want. Hence, the wavelet transform can effectively catch pulse-like events. Figure 5-19 plots the resulting wavelet transform at three different scales. From the second plot, we can clearly see the knocking noise.

Another successful application of the wavelet transform is for denoising [276]. As we know, classical filtering is essentially based on the assumption that either the signal or the noise has a narrow frequency band. When the signal has narrow bandwidth, we can apply a bandpass filter to extract the desired signal. When noise is narrow band, such as harmonics of the power line frequency (60Hz in the USA), we can use a notch filter to reject the unwanted noise. However, classical filtering will not work when both the signal and the noise are wide band, such as chirp signals corrupted by Gaussian white noise or pulse-like "engine knocks." For these kinds of signals, we have to use the wavelet-based denoising technique.

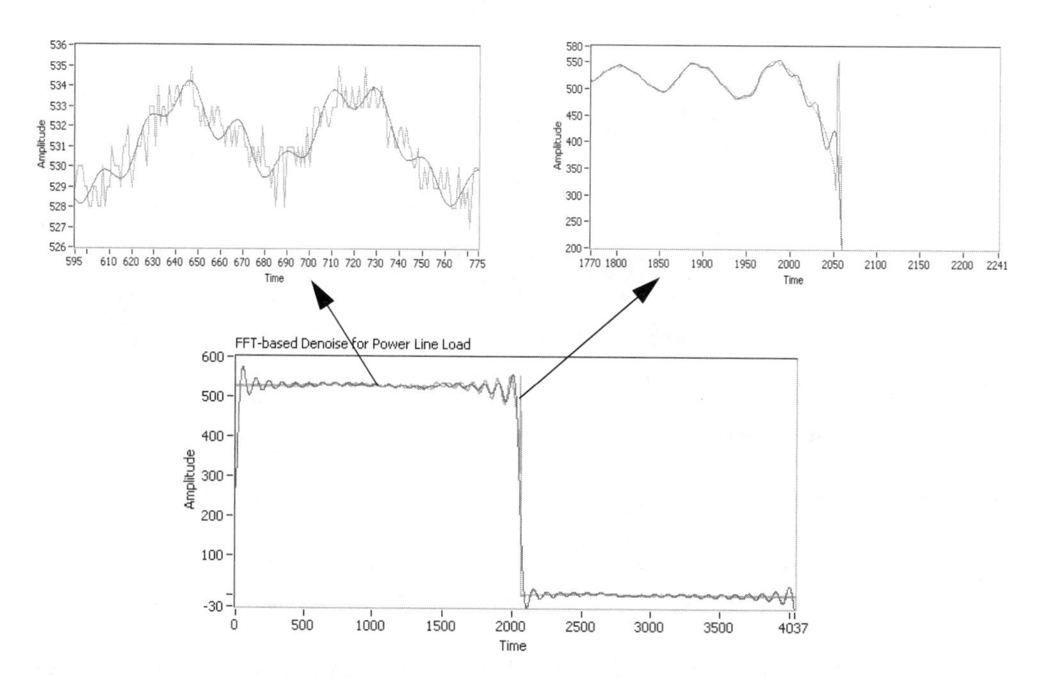

Figure 5-20 With an FFT-based denoising technique, removing unwanted noise is at the cost of badly missing the information at sharp corners (see the plot on the top right).

Figure 5-20 illustrates FFT-based denoising for a power line signal. To apply model-based spectral analysis, such as Prony's method, it is highly desired to first remove unwanted noise. In the meantime, however, we must ensure that the important information, such as sharp changes, is not altered. As shown in Figure 5-20, while removing unwanted noise, the FFT-based denoising method badly misses the information at sharp corners (see the plot on the top right). Moreover, it introduces many artificial ripples.

When the wavelets match the signal of interest, the resulting wavelet transform is sparse; that is, it has many small or zero coefficients. Think of the Fourier transform of a rectangular pulse. It will not be zero for all frequencies. The Fourier series requires an infinite number of Fourier coefficients to square up the edges of the pulse using the smooth and infinitely long sine waves. However, if we compute the wavelet transform of a rectangular pulse with a group of square pulses, such as the Haar wavelet (see Figure 5-10), then except for one coefficient (corresponding to the wavelet that matches the rectangular pulse) all others are zero! If the signal's wavelet transform is indeed sparse, such as in the case of the engine knocking in Figure 5-19, the wavelet-based denoising can effectively remove unwanted background noise.

The concept of wavelet-based denoising is incredibly simple, but has been proven to be statistically optimal [276]. The basic algorithm is simply to apply the wavelet transform, select and apply a threshold to the resulting coefficients, then apply the inverse transform to recover the denoised signal. In this case, coefficients below the threshold are set to zero and the rest are left unchanged. The choice of threshold is based on the standard deviation of the signal — there is a statistically optimal threshold but we can also choose our own based on visual appeal (for images) or any other criteria specific to the particular application.

The algorithm above employs a hard threshold, but we can see some improvement by moving to a soft thresholding approach. If we realize that the "signal" coefficients will still have some noise superimposed on them, we also want to remove that noise. The modified thresholding technique still involves setting coefficients below the threshold to zero, but also requires subtracting the threshold value from the remaining coefficients. This usually gives a superior reconstruction, both visually and from an RMS or maximum error measurement.

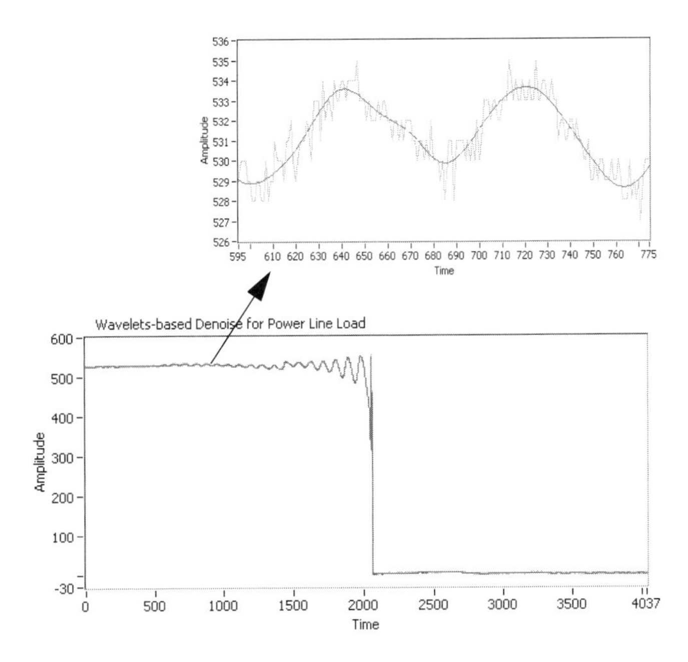

Figure 5-21 While unwanted noise is effectively removed (see the plot on the top), the result of wavelet-based denoising faithfully maintains important information for power engineers to analyze.

Figure 5-21 depicts wavelet-based denoising for the same power line signal as illustrated in Figure 5-20. Obviously, wavelet-based denoising not only removes unwanted noise, but also faithfully preserves important information for further analysis. In this example we used the sixth order Daubechies wavelet and soft thresholding.

The success of wavelet-based denoising hinges on the similarity between the wavelets and the signal. The more the wavelet "resembles" the signal of interest, the larger the good coefficients. At the same time, the less the wavelet "resembles" noise, the smaller the noise coefficients. Thus, it is often possible to apply a threshold to the wavelet coefficients to rid the signal of noise. With the appropriate wavelet, we can denoise a signal that includes noise and covers the entire spectrum with no loss of sharpness at the corners.

The selection of the desired wavelets can be greatly assisted by using commercial wavelet software, such as the Signal Processing Toolset provided by National Instruments. With the aid of the graphical user interface, as provided in the Toolset, the user can immediately see the results of applying different wavelets and thereby readily determine a suitable mother wavelet.

CHAPTER 6

Digital Filter Banks and the Wavelet Transform

*I*n Chapter 5 we investigated the relationship between the wavelet transform and digital filter banks. It turns out that the wavelet transform can be simply achieved by a tree of digital filter banks, with no need of computing mother wavelets.[1] Hence, the filter banks have been playing a central role in the area of wavelet analysis. In this chapter we will introduce the basics of filter banks.

The theory of filter banks was developed a long time ago, before modern wavelet analysis became popular. The reader can find many excellent textbooks in this area, such as *Crochiere* and *Rabiner* [12], *Strang* and *Nguyen* [47], and *Vaidyanathan* [48]. What will be introduced in this chapter are the fundamental concepts and designs of some of the most popular filter banks. They cover the majority of commonly used biorthogonal as well as orthogonal wavelets, except for the coiflets.

The Daubechies wavelets achieve the maximum number of wavelet zero moments, whereas the coiflets are a combination of wavelet zero moments and scaling function zero moments. The digital filters used to compute the coiflet wavelet transform cannot be generated by the structures discussed in this chapter. The design of digital filters for the coiflet wavelet transform needs some special skills and a certain level of mathematical preparation, which are beyond the scope of this book. The reader can find related information in the literature: [6], [13], [270], [247], [288], and [400].

Undoubtedly, experiments with new types of filter banks are proceeding, even as this book is being read. Nevertheless, this chapter provides an adequate tutorial for applied engineers and

1. Although we have only shown the proof for the orthogonal wavelet transform, this conclusion, in fact, is also true for the biorthogonal case.

scientists who want to use wavelet analysis. It also can serve as an introduction for students who are new to the topic. While this chapter assumes some knowledge of the z-transform and digital filters, such knowledge is not essential. The basic principles of filter banks presented here are well developed and easily understood, so that even those with limited experience in wavelet analysis can quickly master this technique.

6.1 Two-Channel Perfect Reconstruction Filter Banks

In Section 5.3 we discussed the relationship between the scaling function, the mother wavelet, and filter banks. Once the lowpass and highpass filters have been determined, we can compute the scaling function and the mother wavelet through the refinement equations (5.40) and (5.65), respectively. Moreover, Section 5.4 further proves that under certain conditions the outputs of the highpass filters are good approximations of the wavelet series. Consequently, the selection of desired scaling functions and mother wavelets reduces to the design of lowpass and highpass filters of two-channel perfect reconstruction (PR) filter banks. The wavelet transform can simply be realized by a tree of two-channel PR filter banks. In this section we will briefly introduce the two-channel PR filter banks.

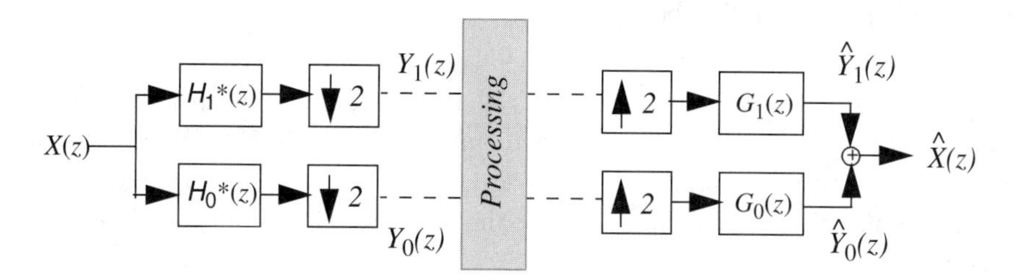

Figure 6-1 Two-channel filter bank. $H_0(z)$ and $H_1(z)$ form an analysis filter bank, whereas $G_0(z)$ and $G_1(z)$ form a synthesis filter bank. Note that $H(z)$ and $G(z)$ can be interchanged.

Figure 6-1 sketches a typical two-channel filter bank system,[2] where the z-transform is defined as

$$H(z) = \sum_{n=0}^{N} h[n]z^{-n} = H(e^{j\omega})\Big|_{z = e^{j\omega}} \equiv H(\omega) \tag{6.1}$$

Clearly, $\omega = 0$ is equivalent to $z = 1$, and $\omega = \pi$ is equivalent to $z = -1$. Hence, $H(0)$ and $H(\pi)$ in the frequency domain correspond to $H(1)$ and $H(-1)$ in the z-domain. In this book, we use N for

2. While Figures 5-16 and 5-17 depict the orthogonal filter banks, the system in Figure 6-1 represents more general cases. As we shall see shortly, the conditions for the orthogonal filter banks obtained in Section 5.3, such as the paraunitary condition and the power complementarity condition, are special cases of the conclusions derived from the system in Figure 6-1.

the filter order. The length of the filter in Eq. (6.1) is equal to $N + 1$.

As shown in Figure 6-1, the signal $X(z)$ is first filtered by a filter bank consisting of $H_0(z)$ and $H_1(z)$. The outputs of $H_0(z)$ and $H_1(z)$ are downsampled by 2 to obtain $Y_0(z)$ and $Y_1(z)$. After some processing, the modified signals are upsampled and filtered by another filter bank consisting of $G_0(z)$ and $G_1(z)$. The downsampling operators are decimetors and the upsampling operators are expanders. If no processing takes place between the two filter banks (in other words, $Y_0(z)$ and $Y_1(z)$ are not altered), the sum of the outputs of $G_0(z)$ and $G_1(z)$ is identical to the original signal $X(z)$, except for a time delay. Such a system is commonly referred to as a two-channel perfect reconstruction filter bank. $H_0(z)$ and $H_1(z)$ form an analysis filter bank, whereas $G_0(z)$ and $G_1(z)$ form a synthesis filter bank. Note that $H(z)$ and $G(z)$ can be interchanged. For instance, we can use $G_0(z)$ and $G_1(z)$ for analysis and $H_0(z)$ and $H_1(z)$ for synthesis. In this book, $H_0(z)$ and $G_0(z)$ denote lowpass filters, while $H_1(z)$ and $G_1(z)$ are highpass filters. The subscripts 0 and 1 represent lowpass and highpass filters, respectively.

Note that for a discrete-time function $x[n]$, the z-transform of its interpolated version is $X(z^M)$. This is because if

$$y[n] = \begin{cases} x\left[\frac{n}{M}\right] & n = 0, \pm M, \pm 2M\ldots \\ zero & otherwise \end{cases} \tag{6.2}$$

then

$$Y[z] = \sum_n y[n]z^{-n} = \sum_{n = 0, \pm M, \pm 2M\ldots} x\left[\frac{n}{M}\right]z^{-n} = \sum_k x[k](z^M)^{-k} = X(z^M) \tag{6.3}$$

Similarly, the reader can verify that for a discrete-time function $x[n]$, the z-transform of its doubly decimated version is

$$Y[z] = \frac{1}{2}[X(z^{1/2}) + X(-z^{1/2})] \tag{6.4}$$

By applying the relations (6.3) and (6.4), we can derive the output of the lower channel in Figure 6-1 as

$$\hat{Y}_0(z) = \frac{1}{2}G_0(z)[H_0(z)X(z) + H_0(-z)X(-z)] \tag{6.5}$$

and the output of the upper channel as

$$\hat{Y}_1(z) = \frac{1}{2}G_1(z)[H_1(z)X(z) + H_1(-z)X(-z)] \tag{6.6}$$

Hence, the output $\hat{X}(z)$ will be

$$\hat{X}(z) = \frac{1}{2}[G_0(z)H_0(z) + G_1(z)H_1(z)]X(z) + \frac{1}{2}[G_0(z)H_0(-z) + G_1(z)H_1(-z)]X(-z) \tag{6.7}$$

where one term involves $X(z)$ and the other involves $X(-z)$. For perfect reconstruction, the term with $X(-z)$, traditionally called the *alias term*, must be zero. To achieve this, we need

$$G_0(z)H_0(-z) + G_1(z)H_1(-z) = 0 \tag{6.8}$$

which reminds us of Eq. (5.62). As a matter of fact, condition (5.62) can be considered as a special case of condition (6.8). While condition (6.8) leads to biorthogonal filter banks, (5.62) is the necessary condition for orthogonal filter banks.

To accomplish Eq. (6.8), we can let

$$G_0(z) = H_1(-z) \qquad \textit{and} \qquad G_1(z) = -H_0(-z) \tag{6.9}$$

which implies that $\gamma_0[n]$ can be obtained by alternating the sign of $\gamma_1[n]$:

$$\gamma_0[n] = (-1)^n h_1[n] \tag{6.10}$$

Similarly,

$$\gamma_1[n] = (-1)^{n+1} h_0[n] \tag{6.11}$$

Therefore, $\gamma_1[n]$ and $h_1[n]$ are the highpass filters if $\gamma_0[n]$ and $h_0[n]$ are the lowpass filters. Once $H_0(z)$ and $H_1(z)$ [or $G_0(z)$ and $G_1(z)$] are determined, we can find the remaining filters with Eq. (6.9).

For perfect reconstruction, we also need the first term in Eq. (6.7), called the *distortion term*, to be a constant or a pure time delay. For example,

$$H_0(z)G_0(z) + H_1(z)G_1(z) = 2z^{-l} \tag{6.12}$$

where l denotes a time delay.[3] If we satisfy both (6.8) and (6.12), the output of the two-channel filter bank in Figure 6-1 is a delayed version of the input signal

$$\hat{X}(z) = z^{-l}X(z) \tag{6.13}$$

Let's rewrite (6.9) as

$$\boxed{H_1(z) = G_0(-z) \qquad \textit{and} \qquad G_1(z) = -H_0(-z)} \tag{6.14}$$

Substituting Eq. (6.14) into Eq. (6.12) yields

$$H_0(z)G_0(z) - H_0(-z)G_0(-z) = P_0(z) - P_0(-z) = 2z^{-l} \tag{6.15}$$

where $P_0(z)$ denotes the product of two lowpass filters, $H_0(z)$ and $G_0(z)$

$$\boxed{P_0(z) = H_0(z)G_0(z)} \tag{6.16}$$

Eq. (6.15) indicates that all odd terms of the product of two lowpass filters, $H_0(z)$ and $G_0(z)$, must be zero, except for order l. But the even order terms are arbitrary. The delay parameter l must be odd, which is usually the center of the filter $P_0(z)$. We can summarize these observations by the following formula:

3. The power complementarity condition (5.63) can be thought of a special case of (6.13). Since the biorthogonal transform generally does not satisfy the power complementarity condition, it usually is not a unitary transform.

$$p_0[n] = \begin{cases} 0 & n \text{ odd and } n \neq l \\ 1 & n = l \\ \text{arbitrary} & n \text{ even} \end{cases} \tag{6.17}$$

Consequently, the design of two-channel PR filter banks reduces to two steps:

1. Design a filter $P_0(z)$ that satisfies Eq. (6.17).
2. Factorize $P_0(z)$ into $H_0(z)$ and $G_0(z)$. Then, use Eq. (6.14) to compute $H_1(z)$ and $G_1(z)$.

Obviously, there are many ways to design $P_0(z)$. And there are many ways to factor it. The choice of four filters given by Eq. (6.9) is also not unique. Experiments are going on even as this book is being written, and undoubtedly they will still be on even as the book is being read. As an introduction to the wavelet transform, however, we will limit our presentation to those cases that are most popular in practical applications. The reader can find a more comprehensive treatment of filter banks in *Strang* and *Nguyen* [47] and *Vaidyanathan* [48].

One of the most well-known selections of the product filter $P_0(z)$ is defined by

$$P_0(z) = (1+z^{-1})^{2k}Q(z) = (1+z^{-1})^{2k}\sum_{m=0}^{2k-2} a_m z^{-m} \tag{6.18}$$

The polynomial $Q(z)$ of degree $2k-2$ is chosen so that Eq. (6.15) is satisfied. In this case, the order of $P_0(z)$ in (6.18) is always an even number. Moreover, it is a type I filter, i.e.,

$$p_0[n] = p_0[N-n] \qquad N \text{ is even} \tag{6.19}$$

which implies that the number of coefficients $p_0[n]$ is odd, $N+1$. Based on Eq. (6.17), there are $2k-1$ odd powers in $P_0(z)$, and $2k-1$ coefficients to choose in $Q(z)$. Thus, $Q(z)$ is unique.

The special factor $(1+z^{-1})^{2k}$ is also known as the *binomial filter*. The binomial itself, without $Q(z)$, represents a spline filter. $Q(z)$ is needed to give perfect reconstruction. It can be proved that $2k$ is the maximum number of zeros that $P_0(z)$ can have while preserving perfect reconstruction. Hence, $P_0(z)$ defined by Eq. (6.18) is traditionally named as the *maxflat filter*.

Example 6.1 A maxflat filter with $k = 2$ is

$$P_0(z) = (1+z^{-1})^4 Q(z) \tag{6.20}$$

Since the order of $Q(z)$ is $2k - 2 = 2$, based on condition (6.17) we can compute

$$P_0(z) = \frac{1}{16}(-1+9z^{-2}+16z^{-3}+9z^{-4}-z^{-6}) \tag{6.21}$$

and

$$Q(z) = -1+4z^{-1}-z^{-2} \tag{6.22}$$

Hence, the two roots from $Q(z)$ are at

$$c = 2 - \sqrt{3} \qquad and \qquad \frac{1}{c} = 2 + \sqrt{3} \tag{6.23}$$

While $|c| < 1$ is inside unit the circle, its reciprocal $|1/c| > 1$ is outside the unit circle. There are several possibilities for factoring $P_0(z)$ in (6.21). For example,

a. $H_0(z) = (1 + z^{-1})^2$ and $G_0(z) = (1 + z^{-1})^2(c - z^{-1})(1/c - z^{-1})$. In this case, $H_0(z)$ has two zeros at $z = -1$ (that is, $\omega = \pi$), which is a typical quadratic spline, as shown in Figure 6-2.

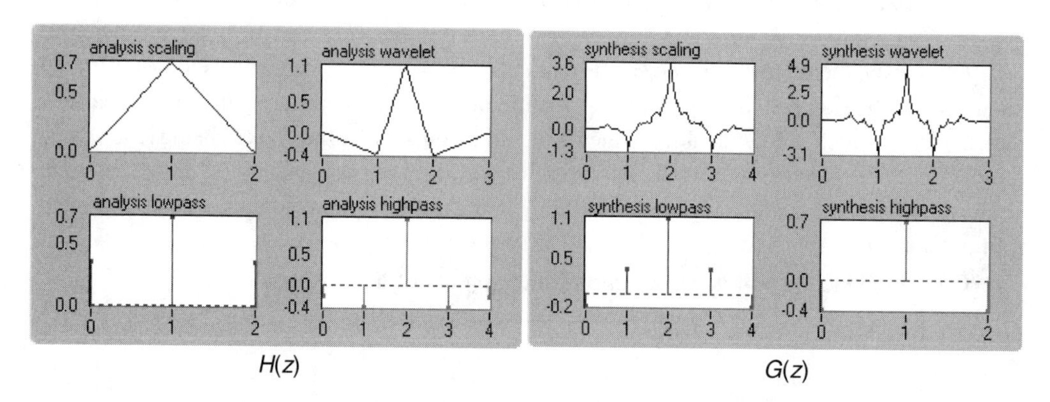

Figure 6-2 Quadratic spline wavelets for case (a)

b. $H_0(z) = (1 + z^{-1})^3$ and $G_0(z) = (1 + z^{-1})(c - z^{-1})(1/c - z^{-1})$. In this case, $H_0(z)$ has three zeros at $z = -1$ (that is, $\omega = \pi$), which is a typical cubic spline. As shown in Figure 6-3, in this case, the analysis mother wavelet is much smoother than that corresponding to the quadratic spline in Figure 6-2. But the synthesis mother wavelet is less desirable. In both cases (a) and (b), all filters are linear phase. The resulting wavelets are both biorthogonal.

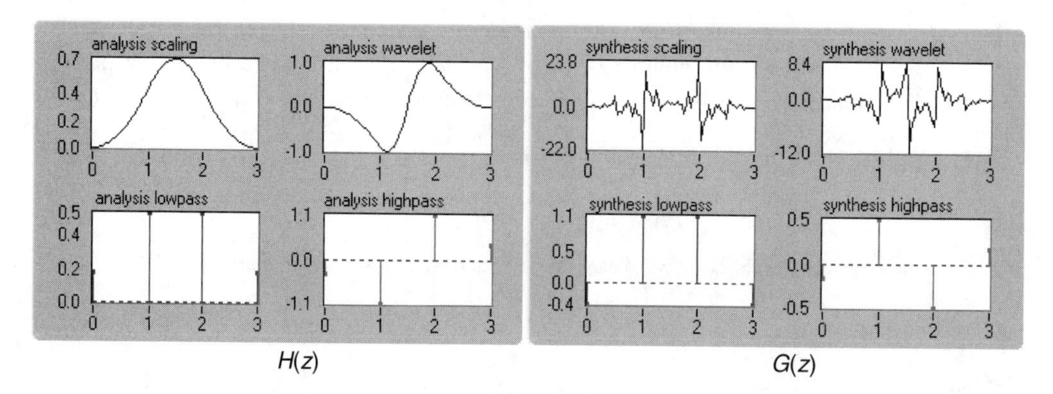

Figure 6-3 Cubic spline wavelets for case (b). Note that the cubic spline analysis mother wavelet is much smoother than that corresponding to the quadratic spline in Figure 6-2. But the corresponding synthesis mother wavelet is less desirable.

c. $H_0(z) = (1 + z^{-1})^2(c - z^{-1})$ and $G_0(z) = (1 + z^{-1})^2(1/c - z^{-1})$. This is the second order Daubechies mother wavelet, as shown in Figure 6-4. Unlike the spline wavelets, filters for the Daubechies mother wavelets are orthogonal. However, they do not have linear phase.

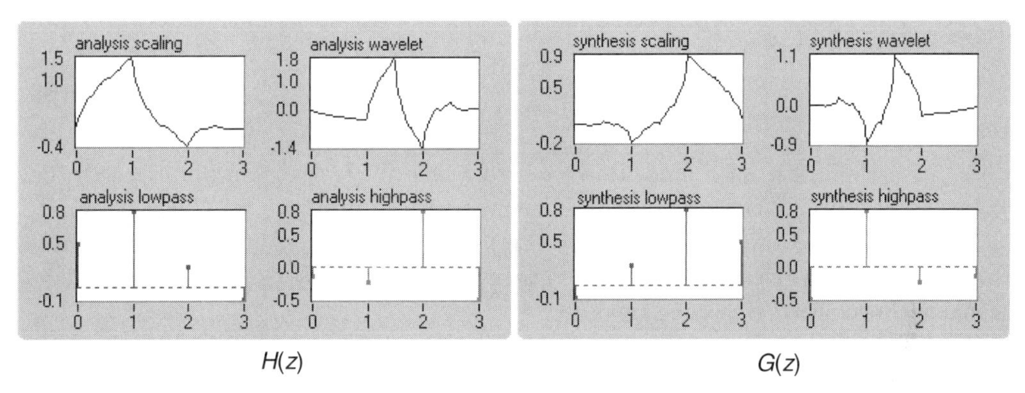

Figure 6-4 Daubechies 2 wavelets for case (c). The Daubechies filters are orthogonal and have the same length. The analysis and synthesis filters have similar behavior.

Example 6.1 demonstrates how to compute the product filter $P_0(z)$ and how to factor it. The common considerations of designing the lowpass FIR filters $H_0(z)$ and $G_0(z)$ include linear phase, minimum phase, and orthogonality. In many applications, such as image processing, linear phase is one of the most fundamental properties. In the time domain, linear phase implies that the coefficients of the causal filters are symmetric or antisymmetric around the central coefficient, i.e.,

$$\begin{array}{lll} h[n] = h[N-n] & & \textit{symmetry} \\ h[n] = -h[N-n] & & \textit{antisymmetry} \end{array} \tag{6.24}$$

In the z-transform domain, if z_k is a zero, then so is $1/z_k$, z_k^*, and $1/z^*_k$ (assume that h[n] is real-valued). In other words, the zeros of a linear phase FIR filter $H(z)$ occur in *reciprocal conjugate pairs*. Spline filters in cases (a) and (b) of Example 6.1 are all linear phase filters.

For FIR filters, *minimum phase* implies that all zeros lie inside the unit circle, i.e.,

$$|z_k| \le 1 \tag{6.25}$$

The minimum phase filter possesses *minimum phase-lag*, but *minimum phase* has historically been the established terminology. Conversely, we name an FIR filter as a maximum phase filter if all its zeros lie outside the unit circle. For the Daubechies wavelets in Eqs. (6.26) and (6.27), while one has minimum phase, the other must have maximum phase. Obviously, no filter can have linear phase and minimum phase simultaneously.

Case (c), in fact, represents one of the most important classes in the wavelet family, the *Daubechies wavelets*. For the kth order Daubechies wavelets (which are commonly named Daubechies k wavelet or simply db$k$),

$$\boxed{H_0(z) = (1+z^{-1})^k \prod_{i=1}^{k-1} (z_i - z^{-1})} \tag{6.26}$$

and

$$\boxed{G_0(z) = (1+z^{-1})^k \prod_{i=1}^{k-1} \left(\frac{1}{z_i} - z^{-1}\right)} \tag{6.27}$$

It has been proved that for a given length of orthogonal filters (or given time support), the Daubechies wavelets filters achieve the maximum number of zeros at $z = -1$ (that is, $\omega = \pi$). Since the zeros of the lowpass filter $H_0(z)$ or $G_0(z)$ at $z = -1$ are related to the zero moments of the wavelet $\psi(t)$ for a given time support, the Daubechies filters in Eqs. (6.26) and (6.27) possess the maximum number of vanishing moments.

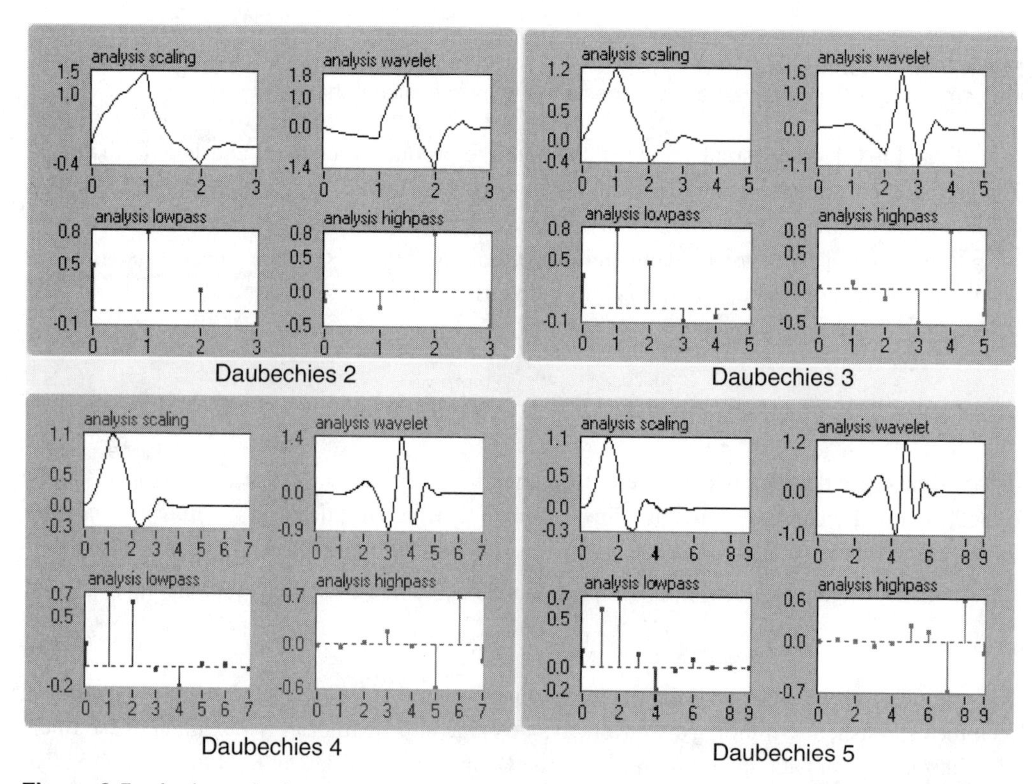

Figure 6-5 As the order increases, the Daubechies wavelets appear more and more smooth and also have more oscillations. On the other hand, as the order increases, the time support becomes wider.

For $k = 1$, the Daubechies wavelet is identical to the Haar wavelet. As the order k in (6.18) increases, the Daubechies wavelets become more and more smooth and also have more number of oscillations. On the other hand, as the order k increases, their time support becomes wider (see Figure 6-5).

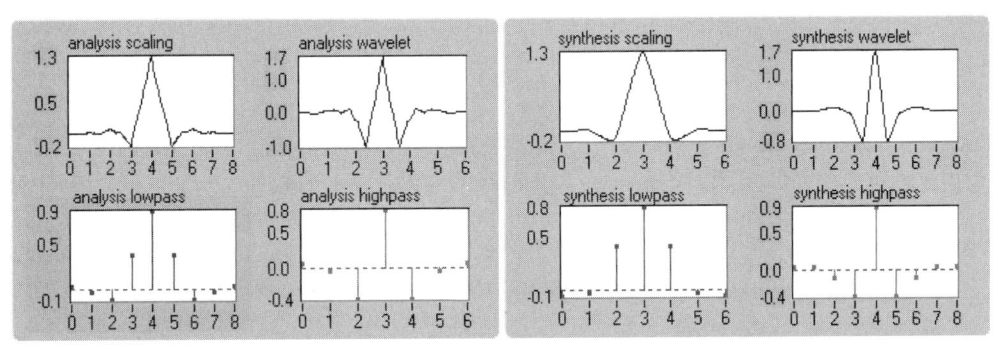

Figure 6-6 Scaling functions and wavelets used for the FBI fingerprint compression

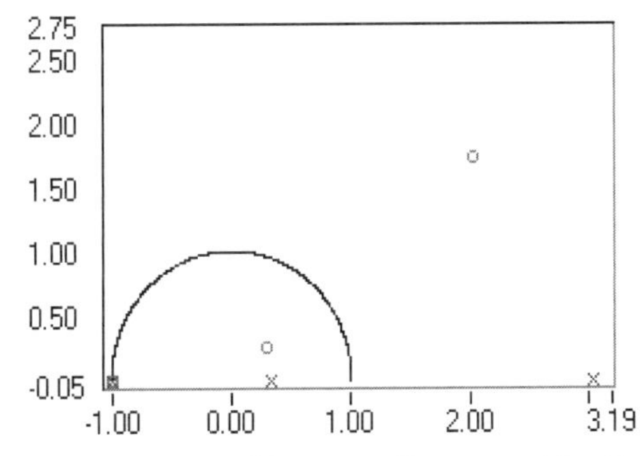

Figure 6-7 Zero distribution of the pair of lowpass filters used for the FBI fingerprint compression. While the symbol "x" represents zeros of the analysis lowpass filter, the symbol "o" indicates zeros of the synthesis lowpass filter. The analysis and synthesis filters both have four zeros at π (that is, $z = -1$), respectively. Since the reciprocal pairs are assigned to the same filter, the resulting filters have linear phase.

Figure 6-6 illustrates another set of popular scaling functions and wavelets that are employed for the FBI (Federal Bureau of Investigation) fingerprint compression. In this case,

$$P_0(z) = (1 + z^{-1})^8 Q(z) \tag{6.28}$$

Figure 6-7 depicts the zero distribution of the corresponding lowpass filters, $H_0(z)$ and $G_0(z)$. Both of them have four zeros at π. Since the pairs of reciprocals, z_i and its reciprocal $1/z_i$ (as well as z_i^* and its reciprocal $1/z_i^*$), are assigned to the same filter, the resulting filters have linear phase.

As mentioned earlier, the derivations in this section are mainly based on the biorthogonal case (see Figure 6-1). The orthogonal transform, in fact, is a special case of its biorthogonal counterpart. In the next section we will investigate conditions, in terms of the product filter $P_0(z)$ and lowpass filters $H_0(z)$ or $G_0(z)$, for orthogonal wavelets.

6.2 Orthogonal Filter Banks

The relationship defined by Eq. (6.9) ensures that the analysis and synthesis filters are orthogonal in the sense of

$$\sum_n h_i[n-2k]\gamma_i[n] = \delta(k) \qquad and \qquad \sum_n h_i[n-2k]\gamma_l[n] = 0 \qquad i \neq l \tag{6.29}$$

However, the condition (6.29) alone does not ensure that the resulting filter banks $\{h_i[n]\}$ or $\{\gamma_i[n]\}$ form orthogonal filter banks and that the resulting wavelets are orthogonal wavelets, such as in the case of the spline wavelets in Example 6.1 a and b. In these cases, there are two sets of scaling functions and mother wavelets. Neither the analysis bank nor the synthesis bank satisfies the power complementarity condition (5.63), i.e.,

$$H_0(\omega)H_0{}^*(\omega) + H_1(\omega)H_1{}^*(\omega) \neq 1 \tag{6.30}$$

and

$$G_0(\omega)G_0{}^*(\omega) + G_1(\omega)G_1{}^*(\omega) \neq 1 \tag{6.31}$$

Figure 6-8 plots Eqs.(6.30) and (6.31) for the quadratic spline wavelets in Example 6.1 a. Although the filters in (6.29) meet the condition of perfect reconstruction, the transformation formed by the set of filters $H_n(\omega)$ or $G_n(\omega)$ is not energy conserving.

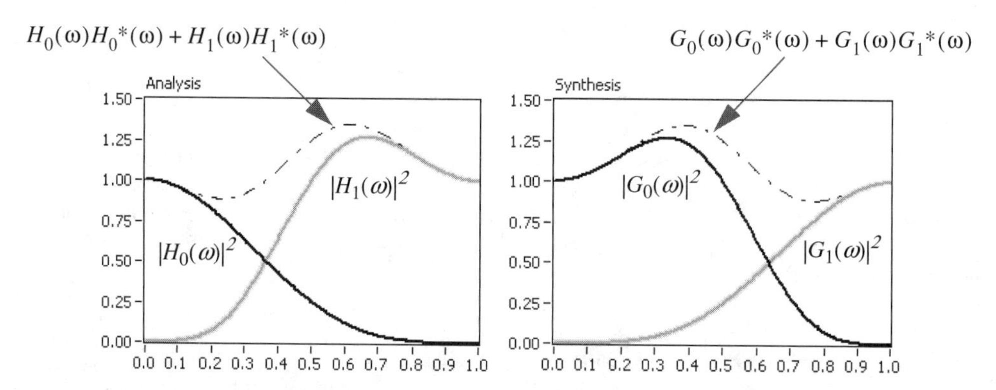

Figure 6-8 In the biorthogonal case, neither the analysis filter bank $H_n(\omega)$ nor the synthesis filter bank $G_n(\omega)$ satisfy the power complementarity condition. The bases of biorthogonal wavelets do not reproduce the signal energy exactly.

For the filter banks in Example 6.1 a and b the relation corresponding to (5.56) has a form

$$H_m(z)G_n{}^*(z) + H_m(-z)G_n{}^*(-z) = 0 \qquad m \neq n \tag{6.32}$$

we name the resulting filter banks the *biorthogonal* filter banks.

In addition to Eq. (6.29), if the filters of the PR filter banks also satisfy the following relationship

$$\sum_n h_i[n-2k]h_i[n] = \delta(k) \qquad and \qquad \sum_n h_i[n-2k]h_l[n] = 0 \qquad i \neq l \tag{6.33}$$

then the resulting filter banks are *orthogonal filter banks*. Obviously, it is a special case of the biorthogonal filter banks. For the orthogonal filter banks, the synthesis and the analysis filters are related by time reversal. The reader can verify that the filter banks for the Daubechies wavelets in Eqs. (6.26) and (6.27) satisfy the condition (6.33).

For orthogonal filter banks, once the product filter $P_0(z)$ is selected, we only need to define $H_0(z)$. With $H_0(z)$, as we will see shortly, we can easily find all the other filters. Many applications demonstrate that the lack of orthogonality complicates quantization and bit allocation between bands, eliminating the principle of conservation of energy. The bases of biorthogonal wavelets do not reproduce the signal energy exactly. Reconstructing a signal from these coefficients may amplify any error introduced in the coefficients. Hence, orthogonal filter banks are often the first choice when we factor the product filter $P_0(z)$. The remaining question is: What constraints does $P_0(z)$ have to meet for orthogonal filter banks?

To achieve Eq. (6.33), we can let

$$\boxed{H_1(z) = -z^{-N}H_0(-z^{-1})} \tag{6.34}$$

which implies that $h_1[n]$ is the *alternating flip* of $h_0[n]$, i.e.,

$$(h_1[0], h_1[1], h_1[2], \ldots) = (h_0[N], -h_0[N-1], h_0[N-2], \ldots) \tag{6.35}$$

From Eqs. (6.9) and (6.34), we can readily compute $G_0(z)$ and $G_1(z)$. For example,

$$\boxed{G_0(z) = z^{-N}H_0(z^{-1})} \tag{6.36}$$

Therefore, $\gamma_0[n]$ is the flip of $h_0[n]$, i.e.,

$$(\gamma_0[0], \gamma_0[1], \gamma_0[2], \ldots) = (h_0[N], h_0[N-1], h_0[N-2], \ldots) \tag{6.37}$$

Substituting Eq. (6.36) into Eq. (6.16), we have

$$P_0(z) = z^{-N}H_0(z)H_0(z^{-1}) \tag{6.38}$$

If we define

$$P(z) = H_0(z)H_0(z^{-1}) \tag{6.39}$$

then

$$P_0(z) = z^{-N}P(z) \tag{6.40}$$

Moreover,

$$P(e^{j\omega}) = \sum_{n=-N}^{N} p[n]e^{-j\omega n} = \left|\sum_{n=0}^{N} h_0[n]e^{-j\omega n}\right|^2 \tag{6.41}$$

which implies that $P(z)$ is non-negative. $P_0(z)$ is the time-shifted non-negative function $P(z)$. It can be shown that the maximally flat filter defined in Eq. (6.18) ensures that this requirement is met. However, special care must be taken when $P_0(z)$ has other forms, such as an equiripple half-band filter [47]. Table 6-1 compares different types of filters. Note that the filter banks cannot be orthogonal and linear phase simultaneously.

Table 6-1 Digital Filters for PR Filter Banks

Filter Type	Location of Zeros	Comments
Real	Complex conjugate symmetrical	
Linear Phase	Each filter must contain both z_i and its reciprocal $1/z_i$. (The pair of reciprocals must be in the same filter.)	Desirable for image processing
Minimum Phase	All zeros have to be on or inside of the unit circle.	Minimum phase lag
Orthogonal	Each filter cannot have z_i and its reciprocal $1/z_i$ simultaneously. z_i and its reciprocal $1/z_i$ have to be in separate filters. This condition is contradictory to that required for linear phase filters.	Analysis and synthesis have the same performance Even length (N odd) Convenient for bit allocation and quantization error control Not linear phase

The discussion in this section has been focused on the two-channel perfect reconstruction filter banks. The relationship of the Fourier transform and the scaling function to the frequency response of the FIR filter is given by the infinite products (5.47). From these connections, we reason that since $H_0(z)$ or $G_0(z)$ is lowpass and, if it has a high order zero at $z = -1$ (i.e., $\omega = \pi$), the Fourier transform of the analysis/synthesis scaling function $\phi(t)$ should drop off rapidly and, therefore, $\phi(t)$ should be smooth. It turns out that this is indeed true. This is related to the fact that the differentiability of a function is tied to how fast the magnitude of its Fourier transform drops off as the frequency goes to infinity.

It can be shown [6] that the number of zeros at $z = -1$ of the lowpass filter $H_0(z)$ or $G_0(z)$ determines the number of zero moments of the wavelets. Table 6-2 lists the discrete and continuous moments of the second (db2, see Example 6.1) and third order (db3) Daubechies scaling function and wavelets. While the continuous moments are defined in (5.79) and (5.80), the discrete moments are defined as

$$\mu_0[k] = \sum_n n^k h_0[n] \tag{6.42}$$

and

$$\mu_1[k] = \sum_n n^k h_1[n] \tag{6.43}$$

The Daubechies filter coefficients ensure the maximum number of zero moments of the wavelets (or maximum vanishing moments), which is weakly related to the number of oscillations. Researchers have also recognized that in some applications, the zero moments of the scaling function are also useful.[4] However, the filters yielding a combination of zero wavelet and zero scaling function moments cannot be directly generated by the filter $P_0(z)$ or $P(z)$ that were introduced earlier. The design of these kinds of filters need to employ other techniques that are beyond the scope of this book. The resulting wavelets are traditionally named *coiflets*. The reader can find related materials in [13], [270], [247], [288], and [400].

Table 6-2 Moments of Daubechies Scaling and Wavelet Functions [6]

	Scaling Functions				Wavelets			
	$\mu_0[k]$		$m_0[k]$		$\mu_1[k]$		$m_1[k]$	
k	db2	db3	db2	db3	db2	db3	db2	db3
0	1.41421	1.41421	1.00000	1.00000	0	0	0	0
1	0.89657	1.15597	0.63439	0.81740	0	0	0	0
2	0.56840	0.94489	0.40192	0.66814	1.22474	0	0.21650	0
3	-0.8643	-0.2243	0.13109	0.44546	6.57201	3.35410	0.78677	0.29646
4	-6.0593	-2.6274	-0.3021	0.11722	25.9598	40.6796	2.01434	2.28246
5	-23.437	5.30559	-1.0658	-0.0466	90.8156	329.323	4.44427	11.4461

The theory of perfect reconstruction filter banks was developed a long time before wavelet analysis became popular, but the original filter banks had no vanishing moments and thus did not always generate finite energy wavelets. "The connection between the number of vanishing moments of a filter and the corresponding wavelet having finite energy is not immediately apparent. But having a conjugate mirror filter $h[n]$ such that its Fourier transform $H(\omega)$ vanishes at $\omega = \pi$ ($z = -1$) is a necessary condition so that the cascade of such filters defines a finite energy scaling function and hence a finite energy wavelet. In addition, the number of vanishing moments of a wavelet is equal to the number of zeros of the Fourier transform of its filter at $\omega = \pi$; saying that a wavelet has one vanishing moment is equivalent to saying that $H(\pi) = 0$. More generally, if a wavelet has k vanishing moments, then $H(\omega)$ and its first $k - 1$ derivatives vanish at $\omega = 0$" [22].

Unlike the Fourier transform in which there is only one set of basis functions, for the wavelet transform one can choose from an infinite number of mother wavelets. The success of the application of wavelet analysis largely hinges on the selection of the mother wavelet. In most applications, such as denoising, the ideal wavelet is one that will encode a signal using the greatest possible number of zero coefficients, or the majority of coefficients closest to zero. Unfortu-

4. Except for the 0th moment. For a valid scaling function, its 0th moment has to be equal to one in order for the scaling function to be of lowpass.

nately, such requirements cannot be described mathematically in most cases. The most effective procedure for selecting a proper mother wavelet may be though trial and error. With the help of computer software, such as the Signal Processing Toolset provided by National Instruments, engineers and scientists now can immediately see the effect on their data samples of selecting between different product filters $P_0(z)$ and factorization schemes.

6.3 General Tree-Structure Filter Banks and Wavelet Packets

Once the two-channel perfect reconstruction filter banks have been determined, based on the result obtained in Section 5.4, we can readily compute the discrete wavelet transform through the tree of filter banks, as shown in Figure 6-9. While the subscript "1" indicates a highpass, the subscript "0" corresponds to a lowpass filter. Hence, the symbol y_{001} represents a sequence after two consecutive stages of lowpass filtering and one stage of highpass filtering. Accordingly, the symbol y_{000} represents a sequence after three consecutive stages of lowpass filtering. To compute the discrete wavelet transform, we repeatedly split, filter, and decimate the lowpass bands. The outputs of the highpass filters, such as y_1, y_{01}, and y_{001}, are the discrete wavelet transform that we need. The computational complexity of such a wavelet transform is $O(N)$.

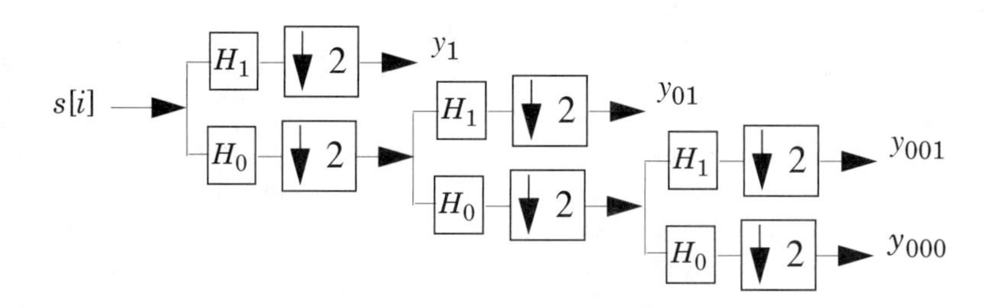

Figure 6-9 The tree of filter banks for computing the discrete wavelet transform

Intuitively, the tree of filter banks illustrated in Figure 6-9 is not the only possible decomposition scheme. For instance, the signal can also be decomposed by the system depicted in Figure 6-10. Particularly, if we split both the lowpass and highpass bands at all stages, the resulting filter bank structure becomes a full binary tree, as shown in Figure 6-11. This type of tree takes $O(N\log N)$ calculations and results in a completely evenly spaced frequency resolution. The result of the full binary tree is similar to that calculated by the STFT algorithm.

At this point, the question arises as to whether one can generate alternative decomposition trees that allow branches on the highpass bands, and whether such a decomposition has an interpretation in terms of basis functions for linear vector spaces of functions. The answer is yes. The reader can find a corresponding mathematical proof in [27] or [40]. The resulting decomposition is called *wavelet packets* by *Coifman* and *Wickerhauser* [261]. The main advantage of the wavelet packets is the flexibility it offers, which allows adaptation to particular signals. The potential of wavelet packets lies in the capacity to offer a rich menu of orthonormal bases from which the "best" one can be chosen ("best" according to a particular criterion).

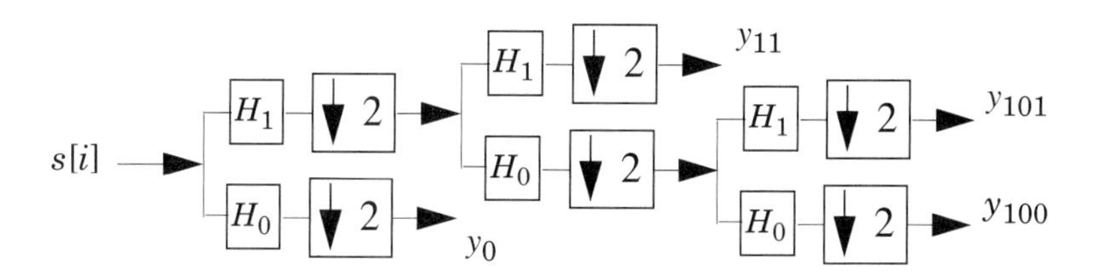

Figure 6-10 An alternative three-stage decomposition

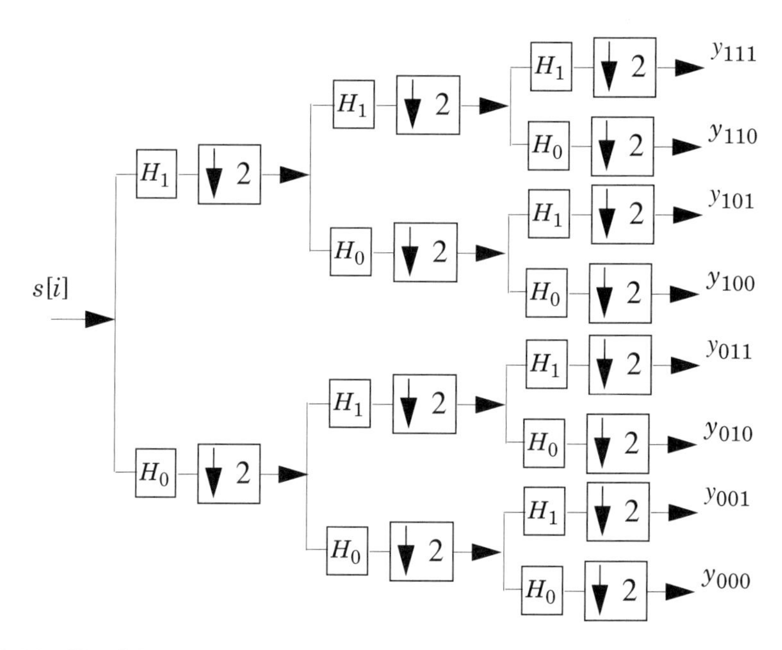

Figure 6-11 The full binary tree results in a completely evenly spaced frequency resolution, which is similar to that calculated by the STFT.

It can be easily verified that for a given number of stages k, the number of different decompositions is

$$P(k) = P(k-1)^2 + 1 \qquad P(1) = 1 \tag{6.44}$$

The number of possible choices grows dramatically as the stage k (or scale) increases. A successful application of the wavelet packets includes the FBI standard for fingerprint image compression (see [248] and [250]).

CHAPTER 7

Wigner-Ville Distribution

The short-time Fourier transform and the wavelet transform introduced in preceding chapters compute correlations between the signal and a family of time-frequency functions. The time-frequency resolution of these transforms is thus determined by a corresponding set of elementary functions. Since the relationship between the time and frequency resolutions of any given function is governed by the uncertainty principle, neither the short-time Fourier transform nor the wavelet transform can achieve arbitrary fine resolutions in both time and frequency simultaneously.

In addition to these correlation-based approaches, there is another type of time-frequency representation that is motivated by the time-frequency energy density. In contrast to the conventional power spectral density, the time-frequency energy density function describes the signal's energy distribution in terms of both time and frequency. Compared to correlation-based methods, such as the short-time Fourier transform and the wavelet transform, the approaches motivated by the time-frequency energy distribution can yield representations with a better time-frequency resolution and with many other properties that are of importance for time-frequency analysis.

One of the most prominent members of time-frequency energy density functions is the Wigner-Ville distribution which is qualitatively different from the STFT spectrogram introduced earlier. The Wigner-Ville distribution first appeared in Wigner's paper in 1932 in the area of quantum mechanics [226]. Wigner's original motivation was to calculate the quantum correlation of the second virial coefficients of a gas. Fifteen years later, a French mathematician, *J. Ville*, derived, by a method based on characteristic functions, a similar result for the signal's time-frequency energy distribution [219]. Although the formulas obtained by Ville and Wigner[1] are similar, their motivations and methods were rather different.

The initial reaction to Ville's work was not enthusiastic. It was not until 1980 when an important set of papers by *Claasen* and *Mecklenbräuker* was published [91]. There, the authors presented a comprehensive approach and originated many new ideas and procedures uniquely suited to time-frequency analysis. Since then, time-frequency analysis has become one of the most active research areas in the field of signal processing. The main obstacle to apply the Wigner-Ville distribution is the cross-term interference that greatly limits its applications. Consequently, the discovery of the strengths and shortcomings of the Wigner-Ville distribution has become a major thrust in the field of time-frequency analysis. This is also the main focus of the rest of this book.

Sections 7.1 and 7.2 introduce the definition and general properties, respectively, of the Wigner-Ville distribution. The Wigner-Ville distribution is computed by correlating the signal with a time and frequency translated version of itself. Unlike the short-time Fourier transform and the wavelet transform, there are no predetermined elementary (or window) functions. So, there is no loss of resolution in the Wigner-Ville distribution. Moreover, the Wigner-Ville distribution possesses many remarkable properties desirable for time-frequency analysis.

Although in many aspects the Wigner-Ville distribution is superior to the STFT spectrogram, its applications are limited by the existence of cross-term interference. The cause and effect of the cross-term is investigated in Section 7.3.

In Section 7.4, we address the relationships between the Wigner-Ville distribution and the square of the short-time Fourier transform (STFT spectrogram) and the square of the wavelet transform (scalogram). It is interesting to note that both the STFT spectrogram and the scalogram can be interpreted as smoothed versions of the Wigner-Ville distribution.

Section 7.5 discusses the effect of the Hilbert transform on the Wigner-Ville distribution. Employing the analytic signal is a popular way of reducing cross-term interference, but it has to be kept in mind that the analytic signal can also bring about distortion, especially in the low frequency band.

The transition from the continuous Wigner-Ville distribution to its discrete counterpart is discussed in Section 7.6. The importance of developing the discrete algorithm is due to the increasing use and capabilities of digital computers. The great flexibility of the digital computer

1. *Eugene P. Wigner* was born on November 17, 1902, in Budapest, Hungary. He attended Budapest Lutheran High School, where he met *John von Neumann* (1903-1957). At age 23, Wigner received his doctorate in chemical engineering. He first came to the United States in 1930 as a lecturer in mathematics at Princeton, where he spent most of his career. He worked on a series of projects during World War II, including the Manhattan Project. On December 10, 1963, *Eugene Wigner*, with *Maria Goeppert Mayer* and *J. Hans D. Jensen*, received the Nobel Prize for their discoveries concerning the theory of the atomic nucleus and elementary particles, which were based on atomic research that had been conducted during the first three decades of the twentieth century. Wigner laid the groundwork for the revision of concepts concerning right-left symmetry by *Chen Ning Yang* and *Tsung-Dao Lee*, both of whom won the Nobel Prize in 1957. Wigner's sister, *Margaret*, married *Paul Dirac*, the brilliant Nobel laureate, whom Wigner always described as "my famous brother-in-law." Wigner was aware that there were other joint densities but chose what has now become the Wigner distribution "because it seems to be the simplest."

has spurred experimentation with the design of increasingly sophisticated discrete-time systems for which no apparent practical implementation using analog equipment exists. For applied engineers and scientists, the job can never be completed without feasible numerical implementations.

The Wigner-Ville distribution has been known for many years, appearing in numerous papers and books. What we introduce in this chapter are the fundamentals for those who want to have a basic understanding of time-frequency analysis and perhaps to use it for their applications. There is no effort to explain how Wigner, Ville, and many other great scientists derived their results, though it is interesting and enlightening. The reader can find more comprehensive treatments of the Wigner-Ville distribution by *Cohen* [10], *Claasen* and *Mecklenbräuker* [91], and *Nuttall* [178].

7.1 Wigner-Ville Distribution

For a signal $s(t)$, its Wigner-Ville distribution is

$$WVD_s(t, \omega) = \int_{-\infty}^{\infty} s\left(t + \frac{\tau}{2}\right) s^*\left(t - \frac{\tau}{2}\right) e^{-j\omega\tau} d\tau \tag{7.1}$$

where the product $s(t+\tau/2)s^*(t-\tau/2)$ is Hermitian symmetry in τ. In contrast to the linear representations, such as the short-time Fourier transform and the wavelet transform, the Wigner-Ville distribution is said to be bilinear in the signal because the signal enters twice in its calculation. Obviously, the Wigner-Ville distribution in (7.1) is phase shift invariant. For example, the phase shift $e^{j\phi}$ does not modify its value.

While (7.1) is considered as an auto-*WVD*, the cross-*WVD* is defined as

$$WVD_{s,g}(t, \omega) = \int_{-\infty}^{\infty} s\left(t + \frac{\tau}{2}\right) g^*\left(t - \frac{\tau}{2}\right) e^{-j\omega\tau} d\tau \tag{7.2}$$

Obviously,

$$WVD_{s,g}(t, \omega) = WVD^*_{g,s}(t, \omega) \tag{7.3}$$

Therefore,

$$WVD_s(t, \omega) = WVD^*_s(t, \omega) \tag{7.4}$$

which implies that the auto-*WVD* is always a real-valued function.

To get a better idea of how the Wigner-Ville distribution overcomes the drawbacks possessed by the Fourier spectra, in what follows, let us compute the Wigner-Ville distributions for a couple of typical time varying signals.

Example 7.1 Wigner-Ville distribution of a Gaussian-type function

$$s(t) = \sqrt[4]{\frac{\alpha}{\pi}} e^{-\alpha t^2/2} \tag{7.5}$$

where $s(t)$ is a normalized Gaussian function that has unit energy. Obviously, in the time domain $s(t)$ is symetrical with respect to $t = 0$. Example 2.8 further tells us that the signal's energy is concentrated around $\omega = 0$. Its Wigner-Ville distribution is

$$WVD_s(t, \omega) = \sqrt{\frac{\alpha}{\pi}}\int_{-\infty}^{\infty} \exp\left\{-\frac{\alpha}{2}\left[\left(t+\frac{\tau}{2}\right)^2 + \left(t-\frac{\tau}{2}\right)^2\right]\right\}e^{-j\omega\tau}d\tau \tag{7.6}$$

$$= e^{-\alpha t^2}\sqrt{\frac{\alpha}{\pi}}\int_{-\infty}^{\infty} \exp\left\{-\frac{\alpha}{4}\tau^2\right\}e^{-j\omega\tau}d\tau$$

$$= 2\exp\left\{-\left[\alpha t^2 + \frac{1}{\alpha}\omega^2\right]\right\}$$

which indicates that the $WVD(t,\omega)$ of the Gaussian function is concentrated at the origin (0,0) – a signal's time and frequency center, as we have observed from the signal's time and frequency representations. The parameter α controls the spread of the $WVD(t,\omega)$ in both the time and frequency domains. A larger value of α leads to less spreading in the time domain but more spreading in the frequency domain, and vice versa.

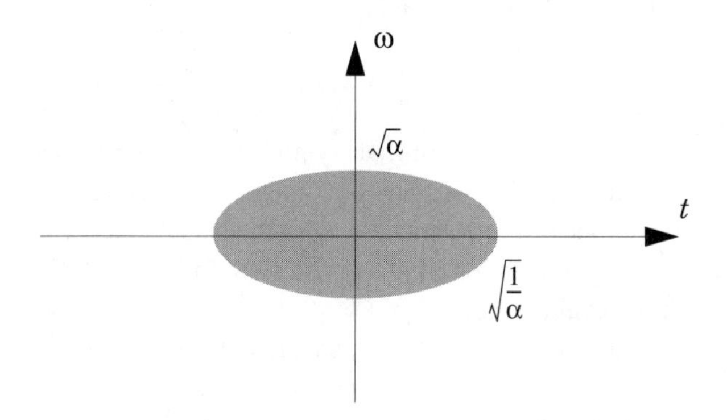

Figure 7-1 The Wigner-Ville distribution of the Gaussian function is concentrated at (0,0). The parameter α controls the spread of the $WVD(t,\omega)$ in both the time and frequency domains. The area enclosed by the ellipse where the levels are down to e^{-1} of their peak value is $A = \pi$, which is half the size of the STFT spectrogram discussed in Example 3.2.

The contour plot of the $WVD(t,\omega)$ consists of concentric ellipses. The contour for the case where the levels are down to e^{-1} of their peak value is the ellipse indicated in Figure 7-1. The area of this particular ellipse is π. Compared to the STFT spectrogram in Example 3.2, the resolution of the $WVD(t,\omega)$ is fixed, and thus there is no window effect. For the STFT spectrogram, the minimum area of the ellipse where the levels are down to e^{-1} of their peak value is $A = 2\pi$, which is twice as big as that of the $WVD(t,\omega)$. In other words, the resolution of the $WVD(t,\omega)$ is twice as good as that of an STFT spectrogram if we use the area A as a measure. Moreover, it is interesting to note that

$$\frac{1}{2\pi}\int_{-\infty}^{\infty} WVD_s(t, \omega)d\omega = \sqrt{\frac{\alpha}{\pi}}e^{-\alpha t^2} = |s(t)|^2 \tag{7.7}$$

It says that the integration of the Wigner-Ville distribution along the frequency axis yields the instantaneous power in the signal. We commonly call (7.7) a *time marginal condition*. Conversely,

$$\int_{-\infty}^{\infty} WVD_s(t, \omega)dt = \sqrt{\frac{4\pi}{\alpha}} e^{-\omega^2/\alpha} = |S(\omega)|^2 \tag{7.8}$$

where $S(\omega)$ denotes the Fourier transform of the signal $s(t)$ in (7.5).

The relationship (7.8) shows that the integration of the Wigner-Ville distribution along the time axis yields the power spectrum of the signal. Accordingly, we name (7.8) as the *frequency marginal condition*. From (7.7) or (7.8) it is easy to see that the Wigner-Ville distribution is energy conserving. Because of the time and frequency marginal conditions, the Wigner-Ville distribution was proposed as a signal's time-frequency density function.

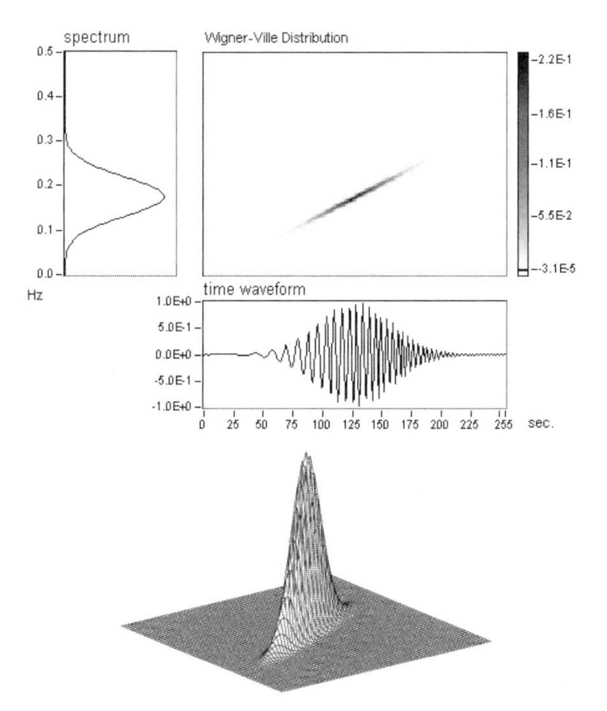

Figure 7-2 While there is no information in the conventional power spectrum (left plot) about how the frequencies of the Gaussian chirplet evolve over time, the Wigner-Ville distribution clearly indicates that the energy of the Gaussian chirplet is concentrated at $\omega = \beta t$.

Example 7.2 Wigner-Ville distribution of a Gaussian chirplet

$$s(t) = \sqrt[4]{\frac{\alpha}{\pi}} \exp\left\{-\frac{\alpha}{2}t^2 + j\frac{\beta}{2}t^2\right\} \tag{7.9}$$

Eq. (7.9) depicts a Gaussian windowed linear chirp signal. Since it is concentrated in a short time period, the signal in (7.9) is traditionally named as a Gaussian chirplet. The power spectrum of the Gaussian chirplet is [36]

$$|S(\omega)|^2 = \sqrt{4\pi\frac{\alpha^2+\beta^2}{\alpha}}\exp\left\{-\frac{\alpha}{\alpha^2+\beta^2}\omega^2\right\} \tag{7.10}$$

From (7.9), it is obvious that frequency of the chirplet linearly increases with time when β is greater than zero, and linearly decreases with time when β is less than zero. However, such important information is not available in its Fourier spectrum (7.10). The Fourier spectrum only tells us what frequencies the Gaussian chirplet contains, not how it changes. Applying Eq. (7.1), we obtain

$$WVD_s(t,\omega) = \exp\{-\alpha t^2\}\sqrt{\frac{\alpha}{\pi}}\int_{-\infty}^{\infty}\exp\left\{-\frac{\alpha}{4}\tau^2\right\}e^{-j(\omega-\beta t)\tau}d\tau \tag{7.11}$$

$$= 2\exp\left\{-\left[\alpha t^2+\frac{1}{\alpha}(\omega-\beta t)^2\right]\right\}$$

which shows that the energy of the Gaussian chirplet is concentrated at $\omega = \beta t$.

Figure 7-2 illustrates the power spectrum of the Gaussian chirplet and the corresponding Wigner-Ville distribution. While the conventional power spectrum (on the left) only tells what frequencies the chirplet contains, the Wigner-Ville distribution further describes how the spectrum evolves with time. The reader can verify that the Wigner-Ville distribution of the Gaussian chirplet in (7.11) also satisfies the marginal conditions defined in (7.7) and (7.8).

If the chirplet (7.9) is written in the form $A(t)^{j\varphi(t)}$, where both amplitude $A(t)$ and phase $\varphi(t)$ are real-valued functions, then the first derivative of the phase $\varphi'(t) = \beta t$. It is interesting to note that

$$\frac{\int_{-\infty}^{\infty}\omega\exp\left\{-\left[\alpha t^2+\frac{1}{\alpha}(\omega-\beta t)^2\right]\right\}d\omega}{\int_{-\infty}^{\infty}\exp\left\{-\left[\alpha t^2+\frac{1}{\alpha}(\omega-\beta t)^2\right]\right\}d\omega} = \beta t = \varphi'(t) \tag{7.12}$$

If we consider the Wigner-Ville distribution $WVD(t,\omega)$ as a signal's time-frequency density function, then the left side of (7.12) can be interpreted as the conditional mean frequency or mean instantaneous frequency. What (7.12) says is that the first derivative of the signal's phase in fact characterizes the signal's mean instantaneous frequency, not the instantaneous frequency as many textbooks suggest.

In Examples 7.1 and 7.2, the resulting Wigner-Ville distributions not only satisfy the marginal conditions, (7.7) and (7.8), but also are non-negative. In such cases, the Wigner-Ville distribution behaves as the signal's time-frequency density function. In general, however, this is not true. Except for Gaussian type functions, the Wigner-Ville distribution of almost all other signals can go negative. We will elaborate this subject shortly.

In addition to the Wigner-Ville distribution, there are a number of infinite time-frequency representations that describe, to a certain degree, the signal's instantaneous power spectra. The Wigner-Ville distribution is only one, the simplest one, of the entire class. The Wigner-Ville distribution of Gaussian type signals, such as (7.5) and (7.9), is mostly concentrated and non-negative. Other representations, such as the Choi-Williams distribution [90] and the cone-shape distribution [236], may preserve some useful properties that the Wigner-Ville distribution has, but their representation for Gaussian type signals will be less concentrated and may go negative.

Besides formulas (7.1) and (7.2), the Wigner-Ville distribution can also be directly computed from the signal's Fourier transform. Let

$$x_t(\tau) = s\left(t + \frac{\tau}{2}\right) \tag{7.13}$$

and

$$y_t(\tau) = g^*\left(t - \frac{\tau}{2}\right) \tag{7.14}$$

Then, the corresponding Fourier transforms of $x_t(\tau)$ and $y_t(\tau)$ are

$$x_t(\tau) \leftrightarrow X_t(\omega) = 2S(2\omega)e^{j2\omega t} \tag{7.15}$$

and

$$y_t(\tau) \leftrightarrow Y_t(\omega) = 2G^*(2\omega)e^{-j2\omega t} \tag{7.16}$$

Based on the convolution theorem (2.58), Eq. (7.2) can be written as

$$\begin{aligned} WVD_{s,g}(t,\omega) &= \int_{-\infty}^{\infty} s\left(t+\frac{\tau}{2}\right)g^*\left(t-\frac{\tau}{2}\right)e^{-j\omega\tau}d\tau \\ &= \int_{-\infty}^{\infty} x_t(\tau)y_t(\tau)e^{-j\omega\tau}d\tau \\ &= X_t(\omega) \otimes Y_t(\omega) \\ &= \frac{4}{2\pi}\int_{-\infty}^{\infty} S(2\alpha)G^*(2\omega-2\alpha)e^{j(4\alpha-2\omega)t}d\alpha \end{aligned} \tag{7.17}$$

Let $2\alpha = \omega + \Omega/2$; then,

$$\boxed{WVD_{s,g}(t,\omega) = \frac{1}{2\pi}\int_{-\infty}^{\infty} S\left(\omega+\frac{\Omega}{2}\right)G^*\left(\omega-\frac{\Omega}{2}\right)e^{j\Omega t}d\Omega} \tag{7.18}$$

where $S(\omega)$ and $G(\omega)$ denote the Fourier transforms of $s(t)$ and $g(t)$, respectively. Accordingly, the corresponding auto-*WVD* is

$$\boxed{WVD_s(t,\omega) = \frac{1}{2\pi}\int_{-\infty}^{\infty} S\left(\omega+\frac{\Omega}{2}\right)S^*\left(\omega-\frac{\Omega}{2}\right)e^{j\Omega t}d\Omega} \tag{7.19}$$

While (7.1) and (7.2) compute the Wigner-Ville distribution from the time waveform of the signal, (7.18) and (7.19) evaluate the Wigner-Ville distribution from the corresponding Fourier transforms. We will frequently employ forms (7.18) and (7.19) in future development.

Finally, if a signal has a finite duration, for instance, $s(t) = 0$ outside $[t_0, t_1]$, then its $WVD_s(t,\omega)$ will also be zero outside $[t_0, t_1]$ for all ω. The same argument is also true for a bandlimited signal. Based on Eq. (7.19), the reader can verify that for a bandlimited signal,

$WVD_s(t,\omega)$ will be zero for frequencies outside the band. Therefore, the Wigner-Ville distribution satisfies the finite support properties in time and frequency.

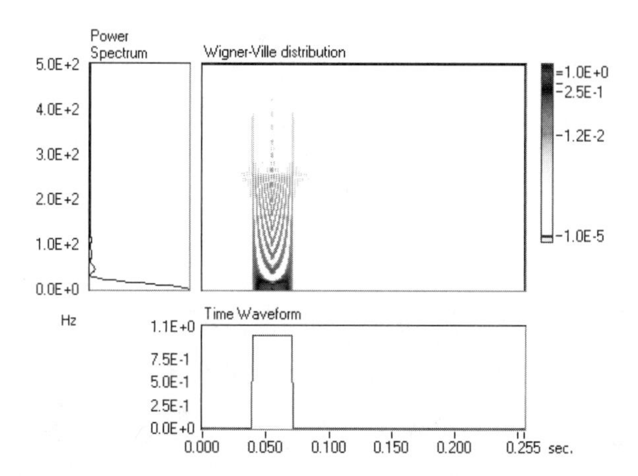

Figure 7-3 The Wigner-Ville distribution has finite support for a signal with one pulse. In this case, $WVD_s(t,\omega) = 0$ whenever $s(t) = 0$. $WVD_s(t,\omega)$ satisfies the time marginal condition.

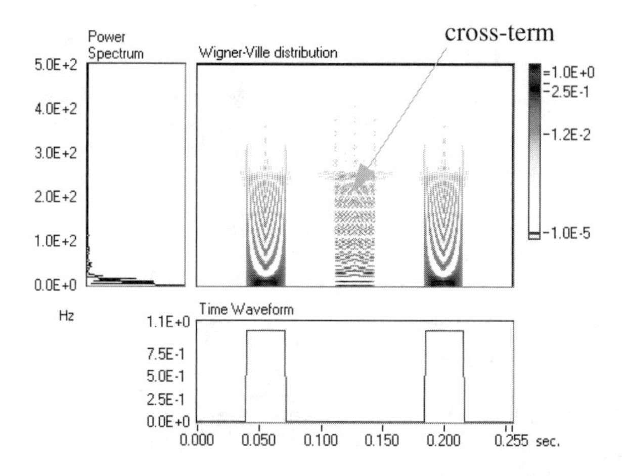

Figure 7-4 The Wigner-Ville distribution does not have finite support for a signal with more than one pulse. In addition to two time limited pulses, there is also a non-zero term between them.

Figure 7-3 depicts the Wigner-Ville distribution of a time pulse. It shows that $WVD_s(t,\omega) = 0$ when $s(t) = 0$. In other words, the Wigner-Ville distribution satisfies the finite support property in time. However, this is not the case when the signal consists of more than one time pulse, as shown in Figure 7-4. In addition to two time limited pulses, there is also a non-zero term between them. The same argument also applies for multiple bandlimited and separated frequency tones. Those non-zero terms are traditionally called cross-terms, and have been the main obstacle preventing the use of the Wigner-Ville distribution for many applications. We will elaborate on this subject in Section 7.3.

7.2 General Properties of the Wigner-Ville Distribution

In the preceding section we introduced the definition of the Wigner-Ville distribution. Since the Wigner-Ville distribution plays a fundamental role in time-frequency analysis, in this section we will discuss it in more detail. Calculating the properties of the Wigner-Ville distribution is fairly straightforward. Hence, many of the proofs will be left as an exercise for the reader. The presentation will be focused on the interpretation of the results.

- **Time Marginal Condition**: In Examples 7.1 and 7.2 we have shown that the Wigner-Ville distributions of Gaussian and chirplet functions satisfy the time (7.7) and frequency (7.8) marginal conditions. As a matter of fact, the Wigner-Ville distribution preserves time and frequency marginal properties for any signal.

$$\frac{1}{2\pi}\int_{-\infty}^{\infty} WVD_s(t,\omega)d\omega = \int_{-\infty}^{\infty} s\left(t+\frac{\tau}{2}\right)s^*\left(t-\frac{\tau}{2}\right)\frac{1}{2\pi}\int_{-\infty}^{\infty} e^{-j\omega\tau}d\omega d\tau \tag{7.20}$$

$$= \int_{-\infty}^{\infty} s\left(t+\frac{\tau}{2}\right)s^*\left(t-\frac{\tau}{2}\right)\delta(\tau)d\tau$$

$$= |s(t)|^2$$

- **Frequency Marginal Condition**:

$$\int_{-\infty}^{\infty} WVD_s(t,\omega)dt = \int_{-\infty}^{\infty}\int_{-\infty}^{\infty} s\left(t+\frac{\tau}{2}\right)s^*\left(t-\frac{\tau}{2}\right)e^{-j\omega\tau}dtd\tau \tag{7.21}$$

$$= \int_{-\infty}^{\infty} e^{-j\omega\tau}\int_{-\infty}^{\infty} s(t)s^*(t-\tau)dtd\tau$$

$$= \int_{-\infty}^{\infty} e^{-j\omega\tau}R(\tau)d\tau$$

$$= |S(\omega)|^2$$

Based on (7.20) and (7.21), we can readily obtain the following relationship

$$\frac{1}{2\pi}\int_{-\infty}^{\infty}\int_{-\infty}^{\infty} s\left(t+\frac{\tau}{2}\right)s^*\left(t-\frac{\tau}{2}\right)e^{-j\omega\tau}dtd\omega = \frac{1}{2\pi}\int_{-\infty}^{\infty}|S(\omega)|^2d\omega = \int_{-\infty}^{\infty}|s(t)|^2dt \tag{7.22}$$

which says that the Wigner-Ville distribution is unitary. The energy contained in $WVD_s(t,\omega)$ is equal to the energy in the original signal $s(t)$. Because of properties (7.20), (7.21), and (7.22), the Wigner-Ville distribution was traditionally treated as a signal's *time-frequency density function*.

- **Mean Instantaneous Frequency**: In Example 7.2 we had pointed out that the conditional mean frequency of the Wigner-Ville distribution of a Gaussian chirplet is equal to the first directive of the phase of the signal. Now, let's prove it for general cases.

Assume that $s(t) = A(t)e^{j\varphi(t)}$, where the magnitude $A(t)$ and the phase $\varphi(t)$ are both real-valued functions. Then, at any time instant t, the conditional mean frequency of the Wigner-Ville distribution is equal to $\varphi'(t)$, the first directive of the phase of the signal, i.e.,

$$\langle\omega\rangle_t = \frac{\int_{-\infty}^{\infty}\omega WVD_s(t,\omega)d\omega}{\int_{-\infty}^{\infty}WVD_s(t,\omega)d\omega} = \frac{1}{2\pi|s(t)|^2}\int_{-\infty}^{\infty}\omega WVD_s(t,\omega)d\omega = \varphi'(t) \tag{7.23}$$

Proof

First, let's look at the numerator

$$\frac{1}{2\pi}\int_{-\infty}^{\infty}\omega WVD_s(t,\omega)d\omega = \left(\frac{1}{2\pi}\right)^2\int_{-\infty}^{\infty}e^{j\Omega t}\int_{-\infty}^{\infty}\omega S\left(\omega+\frac{\Omega}{2}\right)S^*\left(\omega-\frac{\Omega}{2}\right)d\omega d\Omega \tag{7.24}$$

Let

$$H(\omega) = \omega S\left(\omega+\frac{\Omega}{2}\right) \tag{7.25}$$

and

$$G(\omega) = S\left(\omega-\frac{\Omega}{2}\right) \tag{7.26}$$

Based on the derivative property (2.55) and the shifting property (2.51), the time domain representations of (7.25) and (7.26) can be derived as

$$h(t) = -j\frac{d}{dt}(s(t)e^{-j\Omega t/2}) \tag{7.27}$$

and

$$g(t) = s(t)e^{j\Omega t/2} \tag{7.28}$$

respectively. According to Parseval's formula (2.60), (7.24) is equivalent to

$$\frac{1}{2\pi}\int_{-\infty}^{\infty}\omega WVD_s(t,\omega)d\omega = \frac{1}{2\pi}\int_{-\infty}^{\infty}e^{j\Omega t}\int_{-\infty}^{\infty}\left\{-j\frac{d}{da}[s(a)e^{-ja\Omega/2}]s^*(a)e^{-ja\Omega/2}\right\}dad\Omega \tag{7.29}$$

$$= \frac{1}{2\pi}\int_{-\infty}^{\infty}e^{j\Omega t}\int_{-\infty}^{\infty}\left\{-\frac{\Omega}{2}|s(a)|^2e^{-ja\Omega}-je^{-ja\Omega}s^*(a)\frac{d}{da}s(t)\right\}dad\Omega$$

$$= \frac{1}{2\pi}\int_{-\infty}^{\infty}e^{j\Omega t}\int_{-\infty}^{\infty}-\frac{\Omega}{2}|s(a)|^2e^{-ja\Omega}dad\Omega - \frac{j}{2\pi}\int_{-\infty}^{\infty}s(a)\frac{d}{da}s(t)\int_{-\infty}^{\infty}e^{j\Omega(t-a)}d\Omega da$$

$$= -\frac{1}{4\pi}\int_{-\infty}^{\infty}\Omega e^{j\Omega t}\int_{-\infty}^{\infty}|s(a)|^2e^{-ja\Omega}dad\Omega - js^*(a)\frac{d}{da}s(t)$$

Applying the convolution theorem (2.58), the first term becomes

$$-\frac{1}{4\pi}\int_{-\infty}^{\infty}\Omega e^{j\Omega t}\int_{-\infty}^{\infty}|s(a)|^2 e^{-ja\Omega}dad\Omega = -\frac{1}{4\pi}\int_{-\infty}^{\infty}\Omega e^{j\Omega t}[S(\Omega)\oplus S(\Omega)]d\Omega \tag{7.30}$$

$$= \frac{1}{4\pi}\int_{-\infty}^{\infty}S(b)\frac{1}{2\pi}\int_{-\infty}^{\infty}\Omega S^*(\Omega-b)e^{j\Omega t}d\Omega db$$

$$= \frac{-j}{2}\int_{-\infty}^{\infty}S(b)\frac{d}{dt}[s^*(t)e^{jbt}]db$$

$$= \frac{-j}{2}\int_{-\infty}^{\infty}S(b)\left[e^{jbt}\frac{d}{dt}s^*(t)+jbs^*(t)e^{jbt}\right]db$$

$$= \frac{-j}{2}s(t)\frac{d}{dt}s^*(t)+\frac{j}{2}s^*(t)\frac{d}{dt}s(t)$$

Substituting (7.30) into (7.29), we obtain

$$\frac{1}{2\pi}\int_{-\infty}^{\infty}\omega WVD_s(t,\omega)d\omega = \frac{-j}{2}\left\{s(t)\frac{d}{dt}s^*(t)+s^*(t)\frac{d}{dt}s(t)\right\} = |A(t)|^2\varphi'(t) \tag{7.31}$$

Replacing the numerator in (7.23) by (7.31), we can readily obtain the instantaneous frequency property.

- **Group (Time) Delay Property**: Assume that the Fourier transform of a signal $s(t)$ is $S(\omega) = B(\omega)e^{j\psi(\omega)}$. Then, the first derivative of the phase $\psi'(\omega)$ is called the *group delay*, which reflects the signal's time delay. For the Wigner-Ville distribution we can prove that

$$\frac{\int_{-\infty}^{\infty}tWVD_s(t,\omega)dt}{\int_{-\infty}^{\infty}WVD_s(t,\omega)dt} = \frac{\int_{-\infty}^{\infty}tWVD_s(t,\omega)dt}{|S(\omega)|^2} = -2\pi\psi'(\omega) \tag{7.32}$$

which says that the conditional mean time of the Wigner-Ville distribution is equal to the group delay. We leave the proof for the reader to practice.

The mean instantaneous frequency and group delay are the two most important attributes of time-dependent power spectra, which are traditionally used as a figure of merit of prospective time-frequency representations. As a matter of fact, it is not difficult to build a time-frequency function $P_s(t,\omega)$ that meets the time and frequency marginal conditions (7.20) to (7.22). For example, we can let

$$P_s(t,\omega) = \frac{|s(t)|^2}{E}|S(\omega)|^2 \tag{7.33}$$

where

$$E = \int_{-\infty}^{\infty}|s(t)|^2dt = \frac{1}{2\pi}\int_{-\infty}^{\infty}|S(\omega)|^2d\omega \tag{7.34}$$

Obviously, $P_s(t,\omega)$ in (7.33) satisfies all the marginal conditions, but it does not provide any information about how a signal's frequencies change with time. In other words, the time and frequency marginal conditions alone are not enough for time-frequency analysis. What makes the Wigner-Ville distribution so special is that it truly describes how the signal's power spectrum evolves over time. Only those functions $P_s(t,\omega)$ that not only meet the marginal conditions, but also are able to characterize the change of a signal's frequencies, are useful for time-frequency analysis.

At this point, it is interesting to note that the useful properties of the Wigner-Ville distribution, such as (7.20), (7.21), (7.23), and (7.32), are all determined by *averaging* the Wigner-Ville distribution. For instance, the time marginal condition (7.20) is obtained by averaging the Wigner-Ville distribution over frequency. The instantaneous frequency property (7.23) is the average of the frequency at time instant t. This suggests that the properties that are useful to signal processing will be mainly determined by the smooth portions of the Wigner-Ville distribution. Because the average of the highly oscillatory portions is small, these portions have limited influence on the useful properties. This is a very important observation that leads to an improved time-frequency representation – the time-frequency distribution series introduced in Chapter 9.

- **Time-Shift Invariant**: If the Wigner-Ville distribution of a signal $s(t)$ is $WVD_s(t,\omega)$, then the Wigner-Ville distribution of the time-shifted version $s_0(t) = s(t-t_0)$ is the time-shifted Wigner-Ville distribution of $s(t)$, i.e.,

$$WVD_{s0}(t,\omega) = WVD_s(t-t_0,\omega) \tag{7.35}$$

- **Frequency Modulation Invariant**: If $WVD_s(t,\omega)$ is the Wigner-Ville distribution of signal $s(t)$, then the Wigner-Ville distribution of the frequency-modulated version $s_0(t) = s(t)\exp\{j\omega_0 t\}$ is the frequency-shifted Wigner-Ville distribution of $s(t)$, i.e.,

$$WVD_{s0}(t,\omega) = WVD_s(t,\omega-\omega_0) \tag{7.36}$$

Based on (7.35) and (7.36), one can readily obtain the Wigner-Ville distribution of time-shifted and frequency-modulated Gaussian type functions, i.e.,

$$\sqrt[4]{\frac{\alpha}{\pi}}\exp\left\{-\frac{\alpha}{2}(t-t_0)^2\right\}e^{j\omega_0(t-t_0)} \Rightarrow 2\exp\left\{-\alpha(t-t_0)^2-\frac{1}{\alpha}(\omega-\omega_0)^2\right\} \tag{7.37}$$

and

$$\sqrt[4]{\frac{\alpha}{\pi}}\exp\left\{-\frac{\alpha}{2}(t-t_0)^2\right\}e^{j[\omega_0(t-t_0)+\beta(t-t_0)^2/2]} \Rightarrow 2\exp\left\{-\alpha(t-t_0)^2-\frac{1}{\alpha}(\omega-\omega_0-\beta t)^2\right\} \tag{7.38}$$

In (7.37) and (7.38), we employ the results obtained from Examples 7.1 and 7.2, respectively.

Finally, as in the case of the short-time Fourier transform, not all time-frequency functions $P(t,\omega)$ can be Wigner-Ville distributions. For example,

$$P(t, \omega) = \begin{cases} 1 & |t| \le t_0 \ and \ |\omega| \le \omega_0 \\ 0 & otherwise \end{cases} \tag{7.39}$$

will not be a valid Wigner-Ville distribution, because no signal can be time and frequency limited simultaneously.

7.3 Wigner-Ville Distribution for the Sum of Multiple Signals

The Wigner-Ville distribution not only possesses many useful properties, but also has better resolution than the STFT spectrogram. However, its applications are very limited due to the existence of the so-called cross-term interference.

For $s(t) = s_1(t) + s_2(t)$, we can show that the corresponding Wigner-Ville distribution is

$$WVD_s(t, \omega) = WVD_{s1}(t, \omega) + WVD_{s2}(t, \omega) + 2Re\{ WVD_{s1,s2}(t, \omega)\} \tag{7.40}$$

Obviously, the Wigner-Ville distribution of the sum of two signals is not the sum of their corresponding Wigner-Ville distributions only. In addition to two auto-terms, (7.40) contains one cross-term $WVD_{s1,s2}(t,\omega)$. Because the cross-term usually oscillates and its magnitude is twice as large as that of the auto-terms, it often obscures the useful patterns of the time-dependent spectra.

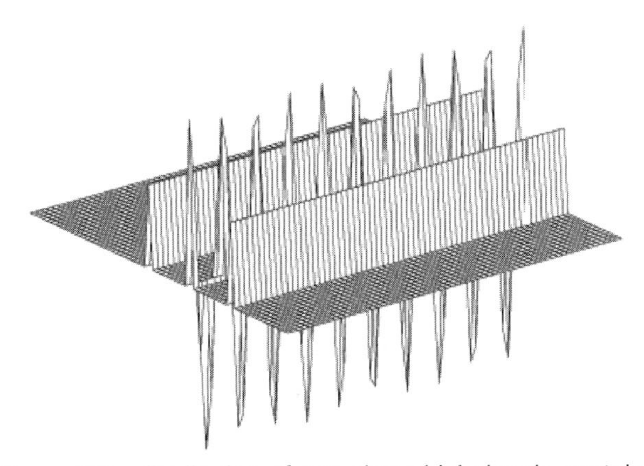

Figure 7-5 The Wigner-Ville distribution of two sinusoidal signals contains two desired auto-terms at frequencies ω_1 and ω_2, and an unwanted cross-term $\cos(\omega_1-\omega_2)t$ midway between the auto-terms. The cross-term oscillates and its amplitude is twice as large as that of the auto-terms.

Example 7.3 Wigner-Ville distribution of complex sinusoidal signals

When $s(t)$ is a single complex sinusoidal signal, for instance, $s(t) = exp(j\omega_0 t)$, we have

$$WVD_s(t, \omega) = \int_{-\infty}^{\infty} \exp\left\{j\omega_0\left(t + \frac{\tau}{2} - t + \frac{\tau}{2}\right)\right\} e^{-j\omega\tau} d\tau = 2\pi\delta(\omega - \omega_0) \tag{7.41}$$

which shows that $WVD_s(t,\omega)$ is concentrated at frequency ω. This is exactly what we expect. When $s(t)$ consists of the sum of two complex sinusoidal signals, such as $s(t)$ =

$\exp(j\omega_1 t)+\exp(j\omega_2 t)$, the conventional power spectrum is

$$|S(\omega)|^2 = 2\pi\delta(\omega_1) + 2\pi\delta(\omega_2) \tag{7.42}$$

The corresponding Wigner-Ville distribution is

$$WVD_s(t, \omega) = 2\pi\delta(\omega - \omega_1) + 2\pi\delta(\omega - \omega_2) + 4\pi\delta(\omega - \omega_\mu)\cos(\omega_d t) \tag{7.43}$$

where ω_μ and ω_d denote the geometric center and the distance between the two complex sinusoidal functions in the frequency domain, i.e.,

$$\omega_\mu = \frac{\omega_1 + \omega_2}{2} \qquad \omega_d = \omega_1 - \omega_2 \tag{7.44}$$

Equation (7.43) is plotted in Figure 7-5. In addition to the two desired auto-terms at frequencies ω_1 and ω_2, we get a large cross-term $\cos(\omega_1\text{-}\omega_2)t$ at ω_μ, midway between the two auto-terms. Unlike the two auto-terms that are non-negative, the cross-term oscillates at frequency ω_d, the frequency difference between the two components. Note that the average of the cross-term is equal to zero, i.e.,

$$\int_{-\infty}^{\infty} 4\pi\delta(\omega - \omega_\mu)\cos\{\omega_d t\}dt = 0 \qquad \omega_d \neq 0 \tag{7.45}$$

but the amplitude of the cross-term is twice as large as that of the auto-terms! Because the conventional power spectrum in (7.43) indicates that there is no signal at ω_μ, and the energy contained in the cross-term is zero, the cross-term is commonly considered as interference.

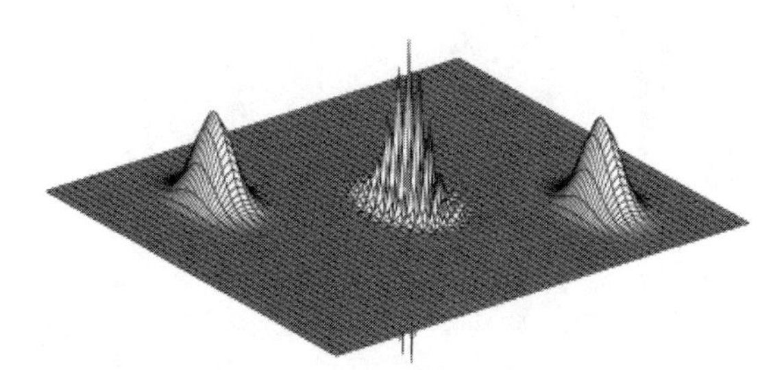

Figure 7-6 The cross-term is at (t_μ, ω_μ), midway between the two auto-terms. It oscillates in both the time and frequency directions.

Example 7.4 Wigner-Ville distribution of two Gaussian functions

$$s(t) = \sum_{i=1}^{2} \sqrt[4]{\frac{\alpha}{\pi}}\exp\left\{-\frac{\alpha}{2}(t - t_i)^2 + j\omega_i t\right\} \tag{7.46}$$

which contains two time-shifted and frequency-modulated Gaussian signals. One is concentrated at (t_1, ω_1). The other is centered at (t_2, ω_2). Then,

$$WVD_s(t, \omega) = 2\sum_{i=1}^{2} \exp\left\{-\alpha(t-t_i)^2 - \frac{1}{\alpha}(\omega-\omega_i)^2\right\} \tag{7.47}$$

$$+ 4\exp\left\{-\alpha(t-t_\mu)^2 - \frac{1}{\alpha}(\omega-\omega_\mu)^2\right\}\cos[(\omega-\omega_\mu)t_d + \omega_d(t-t_\mu) + \omega_d t_\mu]$$

where t_μ and ω_μ denote the geometric time and frequency means between the two individual Gaussian signals. That is,

$$t_\mu = \frac{t_1+t_2}{2} \qquad \omega_\mu = \frac{\omega_1+\omega_2}{2} \tag{7.48}$$

t_d and ω_d are the distances between two individual functions in the time and the frequency domains, i.e.,

$$t_d = t_1 - t_2 \qquad \omega_d = \omega_1 - \omega_2 \tag{7.49}$$

The summation in (7.47) corresponds to the auto-terms, which are non-negative. The last term represents the cross-term at (t_μ, ω_μ), midway between the two auto-terms, as shown in Figure 7-6. The cross-term oscillates in both the time and frequency directions. The rate of oscillation is proportional to t_d and ω_d, the distance between the two auto-terms. Both the auto-terms and the cross-term have the same two-dimensional Gaussian envelope that is concentrated and symmetrical in the joint time-frequency domain.

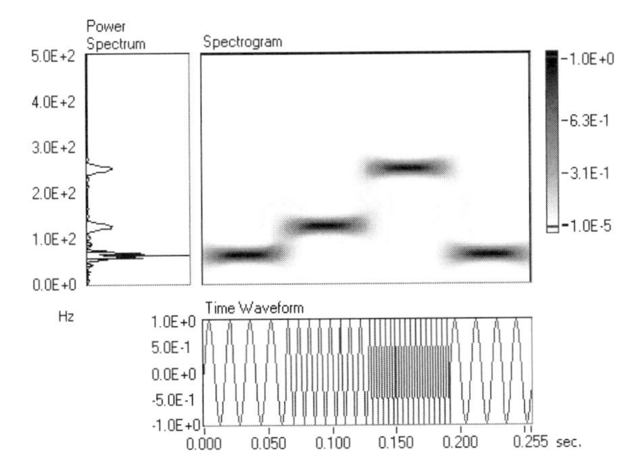

Figure 7-7 As shown in the power spectrum (on the left) and the STFT spectrogram (middle plot), the signal contains four frequency pulses. The STFT spectrogram has lower resolution but it does not have unwanted interference.

Examples 7.3 and 7.4 show that each pair of auto-terms creates one cross-term midway between them. The cross-term oscillates at the rate proportional to the distance between the pair of corresponding auto-terms. While the time domain oscillation is determined by the frequency difference ω_1 - ω_2, the frequency domain oscillation is controlled by the time difference t_1 - t_2.

For N separated components, the total number of cross-terms will be $N(N-1)/2$. For simple cases, such as in Examples 7.3 and 7.4, we may be able to identify the cross-term interference. When the number of separated components gets larger, as shown in Figures 7-7 and 7-8, distinguishing between the auto- and cross-terms can become very challenging. Moreover, except for very simple cases, such as the two sinusoidal signals in Example 7.3 and the two Gaussian functions in Example 7.4, the cross-term in general is not well defined. As we know, the signal can be broken into parts in an infinite number of ways, and the different decompositions will lead to different cross-terms. In most applications, the auto- and cross-terms are mixed together. Often, there is no way to tell which are desired and which are unwanted. It is the presence of cross-term interference that has greatly prevented the Wigner-Ville distribution from being used in many applications, even though it possesses many remarkable properties for time-frequency analysis. Over the years, searching for a method to reduce cross-term interference without destroying the desired properties has been one of the most active research topics in the field of time-frequency analysis.

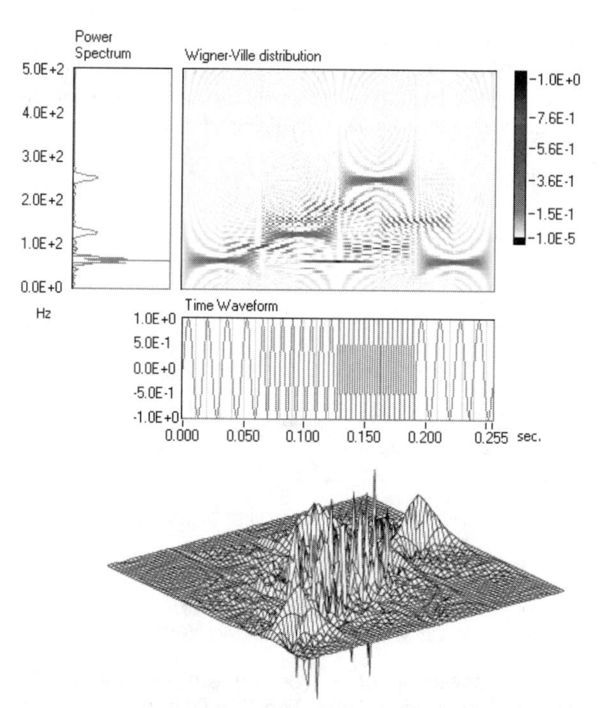

Figure 7-8 Due to the presence of cross-term interference, it is difficult to identify four distinct frequency pulses from the Wigner-Ville distribution.

7.4 Smoothed Wigner-Ville Distribution

Although the Wigner-Ville distribution possesses remarkable properties for signal analysis, its application has been limited due to the presence of cross-term interference. Since the cross-terms usually are strongly oscillating, the most straightforward way of removing cross-term interference is to apply 2D lowpass filtering. The resulting 2D convolution of the Wigner-Ville distribution can be considered as a smoothed Wigner-Ville distribution (SWVD), i.e.,

$$SWVD_s(t,\omega) = \int_{-\infty}^{\infty}\int_{-\infty}^{\infty}\phi(x,y)WVD_s(t-x,\omega-y)dxdy \qquad (7.50)$$

where $\phi(t,\omega)$ is a 2D lowpass filter, such as the 2D separable Gaussian function,

$$\phi(t,\omega) = e^{-\alpha t^2-\beta\omega^2} \qquad \alpha,\beta > 0 \qquad (7.51)$$

where the parameters α and β control the spreading of the function $\phi(t,\omega)$ in the time domain and the frequency domain, respectively. Obviously, the wider the spreading, the more the smoothing. The lowpass filtering can substantially suppress the cross-terms. On the other hand, smoothing will reduce the resolution. A trade-off exists between the degree of smoothing and the resolution.

In general, the Wigner-Ville distribution can have negative values. It can be proved ([85], [179], and [180]), however, that when

$$\alpha\beta \ge 1 \qquad (7.52)$$

the smoothed Wigner-Ville distribution (7.50) will be non-negative. It is interesting to note that when $\alpha\beta = 1$, $\phi(t,\omega)$ in (7.51), in fact, is the Wigner-Ville distribution of a Gaussian function (see Example 7.1). In this case, the smoothed Wigner-Ville distribution (7.50) is nothing more than $|STFT_s(t,\omega)|^2$, the STFT spectrogram. For an arbitrary signal $s(t)$, its STFT spectrogram is equal to the convolution of the Wigner-Ville distributions of the signal $s(t)$ and the analysis function $\gamma(t)$, i.e.,

$$\boxed{|STF_s(t,\omega)|^2 = \int_{-\infty}^{\infty}\int_{-\infty}^{\infty}WVD_\gamma(x,y)WVD_s(t-x,\omega-y)dxdy} \qquad (7.53)$$

where $WVD_s(t,\omega)$ and $WVD_\gamma(t,\omega)$ denote the Wigner-Ville distributions of the analyzed signal $s(t)$ and the analysis function $\gamma(t)$, respectively.

Proof

Let us expand the right side of (7.53).

$$\iiiint_{-\infty}^{\infty}\gamma\left(x+\frac{\mu}{2}\right)\gamma^*\left(x-\frac{\mu}{2}\right)s\left(t-x+\frac{\nu}{2}\right)s^*\left(t-x-\frac{\nu}{2}\right)e^{-jy(\mu-\nu)-j\nu\omega}d\mu d\nu dxdy \qquad (7.54)$$

$$= \int_{-\infty}^{\infty}\int_{-\infty}^{\infty}\int_{-\infty}^{\infty}\gamma\left(x+\frac{\mu}{2}\right)\gamma^*\left(x-\frac{\mu}{2}\right)s\left(t-x+\frac{\nu}{2}\right)s^*\left(t-x-\frac{\nu}{2}\right)e^{-j\nu\omega}\delta(\mu-\nu)d\mu d\nu dx$$

$$= \int_{-\infty}^{\infty}\int_{-\infty}^{\infty}\gamma\left(x+\frac{\nu}{2}\right)\gamma^*\left(x-\frac{\nu}{2}\right)s\left(t-x+\frac{\nu}{2}\right)s^*\left(t-x-\frac{\nu}{2}\right)e^{-j\nu\omega}d\nu dx$$

Let

$$a = x + \frac{\nu}{2} \qquad and \qquad b = x - \frac{\nu}{2} \tag{7.55}$$

Then

$$x = \frac{a+b}{2} \qquad \nu = a - b \tag{7.56}$$

The *Jacobian determinant* is

$$J = \begin{vmatrix} \frac{\partial x}{\partial a} & \frac{\partial x}{\partial b} \\ \frac{\partial \nu}{\partial a} & \frac{\partial \nu}{\partial b} \end{vmatrix} = \begin{vmatrix} 0.5 & 0.5 \\ 1 & -1 \end{vmatrix} = -1 \tag{7.57}$$

Because

$$dxd\nu = |J|\, dadb = dadb \tag{7.58}$$

formula (7.54) becomes

$$\begin{aligned} &\int_{-\infty}^{\infty}\int_{-\infty}^{\infty} \gamma(a)\gamma^*(b)s(t-b)s^*(t-a)e^{-j(a-b)\omega}dadb \\ &= \int_{-\infty}^{\infty} \gamma(a)s^*(t-a)e^{-j\omega a}d\alpha \int_{-\infty}^{\infty} \gamma^*(b)s(t-b)e^{j\omega b}db \\ &= |STFT_s(t,\omega)|^2 \end{aligned} \tag{7.59}$$

When the Wigner-Ville distribution of the analysis function $WVD_\gamma(t,\omega)$ is of a lowpass nature, as is the case for most applications, then the STFT spectrogram is a smoothed version of the Wigner-Ville distribution. This explains why the resolution of the STFT spectrogram is inferior to that of the Wigner-Ville distribution. Figure 7-7 depicts the STFT spectrogram for the frequency hopping signal. In this example the analysis function is a Gaussian, hence $WVD_\gamma(t,\omega)$ is 2D lowpass (see Example 7.1). Compared to the Wigner-Ville distribution plotted in Figure 7-8, the STFT spectrogram does not have interference, but it has poor resolution. Moreover, the STFT spectrogram does not preserve the time marginal, frequency marginal, mean instantaneous frequency, and many other useful properties that are possessed by the Wigner-Ville distribution. The STFT spectrogram improves the cross-term interference at the cost of lower resolution and the loss of other useful properties.

Evidently, the STFT spectrogram (7.53) is a special case of the smoothed Wigner-Ville distribution (7.50) with the function $\phi(t,\omega)$ equal to the Wigner-Ville distribution of the window function. In fact, as we will see in Chapter 8 with different kernel functions $\phi(t,\omega)$, all now known bilinear time-frequency transformations $C_s(t,\omega)$, such as the page distribution [185], the Rihaczek distribution [204], the Choi-Williams distribution [90], and the cone-shape distribution [236], can be written in the form of the smoothed Wigner-Ville distribution, i.e.,

$$C_s(t, \omega) = \int_{-\infty}^{\infty}\int_{-\infty}^{\infty} WVD_s(x, y)\phi_c(t-x, \omega-y)dxdy = SWVD_s(t, \omega) \tag{7.60}$$

The smoothed Wigner-Ville distribution actually is one of the many alternative representations of the general bilinear time-frequency transformation (also known as the *Cohen's class*).

Eq. (7.60) shows bilinear time-frequency representations in terms of 2D convolution of the Wigner-Ville distribution. Conversely, we can also write the Wigner-Ville distribution as the convolution of any member of Cohen's class, i.e.,

$$WVD_s(t, \omega) = \int_{-\infty}^{\infty}\int_{-\infty}^{\infty} C_s(x, y)h(t-x, \omega-y)dxdy \tag{7.61}$$

Note that unless $h(t,\omega)$ is a 2D lowpass filter, $WVD_s(t,\omega)$ will not be a smoothed version of $C_s(t,\omega)$. For example, we usually do not say that a Wigner-Ville distribution is a smoothed Choi-Williams distribution because $h(t,\omega)$ is not lowpass in this case.

Finally, in addition to the STFT spectrogram, the scalogram (the square of the wavelet transform) can also be written in terms of the Wigner-Ville distribution, e.g.,

$$\boxed{SCAL(a, b) = |CWT(a, b)|^2 = \int_{-\infty}^{\infty}\int_{-\infty}^{\infty} WVD_s(x, y)WVD_\psi\left(\frac{x-b}{a}, ay\right)dxdy} \tag{7.62}$$

where $WVD_s(t,\omega)$ and $WVD_\psi(t,\omega)$ denote the Wigner-Ville distribution of the analyzed signal $s(t)$ and the mother wavelet function $\psi(t)$, respectively. The operation involved in (7.62) is known as an *affine correlation*.

Proof

Let's expand the right side of (7.62), i.e.,

$$\iiiint_{-\infty}^{\infty} s\left(x+\frac{m}{2}\right)s^*\left(x-\frac{m}{2}\right)\psi\left(\frac{x-b}{a}+\frac{n}{2}\right)\psi^*\left(\frac{x-b}{a}-\frac{n}{2}\right)e^{-j(m+an)y}dxdydmdn \tag{7.63}$$

$$= \iiint_{-\infty}^{\infty} s\left(x+\frac{m}{2}\right)s^*\left(x-\frac{m}{2}\right)\psi\left(\frac{x-b}{a}+\frac{n}{2}\right)\psi^*\left(\frac{x-b}{a}-\frac{n}{2}\right)\delta(m+an)dxdmdn$$

$$= \int_{-\infty}^{\infty}\int_{-\infty}^{\infty} s\left(x+\frac{an}{2}\right)s^*\left(x-\frac{an}{2}\right)\psi\left(\frac{x-b}{a}+\frac{n}{2}\right)\psi^*\left(\frac{x-b}{a}-\frac{n}{2}\right)dxdn$$

Let $u = x+(\alpha v/2)$ and $v = x-(\alpha v/2)$. Then

$$x = \frac{u+v}{2} \qquad and \qquad n = \frac{u-v}{a} \tag{7.64}$$

The *Jacobian determinant* is

$$J = \begin{vmatrix} \frac{\partial x}{\partial u} & \frac{\partial x}{\partial v} \\ \frac{\partial n}{\partial u} & \frac{\partial n}{\partial v} \end{vmatrix} = \begin{vmatrix} 0.5 & 0.5 \\ \frac{1}{a} & -\frac{1}{a} \end{vmatrix} = -\frac{1}{a} \tag{7.65}$$

Because

$$dxdn = |J|dudv = \frac{dudv}{a} \tag{7.66}$$

Eq. (7.64) becomes

$$\int_{-\infty}^{\infty}\int_{-\infty}^{\infty} s(u)\psi\left(\frac{u-b}{a}\right)s^*(v)\psi^*\left(\frac{v-b}{a}\right)\frac{dudv}{a} = |CWT(a,b)|^2 \tag{7.67}$$

Figure 7-9 illustrates the relationships among the Wigner-Ville distribution $WVD_s(t,\omega)$, the STFT spectrogram $|STFT_s(t,\omega)|^2$, and the scalogram $|CWT_s(a,b)|^2$. As Eqs. (7.53) and (7.62) indicate, both the STFT spectrogram and the scalogram are smoothed versions of the Wigner-Ville distribution, which explains why the Wigner-Ville distribution has the best time-frequency resolution.

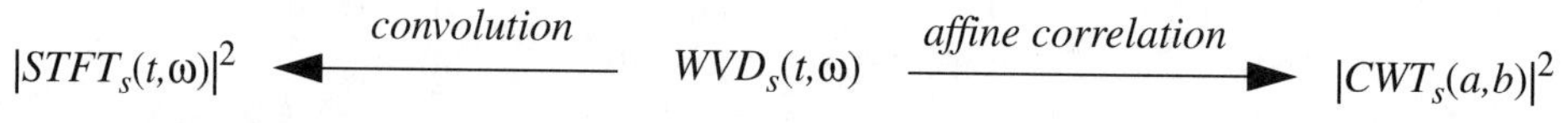

$$|STFT_s(t,\omega)|^2 \xleftarrow{\textit{convolution}} WVD_s(t,\omega) \xrightarrow{\textit{affine correlation}} |CWT_s(a,b)|^2$$

Figure 7-9 Both the spectrogram and the scalogram are smoothed versions of the Wigner-Ville distribution.

7.5 Wigner-Ville Distribution of Analytic Signals

The signals that we deal with under normal circumstances are real. A direct consequence of the realness is that the spectrum of the signal is always symmetric; that is,

$$S(\omega) = S^*(-\omega) \qquad or \qquad |S(\omega)|^2 = |S(-\omega)|^2 \tag{7.68}$$

and

$$WVD_s(t,\omega) = WVD_s(t,-\omega) \tag{7.69}$$

which leads to $\langle\omega\rangle_t = 0$ for all t. Moreover, there will be a huge cross-term in the vicinity of $\omega = 0$. The negative frequency components not only introduce redundancy, but also create cross-terms. To lessen cross-term interference, people usually remove negative frequency components while keeping the total energy of the signal unchanged [80], i.e.,

$$S_a(\omega) = \begin{cases} 2S(\omega) & \omega > 0 \\ S(\omega) & \omega = 0 \\ 0 & \omega < 0 \end{cases} \tag{7.70}$$

which is named as an *analytic signal*. In addition to 7.70, for a given real signal, its corresponding analytic signal can also be obtained by

$$s_a(t) = s(t) + jH\{s(t)\} \tag{7.71}$$

where $H\{\}$ denotes the Hilbert transform. Because the analytic signal is a halfband function, the resulting Wigner-Ville distribution $WVD_a(t,\omega)$ effectively avoids all cross-terms associated with the negative frequency components.

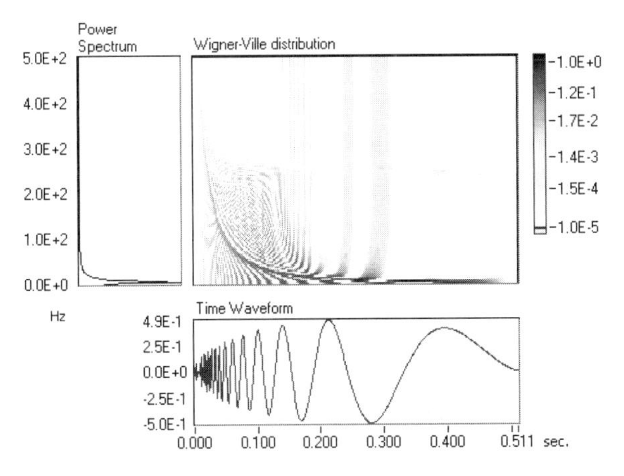

Figure 7-10 Wigner-Ville distribution of a Doppler signal (real-valued)

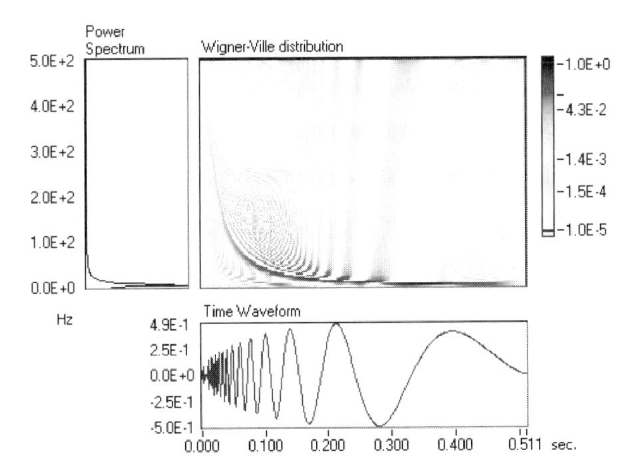

Figure 7-11 Wigner-Ville distribution of a Doppler signal (analytic signal). Compared to Figure 7-10, the analytic signal substantially reduces the cross-term interference.

Figures 7-10 and 7-11 illustrate the Wigner-Ville distribution of a real-valued Doppler signal and the Wigner-Ville distribution of the corresponding analytic signal. Obviously, interference is much less in the Wigner-Ville distribution computed by the analytic signal than that computed by the real signal.

However, it has to be kept in mind that the analytic signal differs from the original signal in several ways. For instance, the analytic signal of a time pulse is no longer time limited, because the analytic signal is band limited. Moreover, instantaneous frequencies of a real signal and its corresponding analytic signal can be very different, particularly in the low frequency band, though they have the same positive power spectrum. Hence, special care has to be taken when the analytic signal is used.

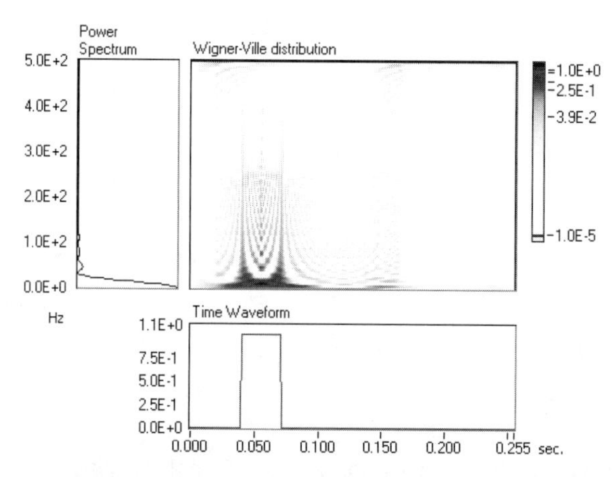

Figure 7-12 The analytic signal of the time-limited signal $s(t)$ is not time limited. The low frequency portion of the Wigner-Ville distribution for the analytic signal significantly spreads out compared to the Wigner-Ville distribution for the real signal in Figure 7-3.

The relationship between the Wigner-Ville distribution of an analytic signal $WVD_a(t,\omega)$ and the Wigner-Ville distribution of a real signal $WVD(t,\omega)$ can best be determined by using Eq. (7.19).

$$WVD_a(t,\omega) = \frac{1}{2\pi}\int_{-\infty}^{\infty} S_a\left(\omega+\frac{\Omega}{2}\right)S_a^*\left(\omega-\frac{\Omega}{2}\right)e^{j\Omega t}d\Omega \tag{7.72}$$

$$= \frac{1}{2\pi}\int_{-2\omega}^{2\omega} S\left(\omega+\frac{\Omega}{2}\right)S^*\left(\omega-\frac{\Omega}{2}\right)e^{j\Omega t}d\Omega$$

which is equivalent to

$$WVD_a(t,\omega) = \frac{1}{2\pi}\int_{-\infty}^{\infty} H(\Omega)S\left(\omega+\frac{\Omega}{2}\right)S^*\left(\omega-\frac{\Omega}{2}\right)e^{j\Omega t}d\Omega \tag{7.73}$$

where $H(\Omega)$ is an ideal lowpass filter with cut-off frequency 2ω. Based on the convolution theorem, we can rewrite Eq. (7.73) as

$$WVD_a(t,\omega) = 2\int_{-\infty}^{\infty}\frac{\sin(2\omega\tau)}{\tau}WVD_s(t-\tau,\omega)d\tau \tag{7.74}$$

This result is a convolution of $WVD_s(t,\omega)$ with an ideal frequency-dependent lowpass filter $\sin(2\omega t)/t$. As shown in Example 2.5, the smaller the quantity ω, the wider the spread. In other words, the spread of the low-frequency portion of $WVD_a(t,\omega)$ (small ω) is wider than that of the high-frequency portion.

Figures 7-3 and 7-12 illustrate $WVD_s(t,\omega)$ and $WVD_a(t,\omega)$ of a time pulse. Figure 7-3 shows that $WVD_s(t,\omega) = 0$ whenever $s(t) = 0$. $WVD_s(t,\omega)$ satisfies the time marginal condition. However, this is not the case for $WVD_a(t,\omega)$ in Figure 7-12. Compared to $WVD_s(t,\omega)$, the low-frequency portion of $WVD_a(t,\omega)$ is significantly altered. Because $WVD_a(t,\omega)$ is smoothed in the time domain, as shown in (7.74), all the time domain properties of $WVD_s(t,\omega)$, such as the time marginal condition and the instantaneous frequency property, are affected. The Wigner-Ville distribution of an analytic signal reduces the cross-term interference at the cost of losing some useful properties.

7.6 Discrete Wigner-Ville Distribution

The continuous-time Wigner-Ville distribution introduced in the previous sections is of great value in analyzing and gaining insight into the properties of continuous-time signals. Because the majority of signals that we deal with are discrete-time signals, in the present section we will address the subject of the discrete Wigner-Ville distribution. The importance of developing the discrete counterpart is due to the increasing use and capabilities of digital computers and the development of design methods for sampled-data systems. The great flexibility of the digital computer has spurred experimentation with the design of increasingly sophisticated discrete-time systems for which no apparent practical implementation using analog equipment exists.

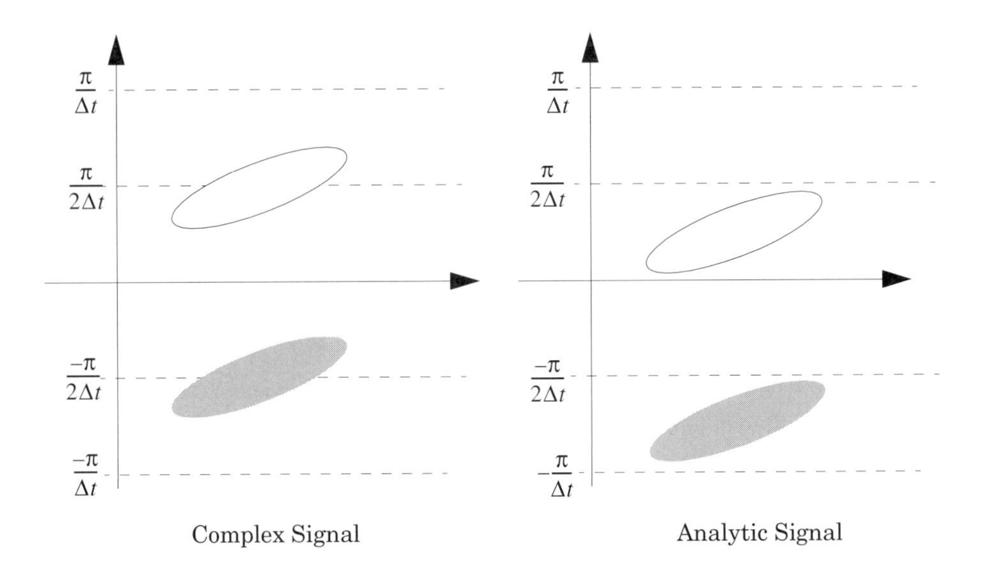

Figure 7-13 Discrete Wigner-Ville distributions for complex (left) and analytic (right) signals

By letting $u = \tau/2$ in (7.1), we have

$$WVD_s(t, \omega) = 2\int_{-\infty}^{\infty} s(t+u)s^*(t-u)e^{-j2\omega u}du \tag{7.75}$$

Assume the interval of integration in (7.75) is Δ. Applying the *Trapezoidal rule*, we have an approximation of the integral by

$$\overline{WVD_s}(t, \omega) = 2\Delta\sum_{n} s(t+n\Delta)s^*(t-n\Delta)e^{-j2\omega n\Delta} \tag{7.76}$$

If the signal $s(t)$ is sampled every Δt seconds, $\Delta t = \Delta$, then we obtain the discrete-time Wigner-Ville distribution as

$$\overline{WVD_s}[m\Delta t, \omega) = 2\Delta t\sum_{n} s[(m+n)\Delta t]s^*[(m-n)\Delta t]e^{-j2\omega n\Delta t} \tag{7.77}$$

Obviously,

$$\overline{WVD_s}[m\Delta t, \omega + \frac{\pi}{\Delta t}) = \overline{WVD_s}[m\Delta t, \omega) \tag{7.78}$$

In other words, the period of $\overline{WVD_s}[m\Delta t, \omega)$ in Eq. (7.78) is $\pi/\Delta t$ rather than $2\pi/\Delta t$. Consequently, for real-valued samples, if the signal's bandwidth is larger than $\pi/(2\Delta t)$, then aliasing will occur.

Without loss of generality, we assume that $\Delta t = 1$, then Eq. (7.77) reduces to

$$\overline{WVD_s}[m, \omega) = 2\sum_{n} s[m+n]s^*[m-n]e^{-j2\omega n} \tag{7.79}$$

Formula (7.79) requires an evaluation from minus infinity to plus infinity, which is not practical in real applications. Usually, we impose a running window $w[n]$, i.e.,

$$\overline{PWVD_s}[m, \omega) = 2\sum_{n=-\infty}^{\infty} w[n]s[m+n]s^*[m-n]e^{-j2\omega n} \tag{7.80}$$

which is the *pseudo Wigner-Ville distribution.* Introducing a running window $w[n]$ not only facilitates the implementation, but also can lessen the cross-term interference at the cost of resolution. While a narrow window will reduce the portion of the cross-term related to time differences, a wide window (narrow frequency bandwidth) will reduce the portion of the cross-term related to frequency differences.

Without loss of generality, let's assume that the running window $w[n]$ is real and symmetrical, that is, $w[n] = w[-n]$, and its length is equal to $2L$-1; then Eq. (7.80) becomes

$$\overline{PWVD_s}[m,\omega) = 2\sum_{n=-(L-1)}^{0} w[n]s[m+n]s^*[m-n]e^{-j2\omega n} \tag{7.81}$$

$$+2\sum_{n=0}^{L-1} w[n]s[m+n]s^*[m-n]e^{-j2\omega n} - 2w[0]s[m]s^*[m]$$

$$= 4Re\left\{\sum_{n=0}^{L-1} w[n]s[m+n]s^*[m-n]e^{-j2\omega n}\right\} - 2w[0]s[m]s^*[m]$$

When the DFT is used, Eq. (7.81) becomes

$$\boxed{DWVD[m,k] = 4Re\left\{\sum_{n=0}^{L-1} w[n]s[m+n]s^*[m-n]\exp\left(-j\frac{4\pi k}{L}n\right)\right\} - 2w[0]s[m]s^*[m]} \tag{7.82}$$

where

$$0 \le n < L \qquad and \qquad 0 \le k \le L \tag{7.83}$$

Because both the time and frequency variables are discrete, Eq. (7.82) is traditionally named as the discrete Wigner-Ville distribution (DWVD). Figure 7-13 illustrates the resulting discrete Wigner-Ville distributions of complex and analytic signals, respectively.

For real-valued data samples, due to aliasing, we need to double the sampling rate. This can be achieved by either applying interpolation filtering as introduced in Example 2.7 or simply padding zeros to the Fourier transform of the real-valued data samples $s[m]$.

Assume that the upsampled sequence is $y[m]$. Then, the corresponding discrete Wigner-Ville distribution is

$$\boxed{DWVD[m,k] = 4Re\left\{\sum_{n=0}^{2L-1} w[n]y[2m+n]y^*[2m-n]\exp\left(-j\frac{4\pi k}{2L}n\right)\right\} - 2w[0]s[2m]s^*[2m]} \tag{7.84}$$

where

$$0 \le n < 2L \qquad and \qquad 0 \le k \le L \tag{7.85}$$

Figure 7-14 depicts the discrete Wigner-Ville distribution for real-valued data samples. The frequency range of $DWVD[m,k]$ in both (7.82) and (7.84) is $0 \le \omega < \pi/\Delta t$, which corresponds to $0 \le k < L$.

Many algorithmic tricks have been proposed to overcome the aliasing problem. However, no matter what kinds of tricks are used, the requirement is usually the same: we must oversample either in the time domain or in the frequency domain. Although there are some variations among the algorithms, the improvements in terms of computational complexity as well as memory usage are marginal. There are two advantages of the method introduced in this section. Firstly, it is closer to the continuous-time Wigner-Ville distribution. Consequently, many properties possessed by the continuous-time Wigner-Ville distribution are carried over to its discrete counterpart. Secondly, the interpolation is very easy to implement.

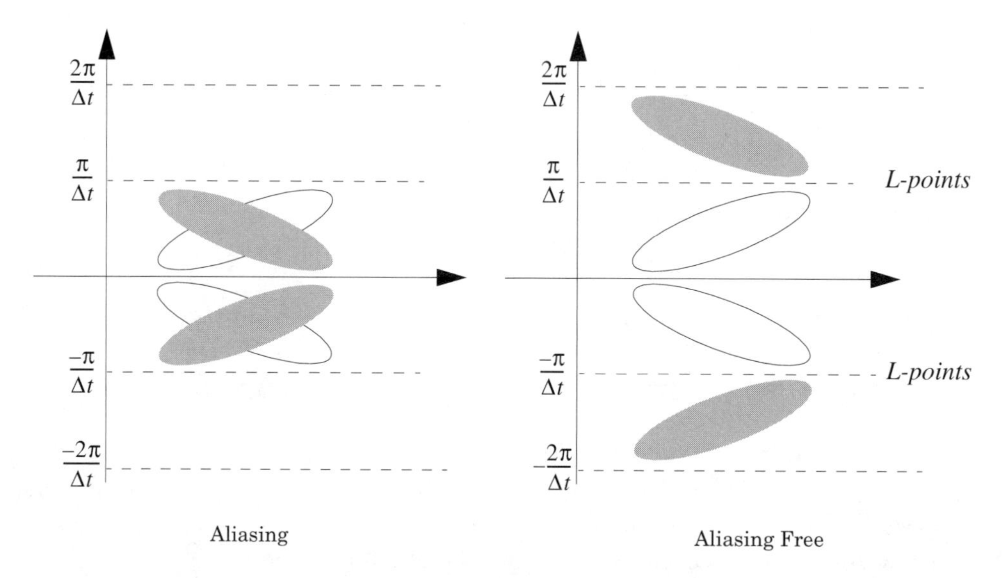

Figure 7-14 Discrete Wigner-Ville distributions of a real-valued signal with aliasing (left) and without aliasing (right)

CHAPTER 8

Other Time-Dependent Power Spectra

In addition to the STFT spectrogram and the Wigner-Ville distribution introduced in the preceding chapters, there are dozens of other time-frequency representations that characterize the time-frequency energy distribution of a signal. A major motivation in the development of these different representations has been the suppression of the cross-term interference appearing in the Wigner-Ville distribution. Since the cross-term is almost always strongly oscillating, the most straightforward technique of reducing the cross-term is to apply 2D lowpass filtering. Different filters then lead to different distributions. Based on the convolution theorem, such a 2D convolution of Wigner-Ville distributions can also be achieved by multiplication of the signal's ambiguity function and a kernel function. More interestingly, both of these engineering approaches are equivalent to the general representation initially developed in the area of quantum mechanics, now known as Cohen's class.

The presentation in this chapter starts with the symmetric ambiguity function. The relationship between the ambiguity function and the Wigner-Ville distribution has revealed a great deal about the mechanism of cross-term interference. In Section 8.2 Cohen's class is introduced, particularly, the relationship between the kernel function and the properties of the resulting representation. "By the facility of its parameterization and the large number of possible interpretations, it provides the theoretical framework within which most of the work in the 1980s was performed"[16]. With Cohen's class, the search for a desired representation reduces to the design of a single kernel function. In Section 8.3 three popular members of Cohen's class, the Choi-Williams distribution, the cone-shape distribution, and the signal-dependent adaptive representation, are reviewed. Finally, in Section 8.4 a recently popularized representation, the reassignment method, is introduced. Although the reassignment method does not belong to Cohen's class, its results, in terms of time-frequency resolution and computational complexity, are rather promising.

As there is no intention of writing an encyclopedic treatise on the very general theme of time-frequency analysis, the materials presented in this chapter have been limited to those that I feel are the basics for a beginner in the area of time-frequency analysis. Many theoretically excellent results, which are not practical for digital implementation, have been omitted. Readers can find a more comprehensive survey of time-frequency transforms in *Hlawatsch* and *Boudreaux-Bartels* [124], *Cohen* [10], and *Flandrin* [16].

8.1 Ambiguity Function

As mentioned in Section 2.2, the time-dependent spectrum can also be written in a form similar to the classical power spectrum, that is, the Fourier transform of the auto-correlation function,

$$P(t, \omega) = \int_{-\infty}^{\infty} R_t(\tau) e^{-j\omega\tau} d\tau \tag{8.1}$$

However, unlike the classical power spectrum in which the auto-correlation function is independent of time, the auto-correlation function $R_t(\tau)$ in (8.1) is time-dependent and is thereby called the time-dependent auto-correlation. Obviously, different time-dependent auto-correlations will lead to different time-dependent power spectra. If

$$R_t(t, \tau) = s\left(t + \frac{\tau}{2}\right) s^*\left(t - \frac{\tau}{2}\right) \tag{8.2}$$

then the resulting time-dependent power spectrum (8.1) becomes the Wigner-Ville distribution (7.1). Taking the Fourier transform with respect to the variable t instead of τ in $R_t(\tau)$ in (8.2), we obtain another popular time-frequency representation, the *symmetric ambiguity function* (AF), i.e.,

$$\boxed{AF_s(\theta, \tau) = \frac{1}{2\pi} \int_{-\infty}^{\infty} s\left(t + \frac{\tau}{2}\right) s^*\left(t - \frac{\tau}{2}\right) e^{j\theta t} dt} \tag{8.3}$$

While (8.3) is called the *auto-AF*, the *cross-AF* is defined as

$$AF_{s,g}(\theta, \tau) = \frac{1}{2\pi} \int_{-\infty}^{\infty} s\left(t + \frac{\tau}{2}\right) g^*\left(t - \frac{\tau}{2}\right) e^{j\theta t} dt \tag{8.4}$$

Unlike the auto-WVD, which is real in all cases, the symmetric ambiguity function generally is complex, i.e.,

$$AF_{s,g}(\theta, \tau) \neq AF^*_{g,s}(\theta, \tau) \tag{8.5}$$

The symmetric ambiguity function was first introduced by *Ville* and *Moyal*. Its relation to matched filters was thoroughly investigated by Woodward [52]. The ambiguity function reflects the correlation of signals in time and phase, and has wide applications in the context of radar and sonar. The discussion in this section will be focused only on those aspects that are important for time-frequency analysis.

Based upon the Fourier theorem, for a given ambiguity function $AF_s(\theta,\tau)$, we can compute the time-dependent auto-correlation function through the Fourier transform, i.e.,

$$\int_{-\infty}^{\infty} AF_s(\theta, \tau)\, e^{-j\theta t} d\theta = s\left(t + \frac{\tau}{2}\right) s^*\left(t - \frac{\tau}{2}\right) \tag{8.6}$$

Substituting (8.6) into (7.1) yields

$$\boxed{WVD_s(t, \omega) = \int_{-\infty}^{\infty}\int_{-\infty}^{\infty} AF_s(\theta, \tau)\, e^{-j(\omega\tau + \theta t)} d\theta d\tau} \tag{8.7}$$

which indicates that the Wigner-Ville distribution is the double Fourier transform of the symmetric ambiguity function. Figure 8-1 illustrates the relationship between the Wigner-Ville distribution and the symmetric ambiguity function.

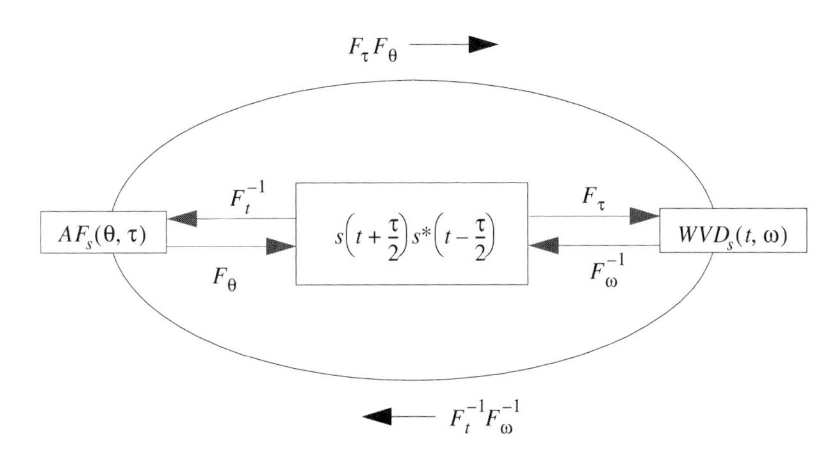

Figure 8-1 Relationship between the Wigner-Ville distribution and the ambiguity function

Example 8.1 Ambiguity function of a Gaussian signal

$$s(t) = \sqrt[4]{\frac{\alpha}{\pi}}\, e^{-\frac{\alpha}{2}(t - t_0)^2 + j\omega_0 t} \tag{8.8}$$

The time and frequency centers of $s(t)$ are t_0 and ω_0, respectively. The corresponding symmetric ambiguity function is

$$AF_s(\theta, \tau) = \exp\left\{-\left(\frac{1}{4\alpha}\theta^2 + \frac{\alpha}{4}\tau^2\right)\right\} e^{j(\omega_0\tau + \theta t_0)} \tag{8.9}$$

Figure 8-2 illustrates the real part of (8.9), which is centered at the origin (0,0) and oscillates. The phase $\omega_0 t + \theta t_0$ is related to the signal's time shift t_0 and frequency modulation ω_0.
In contrast, the Wigner-Ville distribution of the Gaussian function (7.5) is

$$WVD_s(t, \omega) = 2\exp\left\{-\frac{1}{\alpha}(t - t_0)^2 - \alpha(\omega - \omega_0)^2\right\} \tag{8.10}$$

which is centered at (t_0, ω_0). In other words, the signal's time shift and frequency modulation are associated with the geometric center of the WVD.

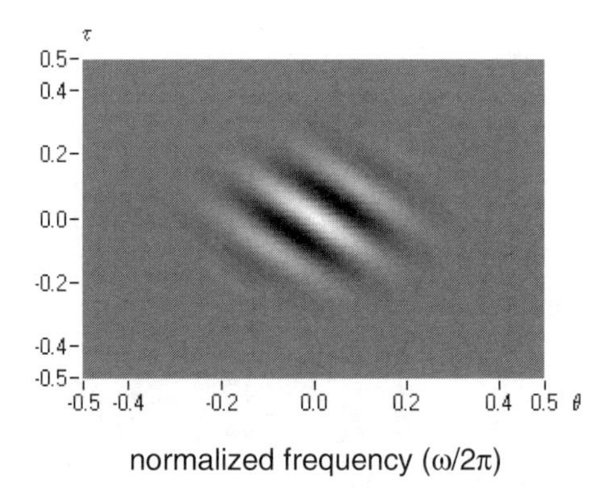

Figure 8-2 Ambiguity function of the Gaussian signal (real parts)

Example 8.2 Ambiguity function of the sum of two Gaussian functions

$$s(t) = \sum_{i=1}^{2} s_i(t) = \sum_{i=1}^{2} \sqrt[4]{\frac{\alpha}{\pi}} e^{-\frac{\alpha}{2}(t-t_i)^2 + j\omega_i t} \tag{8.11}$$

The two Gaussian functions are concentrated in (t_1,ω_1) and (t_2,ω_2), respectively. The corresponding symmetric ambiguity function is

$$AF_s(\theta,\tau) = \sum_{i=1}^{2} AF_{s_i}(\theta,\tau) + AF_{s1,s2}(\theta,\tau) + AF_{s2,s1}(\theta,\tau) \tag{8.12}$$

where the auto $AF_{s1}(\theta,\tau)$ are described by (8.9), and which are all concentrated around the origin (0,0). The cross $AF_{s1,s2}(\theta,\tau)$ is equal to

$$AF_{s1,s2}(\theta,\tau) = e^{-\frac{1}{4\alpha}(\theta-\omega_d)^2 + \frac{\alpha}{4}(\tau-t_d)^2} e^{j(\omega_\mu\tau - \theta t_\mu + \omega_d t_\mu)} \tag{8.13}$$

where

$$t_\mu = \frac{t_1+t_2}{2} \qquad \omega_\mu = \frac{\omega_1+\omega_2}{2} \qquad t_d = t_1 - t_2 \qquad \omega_d = \omega_1 - \omega_2 \tag{8.14}$$

Equation (8.13) indicates that $AF_{s1,s2}(\theta,\tau)$ is concentrated in $(t_1-t_2,\omega_1-\omega_2)$, away from the origin. Moreover, $AF_{s2,s1}(\theta,\tau)$ has a form similar to $AF_{s1,s2}(\theta,\tau)$, except that the center is as $(t_2-t_1,\omega_2-\omega_1)$. Figure 8-3 sketches the locations of each individual term in (8.12).

Example 8.2 reveals an important fact; that is, in the ambiguity domain the auto-term is always concentrated at the origin (0,0), whereas the cross-term is always away from the origin. Such an observation suggests that one can apply the 2D lowpass filter in the ambiguity domain to suppress the cross-term interference, i.e.,

$$\int_{-\infty}^{\infty}\int_{-\infty}^{\infty} AF_s(\theta,\tau)\Phi(\theta,\tau)e^{-j(\vartheta t+\omega\tau)}d\theta d\tau \tag{8.15}$$

where the function $\Phi(\theta,\tau)$ denotes a 2D lowpass filter. If

$$\int_{-\infty}^{\infty}\int_{-\infty}^{\infty} \Phi(\theta,\tau)d\theta d\tau = \phi(t,\omega) \tag{8.16}$$

then we can, based on the convolution theorem, write (8.15) as the 2D convolution of the Wigner-Ville distribution and a 2D filter, i.e.,

$$\begin{aligned}&\int_{-\infty}^{\infty}\int_{-\infty}^{\infty} AF_s(\theta,\tau)\Phi(\theta,\tau)e^{-j(\theta t+\omega\tau)}d\theta d\tau \\ &= \int_{-\infty}^{\infty}\int_{-\infty}^{\infty} \phi(x,y)WVD_s(t-x,\omega-y)dxdy \\ &= SWVD(t,\omega)\end{aligned} \tag{8.17}$$

As introduced in Section 7.4, if $\phi(t,\omega)$ happens to be the Wigner-Ville distribution of a window function $\gamma(t)$, then (8.17) is equal to the square of the STFT.

Equations (8.15) and (8.17) in fact are the two different representations of a generalized time-frequency representation, Cohen's class [93]. Usually, the function $\Phi(\theta,\tau)$ (and $\phi(t,\omega)$) is considered as a kernel function that controls the properties of the resulting smoothed Wigner-Ville distribution. We will study the kernel function in more detail in Section 8.2.

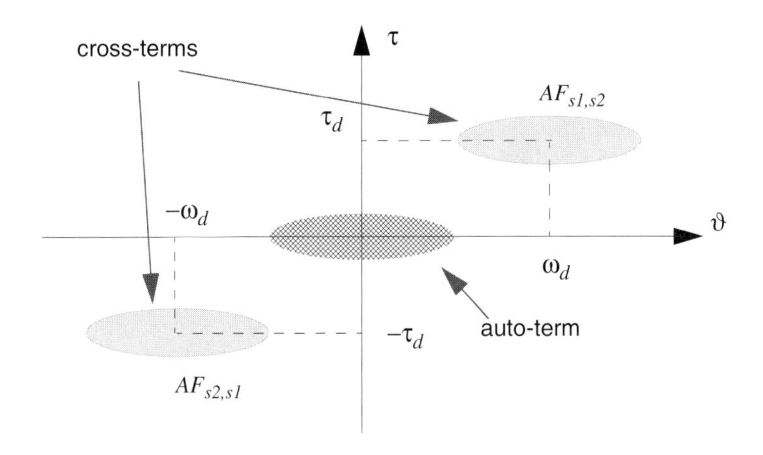

Figure 8-3 Ambiguity function of the sum of two Gaussian signals. While the auto-term is always concentrated at the origin (0,0), the cross-term is always away from the origin

Recall that the cross-WVD has a form

$$WVD_{s1,s2}(t,\omega) = 2e^{-\alpha(t-t_\mu)^2-\frac{1}{\alpha}(\omega-\omega_\mu)^2}e^{j[(\omega-\omega_\mu)t_d+\omega_d t]} \tag{8.18}$$

If we write both the Wigner-Ville distribution and the symmetric ambiguity function in terms of magnitude and phase, i.e.,

$$WVD_{s1,s2}(t,\omega) = A_{WVD}(t,\omega)e^{j\varphi_{WVD}(t,\omega)} \tag{8.19}$$

and

$$AF_{s1,s2}(\theta,\tau) = A_{AF}(\theta,\tau)e^{j\varphi_{AF}(\theta,\tau)} \tag{8.20}$$

then

$$\frac{\partial}{\partial\theta}\varphi_{AF}(\theta,\tau) = -t_{\mu} \qquad \frac{\partial}{\partial\tau}\varphi_{AF}(\theta,\tau) = \omega_{\mu} \tag{8.21}$$

which says that the partial derivatives of the phase of the symmetric ambiguity function are equal to the time-frequency center of the Wigner-Ville distribution.

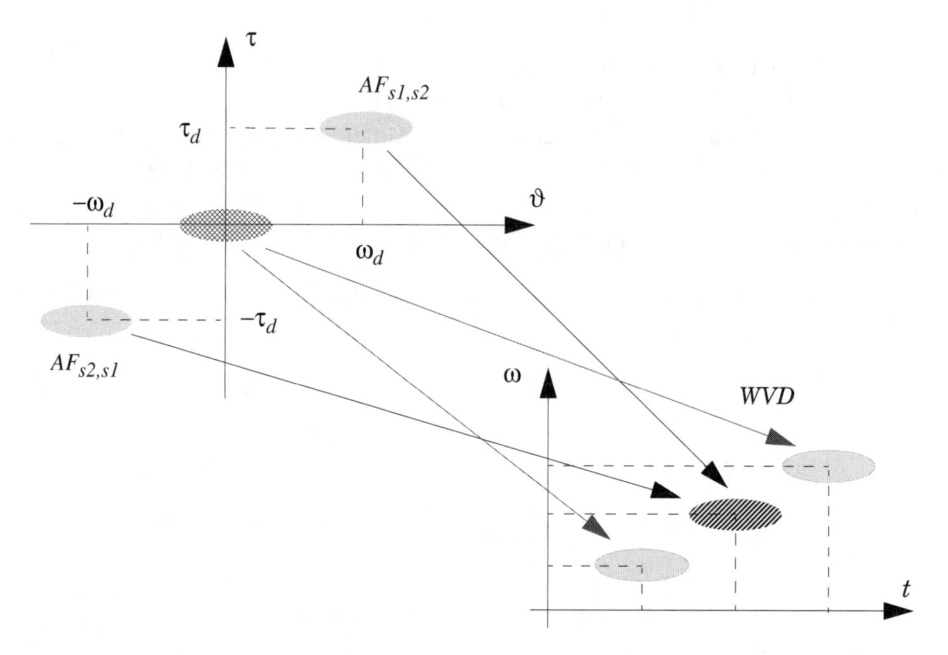

Figure 8-4 Mapping between the ambiguity function and the Wigner-Ville distribution

Conversely,

$$\frac{\partial}{\partial\omega}\varphi_{WVD}(t,\omega) = t_d \qquad \frac{\partial}{\partial t}\varphi_{WVD}(t,\omega) = \omega_d \tag{8.22}$$

which says that the partial derivatives of the phase of the Wigner-Ville distribution are equal to the center of the symmetric ambiguity function. If we consider the derivative of the phase as the frequency, then the location of the symmetric ambiguity function directly relates to the rate of oscillation of the Wigner-Ville distribution. When the cross-term in the Wigner-Ville domain is strongly oscillating, the partial derivative of the phase of the Wigner-Ville distribution must be large. In the ambiguity domain, this is equivalent to the corresponding ambiguity function being away from the origin. The further away $AF_{s1,s2}(\theta,\tau)$ is from the origin, the higher the oscillation of $WVD_{s1,s2}(t,\omega)$.

Because the highly oscillated $WVD_{s1,s2}(t,\omega)$ has a very small average (thereby having negligible contributions to the useful properties), the $AF_{s1,s2}(\theta,\tau)$ that is away from the origin can often be ignored. Figure 8-4 demonstrates the mapping between the ambiguity and Wigner-Ville domains.

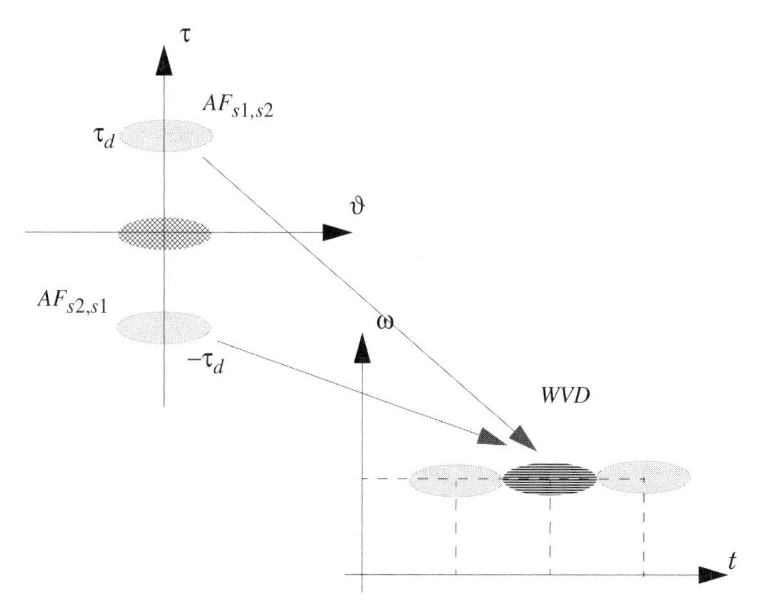

Figure 8-5 The cross-AFs centered in the τ-axis correspond to the cross-term in the WVD, which oscillates in frequency.

If $\omega_1 = \omega_2 = \omega_0$, that is, the two Gaussian functions have the same frequency center, then $AF_{s1,s2}(\theta,\tau)$ in (8.13) reduces to

$$AF_{s1,s2}(\theta,\tau) = e^{-\left[\frac{1}{4\alpha}\theta^2 + \frac{\alpha}{4}(\tau - t_d)^2\right]} e^{j(\omega_0\tau - \theta t_\mu)} \tag{8.23}$$

which is concentrated in the τ-axis. The distance between the center of $AF_{s1,s2}(\theta,\tau)$ to the origin is $|t_d|$, as shown in Figure 8-5. If $|t_d| >> 0$, the $AF_{s1,s2}(\theta,\tau)$ corresponds to the $WVD_{s1,s2}(t,\omega)$ that strongly oscillates in the frequency domain and thereby has a negligible average.

Similarly, if $t_1 = t_2 = t_0$, that is, the two Gaussian functions have the same time center, then $AF_{s1,s2}(\theta,\tau)$ in (8.13) reduces to

$$AF_{s1,s2}(\theta,\tau) = e^{-\left[\frac{1}{4\alpha}(\theta - \omega_d)^2 + \frac{\alpha}{4}\tau^2\right]} e^{j[\omega_\mu\tau - (\theta - \omega_d)t_0]} \tag{8.24}$$

which is concentrated in the θ-axis, as illustrated in Figure 8-6. The distance between the center of $AF_{s1,s2}(\theta,\tau)$ to the origin is $|\omega_d|$. If $|\omega_d| >> 0$, the $AF_{s1,s2}(\theta,\tau)$ corresponds to the $WVD_{s1,s2}(t,\omega)$ that strongly oscillates in time with the frequency ω_d. Because of the relatively small average, such a $WVD_{s1,s2}(t,\omega)$ has negligible contributions to the useful properties.

In order to maintain the useful properties of the WVD, many time-frequency representations tend to retain all portions of $AF(\theta,\tau)$ that are in the τ-axis or θ-axis. As studied earlier, however, the portions of $AF(\theta,\tau)$ in the τ-axis or θ-axis could cause significant undesired terms in the Wigner-Ville domain. On the other hand, when $AF(\theta,0)$ and $AF(0,\tau)$ are far away from the origin, their contribution to the useful properties is limited.

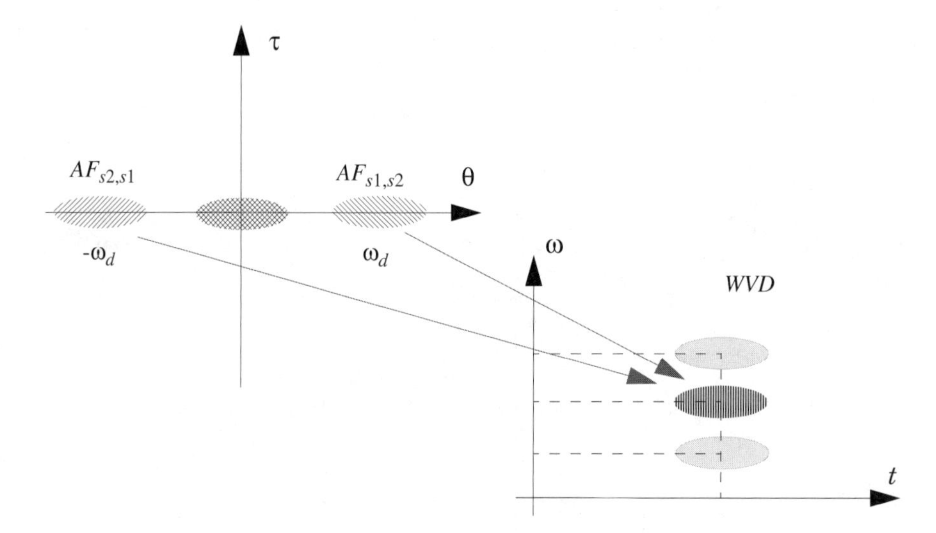

Figure 8-6 The cross-AFs centered in the θ-axis correspond to the cross-term in the WVD, which oscillates in time.

Finally, like the Wigner-Ville distribution, the ambiguity function is also closely related to the wavelet transform, e.g.,

$$CWT(a,b)CWT^*(a,-b) = \frac{1}{a}\int_{-\infty}^{\infty}\int_{-\infty}^{\infty} AF_s(\tau,\nu)AF_\psi\left(\frac{\tau-2b}{a}, a\nu\right)d\tau d\nu \tag{8.25}$$

Proof

The right side of (8.25) is equal to

$$\begin{aligned}
&\frac{1}{a}\iiint_{-\infty}^{\infty} s\left(x+\frac{\tau}{2}\right)s^*\left(x-\frac{\tau}{2}\right)e^{-j\nu x}dx\int_{-\infty}^{\infty}\psi^*\left(y+\frac{\tau-2b}{2a}\right)\psi\left(y-\frac{\tau-2b}{2a}\right)e^{ja\nu y}dyd\tau d\nu \qquad (8.26)\\
&= \frac{1}{a}\iint_{-\infty}^{\infty} s\left(ay+\frac{\tau}{2}\right)s^*\left(ay-\frac{\tau}{2}\right)\psi^*\left(y+\frac{\tau-2b}{2a}\right)\psi\left(y-\frac{\tau-2b}{2a}\right)dyd\tau
\end{aligned}$$

Substitute

$$y' = ay + \frac{\tau}{2} \qquad and \qquad \tau' = ay - \frac{\tau}{2} \tag{8.27}$$

into (8.26). Because

$$dyd\tau = |J|dy'd\tau' = dy'd\tau' \tag{8.28}$$

where $|J|$ denotes the Jacobian determinant (see more details about multivariable calculus in Section 7.4), (8.26) becomes

$$\frac{1}{a}\int_{-\infty}^{\infty}\int_{-\infty}^{\infty} s(y')s^*(\tau')\psi^*\left(\frac{y'-b}{a}\right)\psi\left(\frac{y'+b}{a}\right)dy'd\tau' = CWT(a,b)CWT^*(a,-b) \tag{8.29}$$

8.2 Cohen's Class

So far, we have introduced the STFT spectrogram, the Wigner-Ville distribution, and the symmetric ambiguity function. In addition to these three bilinear time-frequency representations, we can list at least a dozen other counterparts, such as the page distribution [185], the Rihaczek distribution [204], the Choi-Williams distribution [90], and the cone-shape distribution [236]. It is interesting to note that all these bilinear time-frequency representations can be written in a general form,

$$\boxed{C(t,\omega) = \frac{1}{2\pi}\iiint_{-\infty}^{\infty} s\left(u+\frac{\tau}{2}\right)s^*\left(u-\frac{\tau}{2}\right)\Phi(\theta,\tau)e^{-j(\theta t + \omega\tau - \theta u)}dud\tau d\theta} \tag{8.30}$$

with different kernel functions $\Phi(\theta,\tau)$. Equation (8.30) is commonly named as *Cohen's class*, and first appeared in the area of quantum mechanics in 1966 [93]. The original objective of (8.30) was to define joint representations of position-momentum space. *Escudie* and *Grea* in 1976 [104] and later *Classen* and *Mecklenbrauker* in 1980 [91] introduced it into the field of signal processing. "By the facility of its parameterization and the large number of possible interpretations, it provides the theoretical framework within which most of the work in the 1980s was performed" [16]. The significance of the general time-frequency representation (8.30) is to reduce the problem of searching a desired time-dependent spectrum to the selection of the kernel function $\Phi(\theta,\tau)$.

By the definition of the ambiguity function (8.3), we can rewrite (8.30) as

$$C_s(t,\omega) = \int_{-\infty}^{\infty}\int_{-\infty}^{\infty} AF_s(\theta,\iota)\Phi(\theta,\tau)e^{-j(\theta t + \omega\tau)}d\tau d\theta \tag{8.31}$$

where $AF_s(\theta,\tau)$ is the ambiguity function of the signal $s(t)$. Note that (8.31) has exactly the same form as the smoothed Wigner-Ville distribution (8.15). Obviously, if $\Phi(\theta,\tau)$ is allpass, that is, $\Phi(\theta,\tau) = 1$ for all θ and τ, (8.31) reduces to the Wigner-Ville distribution (8.15). The reader can also verify that if the kernel function $\Phi(\theta,\tau)$ is an ambiguity function of a window $\gamma(t)$, then (8.31) is equal to the square of the STFT.

Table 8-1 lists the relationships between the kernel function $\Phi(\theta,\tau)$ in (8.30) and the useful properties without proof. Readers interested in derivations are encouraged to consult [10] and [91]. Moreover, all relationships listed in Table 8-1 are based on the continuous-time cases and, therefore, all conclusions only hold for continuous-time signals. The discrete counterparts were studied by *Jeong* and *Williams* [146], *Morris* and *Wu* [175], and *Kootsookos* et al. [155].

It is important to note that the suppression of cross-term interference and properties 4 – 7 listed in Table 8-1 generally cannot be achieved simultaneously. For example, to preserve these properties both $\Phi(\theta,0)$ and $\Phi(0,\tau)$ have to be equal to one. In other words, all portions of $AF(\theta,\tau)$ that are in the τ- and θ-axis will be retained. As discussed in Section 8.1, however, portions of $AF(\theta,\tau)$ that are in the τ-axis or θ-axis will cause significant undesired cross-terms when they are far away from the origin. Therefore, strictly maintaining properties 4 – 7 will sacrifice the readability (in terms of less cross-term interference and good time-frequency resolution) of the resulting time-frequency representation.

Table 8-1 Relationship between Properties and Kernels

	Properties	Kernel $\Phi(\theta,\tau)$
1	time-shift invariant	independent of time variable t
2	frequency-shift invariant	independent of frequency variable ω
3	realness	$\Phi(\theta,\tau) = \Phi(-\theta,-\tau)$
4	time marginal	$\Phi(\theta,0) = 1$
5	frequency marginal	$\Phi(0,\tau) = 1$
6	mean instantaneous frequency property	$\Phi(\theta,0) = 1$ and $\frac{\partial}{\partial\tau}\Phi(\theta,\tau)\Big\vert_{\tau=0} = 0$
7	group delay property	$\Phi(0,\tau) = 1$ and $\frac{\partial}{\partial\theta}\Phi(\theta,\tau)\Big\vert_{\theta=0} = 0$
8	positivity	$\Phi(\theta,\tau)$ is an ambiguity function of $\gamma(t)$ or signal-dependent

On the other hand, for those $AF(\theta,0)$ and $AF(0,\tau)$ that are far away from the origin, $\theta, \tau \gg 0$, their contribution to properties 4 – 7 is limited. Hence, it seems wiser to selectively retain the portion away from the origin. The result, such as that computed by the low order time-frequency distribution series introduced in Chapter 9, can have good readability but negligible aberrations from useful properties. There is a tradeoff between strictly maintaining properties 4 – 7 and the readability of the resulting time-frequency representation. Table 8-2 lists some popular kernel functions and their corresponding bilinear time-frequency distributions. The distribution associated with

$$\Phi(\theta, \tau) = \frac{\sin\frac{\theta\tau}{2}}{\frac{\theta\tau}{2}} \tag{8.32}$$

has the form

$$\int_{-\infty}^{\infty} \frac{1}{|\tau|} e^{-j\omega\tau} \int_{t-|\tau|/2}^{t+|\tau|/2} s\left(u+\frac{\tau}{2}\right) s^*\left(u-\frac{\tau}{2}\right) du d\tau \tag{8.33}$$

which was first derived by *Cohen* [93], though in the literature it also carries the name *Born-Jordan* [81].

Table 8-2 Some Distributions and Their Kernels

Name	Kernel $\Phi(\theta,\tau)$	Distribution
Born-Jordan [81] Cohen [93]	$\frac{\sin(\theta\tau/2)}{\theta\tau/2}$	Eq. (8.33)
Choi-Williams [90]	$e^{-\alpha(\theta\tau)^2}$	Eq. (8.41)
Cone-shape [236]	$\frac{\sin(\theta\tau/2)}{\theta\tau/2} e^{-\alpha\tau^2}$	Eq. (8.43)
Margenau-Hill [173]	$\cos(\theta\tau/2)$	$Re\frac{1}{\sqrt{2\pi}} s(t)S(\omega)e^{-j\omega t}$
Page [185]	$e^{-j\theta\|\tau\|}$	$\frac{\partial}{\partial t}\left\|\frac{1}{\sqrt{2\pi}}\int_{-\infty}^{t} s(t')e^{-j\omega t'}dt'\right\|^2$
Kirwood [152] Rihaczek [204]	$e^{j\theta\tau/2}$	$\frac{1}{\sqrt{2\pi}} s(t)S(\omega)e^{-j\omega t}$
spectrogram	ambiguity function of $\gamma(t)$	$\left\|\int_{-\infty}^{\infty} s(\tau)\gamma^*(\tau-t)e^{-j\omega\tau}d\tau\right\|^2$
Wigner-Ville [226] [220]	1	Eq. (7.1)

Generally speaking, Cohen's class can be negative unless the kernel function is signal-dependent or $\Phi(\theta,\tau)$ is an ambiguity function of a function $\gamma(t)$. In this case, $C(t,\omega)$ is equivalent to the square of the STFT. The resulting $C(t,\omega)$ does not satisfy properties 4, 5, 6, and 7. From the classical energy concept, the signal's energy distribution (or energy density function) should be non-negative. Wigner showed, however, that the bilinear transform cannot satisfy marginal conditions and be non-negative simultaneously [227]. A natural question at this point is, does a non-negative time-frequency distribution (or time-frequency density function) exist? The answer is *yes*. In fact, if we do not limit ourselves to bilinear transformations, such as Cohen's class (8.30), then we could easily make non-negative time-frequency functions with the right time and frequency marginal. For example, as mentioned in Section 7.2, we can let

$$P(t, \omega) = \frac{|s(t)|^2}{\|s(t)\|^2}|S(\omega)|^2 \tag{8.34}$$

where

$$\|s(t)\|^2 = \int |s(t)|^2 dt = \frac{1}{2\pi}\int |S(\omega)|^2 d\omega \tag{8.35}$$

Evidently, $P(t,\omega)$ in (8.34) is non-negative and satisfies the time/frequency marginal conditions. With normalization, it could be formulated as a joint density function. But such a time-frequency density function is *meaningless* for time-frequency analysis, because it does not convey any information regarding local behavior of the signal.

Then, the next question will be, does a *meaningful* non-negative time-frequency distribution exist? Unfortunately, the answer to this question is, "*we don't know*." Although there are many ways of creating non-negative time-frequency functions, none of them has been proved to truly reflect a signal's time-varying nature, for instance, satisfying the mean instantaneous frequency property. The subject of the existence of a non-negative time-frequency distribution that also reflects the *signal's time-varying nature* so far remains a research topic. The reader interested in this topic is encouraged to consult *Cohen* [10], *Flandrin* [16], *Janssen* ([137], [139], and [142]), *Louphlin* et al. [169], and *Pedersen* [189].

In addition to (7.50) and (8.31), Cohen's class (8.30) can also be written as

$$C_s(t, \omega) = \int_{-\infty}^{\infty}\int_{-\infty}^{\infty} s\left(u+\frac{\tau}{2}\right)s^*\left(u-\frac{\tau}{2}\right)f(t-u, \tau)du e^{-j\omega\tau}d\tau \tag{8.36}$$

where

$$f(t-u, \tau) = \frac{1}{2\pi}\int_{-\infty}^{\infty} \Phi(\theta, \tau)e^{-j(\theta t - \theta u)}d\theta \tag{8.37}$$

Compared to (7.50) and (8.31), formula (8.36) is more suitable for real time implementations. This is because, to evaluate (7.50) or (8.31), we need to have the entire time record for computing the Wigner-Ville distribution or the ambiguity function. However, there is no such need in (8.36) if the duration of the function $f(t, \tau)$ is limited.

8.3 Some Members of Cohen's Class

One of the major activities in the area of time-frequency analysis over the past ten years has been to seek a time-dependent spectrum that not only possesses useful properties, but also has reduced cross-term interference. As discussed in Section 8.1, the portion of the ambiguity function that corresponds to auto-terms is always connected to the origin, whereas the part of the ambiguity function that is related to the cross-terms tends to spread somewhere else. This observation inspired the search for a kernel function $\Phi(\theta,\tau)$ such that the product $|\Phi(\theta,\tau)AF(\theta,\tau)|$ is enhanced in the vicinity of the origin and suppressed everywhere else. At the same time, $\Phi(\theta,\tau)$ should satisfy as many as possible of the properties listed in Table 8-1.

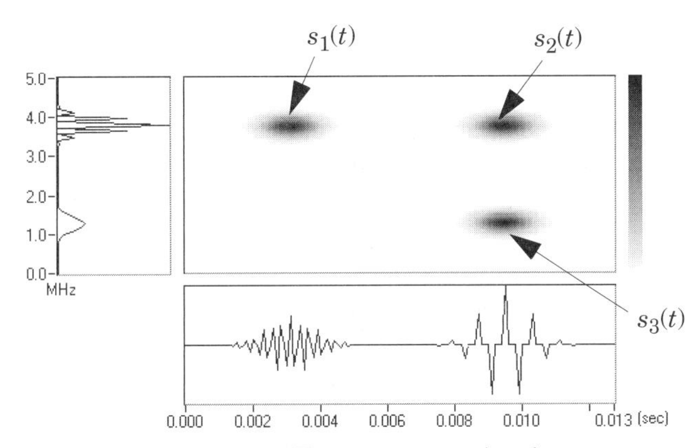

Figure 8-7 Three-tone test signal

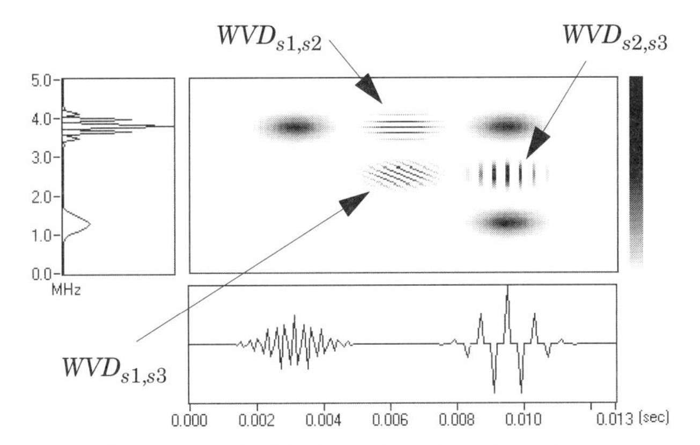

Figure 8-8 WVD of the three-tone test signal

Figure 8-7 illustrates a three-tone test signal. The components $s_1(t)$ and $s_2(t)$ have the same frequency center. The components $s_3(t)$ and $s_2(t)$ have the same time center. Figure 8-8 plots the corresponding Wigner-Ville distribution, which indicates that each pair of auto-terms creates one cross-term.

Figure 8-9 depicts the three-tone test signal in the ambiguity domain. It shows that, except for the cluster centered at (0,0), all other signals are caused by cross-terms. As discussed previously, many useful properties and cross-term suppression cannot be accomplished simultaneously. For example, to reduce the cross-term interference, the product $|\Phi(\theta,\tau)AF(\theta,\tau)|$ has to vanish for larger θ and τ. On the other hand, to preserve the time and frequency marginal conditions, the following must hold:

$$\Phi(\theta, 0)AF(\theta, 0) = AF(\theta, 0) \qquad and \qquad \Phi(0, \tau)AF(0, \tau) = AF(0, \tau) \tag{8.38}$$

which implies that all portions of $AF(\theta, \tau)$ in both the τ-axis and θ-axis have to be kept, no matter how far they are from the origin. The direct consequence is that the resulting representation will preserve all cross-terms that are created by two functions that have either the same time center or frequency center.

In what follows, we will investigate the performances of three popular members of Cohen's class by the test signal introduced above.

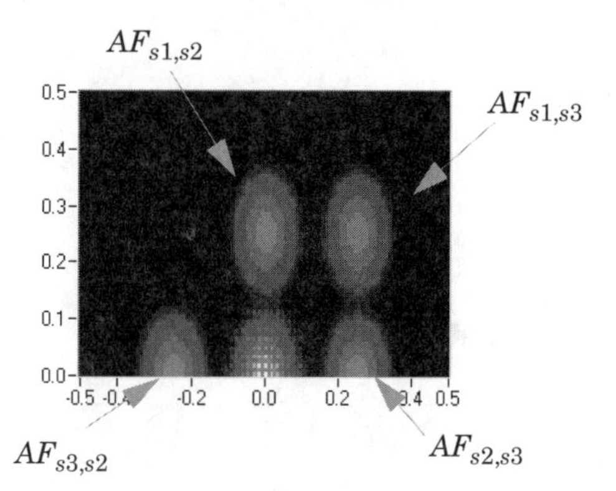

Figure 8-9 Ambiguity function of the three-tone test signal

Choi-Williams Distribution

To suppress the portions of the AF that are away from the origin, Choi and Williams introduced the exponential kernel as [90]

$$\Phi(\theta, \tau) = e^{-\alpha(\theta\tau)^2} \tag{8.39}$$

It is rather easy to check that the exponential kernel in (8.39) satisfies all properties listed in Table 8-1 except for positivity. Moreover, $\Phi(0,0) = 1$ and $\Phi(\theta,\tau) < 1$ for $\theta \neq 0$ and $\tau \neq 0$, which imply that the exponential kernel will suppress the cross-terms created by two functions that have both different time and frequency centers. The parameter α controls the decay speed, as shown in Figure 8-10. The bigger the α, the more the cross-terms are suppressed. On the other hand, the bigger the α, the more the auto-terms are affected. Therefore, there is a trade-off for the selection of the parameter α. When α goes to zero, the corresponding distribution converges to the Wigner-Ville distribution.

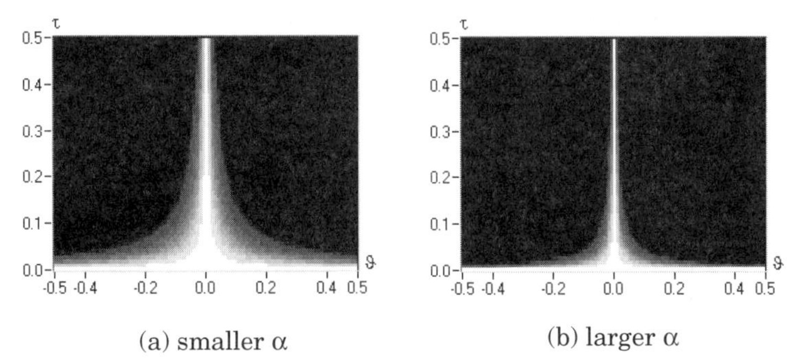

Figure 8-10 Exponential kernel with smaller α (left) and bigger α (right). The bigger the α, the more the cross-terms are suppressed. On the other hand, the bigger the α, the more the auto-terms are affected.

The Fourier transform of the exponential kernel $\Phi(\theta,\tau)$ in (8.39), with respect to the parameter θ, is

$$f(t,\tau) = \frac{1}{\sqrt{4\pi\alpha\tau^2}}\exp\left\{-\frac{1}{4\alpha\tau^2}t^2\right\} \tag{8.40}$$

Substituting (8.40) into (8.36) yields

$$CWD(t,\omega) = \int_{-\infty}^{\infty}\frac{1}{\sqrt{4\pi\alpha\tau^2}}e^{-j\omega\tau}\int_{-\infty}^{\infty}s\left(u+\frac{\tau}{2}\right)s^*\left(u-\frac{\tau}{2}\right)\exp\left\{-\frac{(t-u)^2}{4\alpha\tau^2}\right\}du\,d\tau \tag{8.41}$$

which is commonly named as the *Choi-Williams distribution* (CWD) [90].

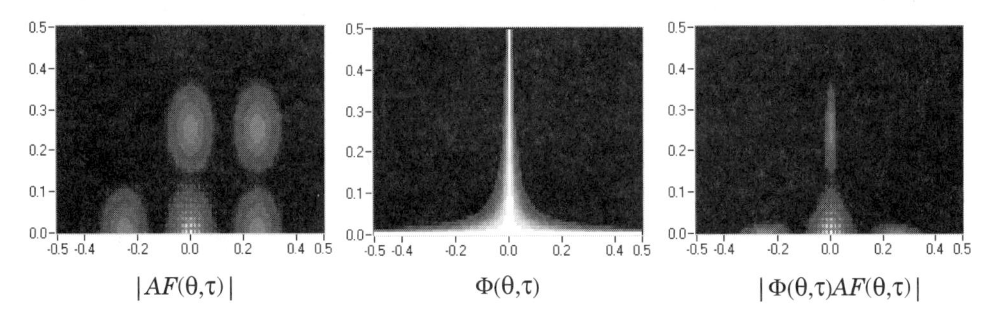

Figure 8-11 Although the exponential kernel suppresses the portion that is away from the origin, it preserves all cross-terms that are in the θ-axis and τ-axis.

Figure 8-11 depicts the process of cross-term suppression by the exponential kernel function for the three-tone test signal. Although the exponential kernel suppresses the portion that is away from the origin, it preserves all cross-terms that are in the θ-axis or τ-axis. Consequently, the Choi-Williams distribution will contain strong horizontal and vertical ripples. While the horizontal ripples are caused by the auto-terms that have the same frequency center, the vertical ripples correspond to the auto-terms that have the same time center. Figure 8-12 illustrates the Choi-Williams distribution of the three-tone test signal. The reader should also bear in mind that the continuous Choi-Williams distribution preserves the properties of the Wigner-Ville distribution, but this is not the case for the discrete-time signal.

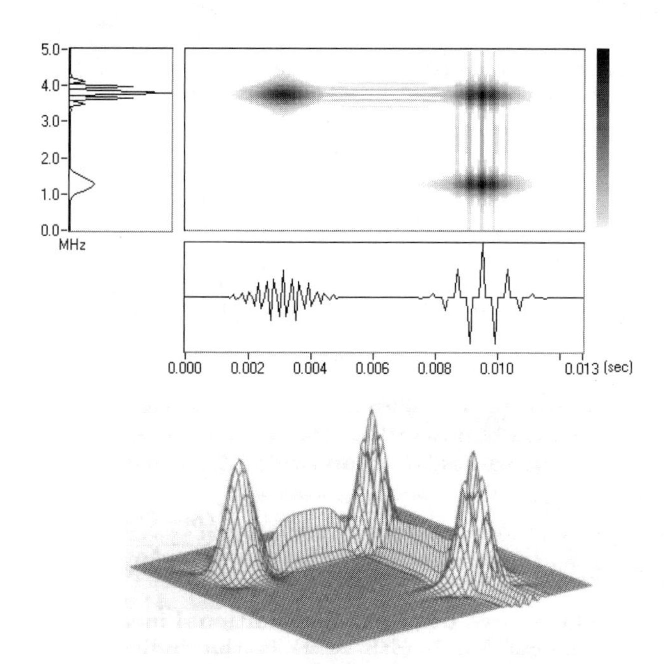

Figure 8-12 The Choi-Williams distribution contains strong horizontal and vertical ripples. While the horizontal ripples are caused by the auto-terms that have the same frequency center, the vertical ripples correspond to the auto-terms that have the same time center.

Cone-Shape Distribution

Another popular representation is the cone-shape distribution introduced by Zhao, et al. [236]. The cone-shape function can be expressed as

$$f(t, \tau) = \begin{cases} g(\tau) & |\tau| \geq 2|t| \\ 0 & otherwise \end{cases} \tag{8.42}$$

which is plotted in Figure 8-13. Substituting it into (8.36) yields the cone-shape distribution (CSD),

$$CSD_s(t, \omega) = \int_{-2t}^{2t} e^{-j\omega\tau} \int_{-\infty}^{\infty} s\left(u + \frac{\tau}{2}\right) s^*\left(u - \frac{\tau}{2}\right) f(t-u, \tau) du d\tau \tag{8.43}$$

In the ambiguity domain, the cone-shape function has the form

$$\Phi(\theta, \tau) = \int_{-\infty}^{\infty} f(t, \tau)e^{-j\theta t}dt = g(\tau)\int_{-\tau/2}^{\tau/2} e^{-j\theta t}dt = 2g(\tau)\frac{\sin(\theta\tau/2)}{\theta} \tag{8.44}$$

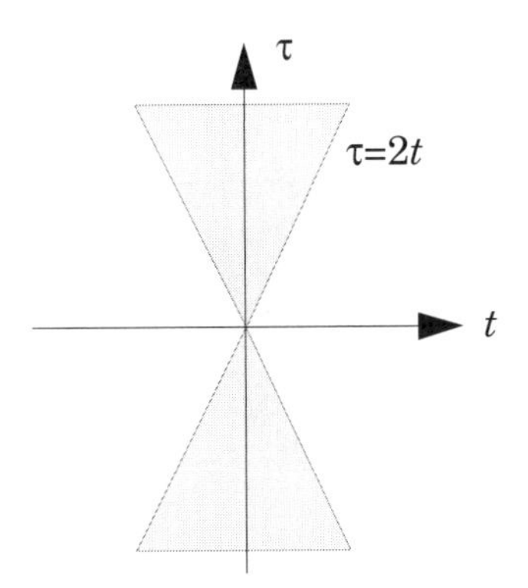

Figure 8-13 Cone-shape function

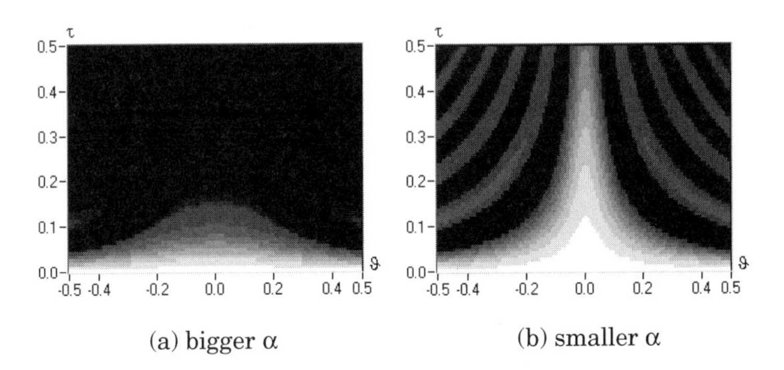

Figure 8-14 Cone-shape kernel with bigger α (left) and smaller α (right). The bigger the α, the more the cross-terms are suppressed. On the other hand, the bigger the α, the more the auto-terms are affected.

Let

$$g(\tau) = \frac{1}{\tau}e^{-\alpha\tau^2} \tag{8.45}$$

Then (8.44) becomes

$$\Phi(\theta, \tau) = \frac{\sin(\theta\tau/2)}{\theta\tau/2}e^{-\alpha\tau^2} \qquad \alpha > 0 \tag{8.46}$$

The parameter α controls the degree of suppression, as shown in Figure 8-14. The bigger the parameter α, the more the cross-terms are suppressed (at the expense of more disturbed auto-terms). Obviously,

$$\Phi(\theta, \tau) = \begin{cases} 1 & \tau = 0 \\ e^{-\alpha\tau^2} & \theta = 0 \end{cases} \tag{8.47}$$

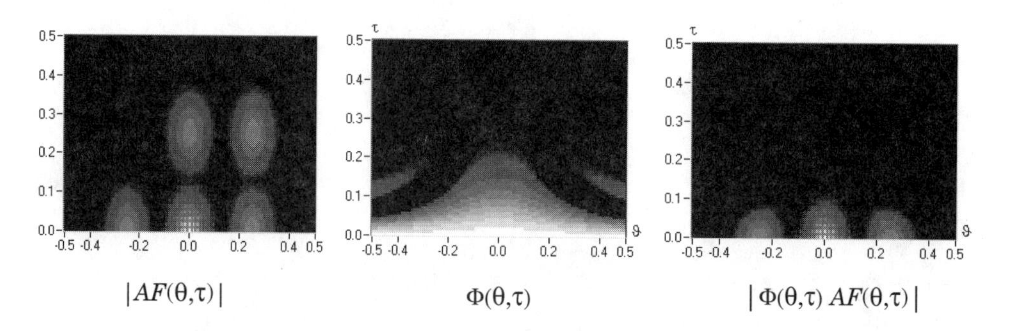

Figure 8-15 The cone-shape kernel preserves cross-terms in the θ-axis, but it suppresses the cross-terms in the τ-axis.

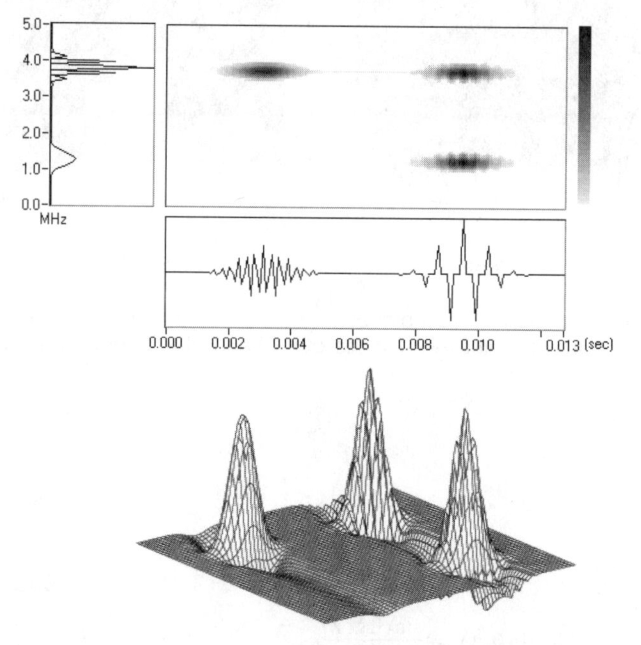

Figure 8-16 Compared to the Wigner-Ville distribution in Figure 8-8 and the Choi-Williams distribution in Figure 8-12, the cross-term interference is significantly reduced in the cone-shape distribution. In particular, the cross-term between two functions that have the same frequency center is completely removed. However, the interference created by two functions with the same time center still exists.

Unlike the exponential kernel function (8.39) that preserves all portions of $AF(\theta,\tau)$ in both the θ-axis as well as τ-axis, the cone-shape kernel (8.46) suppresses those $AF(\theta,\tau)$ that are in the τ-axis. Consequently, the cone-shape distribution is able to effectively attenuate the cross-terms created by two functions that have the same frequency center. Figure 8-15 plots $|AF(\theta,\tau)|$ of the three-tone test signal, $\Phi(\theta,\tau)$ of the cone-shape function, and $|\Phi(\theta,\tau)\ AF(\theta,\tau)|$. Although the cross-terms in the θ-axis are preserved, the cross-term in the τ-axis is suppressed.

Figure 8-16 depicts the cone-shape distribution for the three-tone test signal. Compared to the Wigner-Ville distribution in Figure 8-8 and the Choi-Williams distribution in Figure 8-12, the cross-term interference is significantly reduced in the cone-shape distribution. In particular, the cross-term between two functions that have the same frequency center is completely removed. However, the interference created by two functions with the same time center still exists.

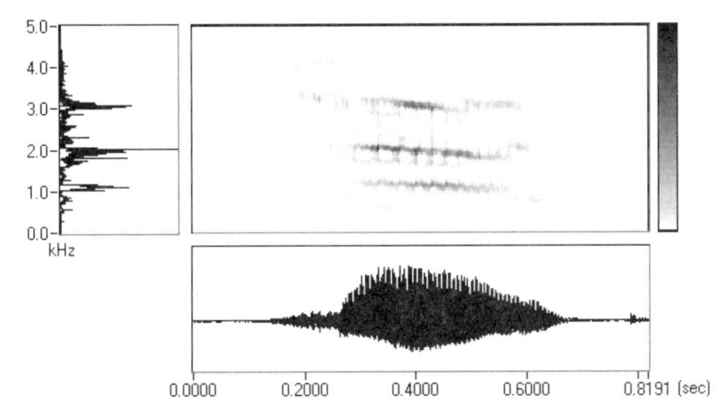

Figure 8-17 Cone-shape distribution of "wood" spoken by a five-year-old boy (Data provided by Y. Zhao of the University of Missouri).

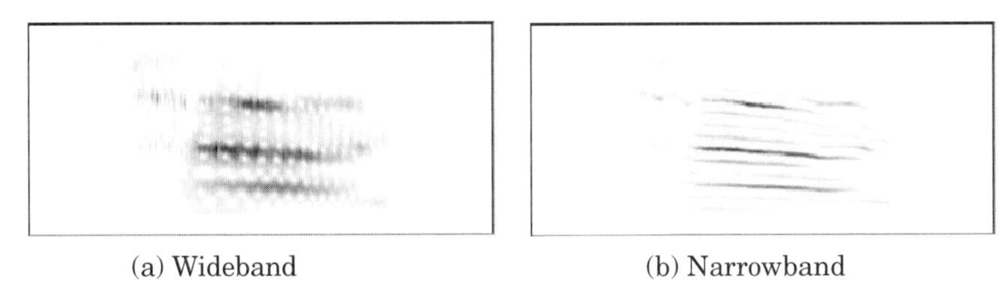

(a) Wideband (b) Narrowband

Figure 8-18 While the formants in the wideband STFT spectrogram spread widely, they are mixed with harmonics in the narrowband STFT spectrogram.

In the area of speech analysis, the cone-shape distribution has been found to be a favorite alternative to the conventional STFT spectrogram. Figure 8-17 illustrates the analysis results of an utterance by a five-year-old boy. The top plot is computed by the cone-shape distribution, which not only clearly demonstrates the formant patterns, but also depicts the vertical structures well. Figure 8-18 plots the same signal computed by both the wideband and the narrowband

STFT spectrogram. While the formants in the wideband STFT spectrogram spread widely, they are mixed with harmonics in the narrowband STFT spectrogram.

Signal-Dependent Time-Frequency Representations[1]

Unlike the previous kernels that emphasize preserving the properties of the Wigner-Ville distribution over matching the shape of auto-components, the signal-dependent kernels aim to optimally pass the auto-components while suppressing cross-components. Since a fixed kernel acts on the ambiguity function as a mask or filter, it is limited in its ability to perform this function. The locations of the auto- and cross-components depend on the signal to be analyzed; hence, we expect to obtain good performance for a broad class of signals only by using a *signal-dependent* kernel. To find the bilinear time-frequency distribution that provides in some sense the "best" time-frequency representations for a given signal, Baraniuk and Jones have formulated the signal-dependent kernel design procedure as an optimization problem ([69] and [70]).

The 1/0 kernel method [70]

Given a signal and its ambiguity function, the optimal 1/0 kernel is defined as the real, non-negative function Φ_{opt} that solves the following optimization problem

$$max_{\Phi}\int_{-\infty}^{\infty}\int_{-\infty}^{\infty}|AF(\theta,\tau)\Phi(\theta,\tau)|^2 d\theta d\tau \tag{8.48}$$

subject to

$$\Phi(0,0) = 1 \tag{8.49}$$

$$\Phi(\theta,\tau) \text{ is radially non-increasing} \tag{8.50}$$

$$\int_{-\infty}^{\infty}\int_{-\infty}^{\infty}|\Phi(\theta,\tau)|^2 d\tau d\theta \le \alpha \qquad \alpha \ge 0 \tag{8.51}$$

The radially non-increasing constraint (8.50) can be expressed explicitly as

$$\Phi(\gamma_1,\psi) \ge \Phi(\gamma_2,\psi) \qquad \forall\, \gamma_1 \le \gamma_2 \qquad \forall\, \psi \tag{8.52}$$

where γ_i and ψ correspond to the polar coordinates radius and angle, respectively.

The constraints (8.49) to (8.51) and performance measure (8.48) are formulated so that the optimal kernel (OK) passes auto-components and suppresses cross-components. The constraints force the optimal kernel to be a lowpass filter of fixed volume α; maximizing the performance measure encourages the passband of the kernel to lie over the auto-components. Both the performance measure and the constraints are insensitive to the orientation angle and aspect ratio (scaling) of the signal components in the (θ,τ) plane.

By controlling the volume under the optimal kernel, the parameter α controls the trade-off between cross-component suppression and smearing of the auto-components. Reasonable

1. Contributed by Richard G. Baraniuk, Department of Electrical and Computer Engineering, Rice University, Houston.

bounds are $1 \leq \alpha \leq 5$. At the lower bound, the optimal kernel shares the same volume as an STFT spectrogram kernel, while at the upper bound, the optimal kernel smooths only slightly. In fact, as α gets larger, the optimal kernel distribution converges to the Wigner-Ville distribution of the signal.

Analyzing a signal with an optimal-kernel distribution requires a three-step procedure: (1) compute the ambiguity function of the signal; (2) solve the linear program (8.48) to (8.52) in variables $|\Phi(\theta,\tau)|^2$ (a fast algorithm is given in [68]); (3) Fourier transform the AF-kernel product $AF(\theta,\tau)\Phi(\theta,\tau)$.

The radially Gaussian kernel method [69]

Although the 1/0 kernel is optimal according to criteria (8.48) to (8.51), its sharp cutoff may introduce ringing into the OK distribution, especially for small values of the kernel volume parameter α. For an alternative, direct approach to smooth optimal kernels, explicit smoothness constraints can be appended to the kernel optimization formulation (8.48) to (8.51). In [69] the kernel is constrained to be Gaussian along radial profiles

$$\Phi(\theta, \tau) = \exp\left\{-\frac{\theta^2 + \tau^2}{2\sigma^2(\psi)}\right\} \tag{8.53}$$

The term $\sigma(\psi)$ represents the dependence of the Gaussian spread on radial angle $\psi = \arctan(\tau/\theta)$. Any kernel of the form (8.53) is bounded and radially non-increasing and, furthermore, smooth if σ is smooth. Since the shape of a radially Gaussian kernel is completely parameterized by this function, finding the optimal, radially Gaussian kernel for a signal is equivalent to finding the optimal function σ_{opt} for the signal. A gradient ascent/Newton algorithm solving the (non-linear) system (described by (8.48), (8.51), and (8.53)) is detailed in [69].

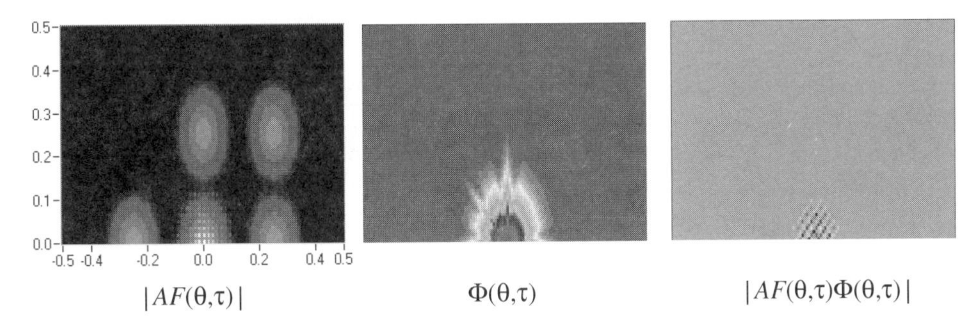

Figure 8-19 Optimal kernel and $|AF(\theta,\tau)\Phi(\theta,\tau)|$ of the three-tone test signal

Figure 8-19 shows the AF domain optimal kernel and product $|AF(\theta,\tau)\Phi(\theta,\tau)|$ for the three-tone test signal. Figure 8-20 shows the optimal-kernel distribution. As another example, Figure 8-21 shows a time-frequency analysis of 2.5 ms of an echo-location pulse emitted by the large brown bat, *Eptesicus fuscus*.

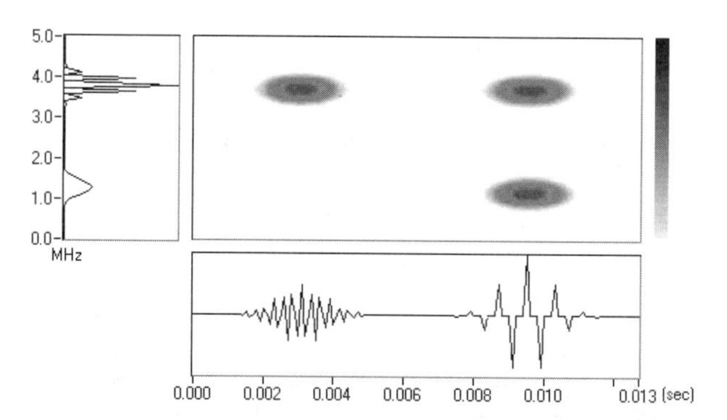

Figure 8-20 Optimal-kernel distribution for the three-tone test signal

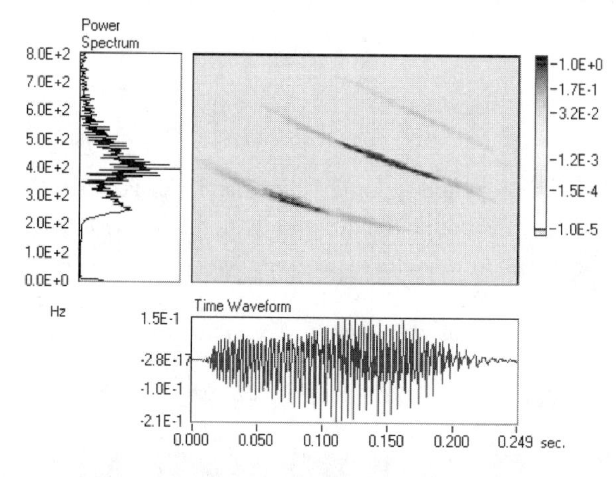

Figure 8-21 Bat sound example (Bat data provided by Curtis Condon, Ken White, and Al Feng of the Beckman Institute at the University of Illinois).

Extensions

Additional constraints: The goal of the optimization problems (8.48) to (8.51) and (8.53) is strictly to find the kernels that optimally pass auto-components and suppress cross-components. However, additional constraints encouraging additional kernel properties can also be considered. An attractive feature of the optimal-kernel design formulation is the ease with which it can be customized to incorporate application-specific knowledge into the design process. For example, constraints have been developed to coerce OK time-frequency distributions to satisfy the marginal distributions and/or preserve the time support of the signal ([69] and [70]).

Adaptive formulations: While the 1/0 and radially Gaussian distributions generally perform well, they are block-oriented techniques that design only one kernel for the entire signal. For analyzing signals with characteristics that change over time, for real-time, on-line operation, or for very long signals, adaptive signal-dependent time-frequency distributions are required.

Adaptation of the kernel to track the local signal characteristics over time requires that the kernel optimization procedure consider only the local signal characteristics. An ambiguity-domain design procedure, such as the radially Gaussian kernel optimization technique described above, does not immediately admit such time localization because the ambiguity function includes information from all times and frequencies in the signal. This difficulty has been surmounted, however, by the development of a time-localized, or short-time, ambiguity function. Application of the radially Gaussian kernel optimization procedure to the short-time ambiguity function localized at time t_0 produces an optimal kernel $\Phi_{opt}(\theta,\tau;\ t_0)$ and an optimal-kernel distribution frequency slice $C_{opt}(t_0, f)$ at t_0. Since the algorithm alters the kernel at each time to achieve optimal local performance, it results in better tracking of the signal changes. A simpler adaptive algorithm based on the short-time Fourier transform is derived in [148].

8.4 Reassignment

The original idea of reassignment was introduced to the STFT spectrogram by *Kodera* et al. in 1976 ([153] and [154]). In addition to preserving all properties of the STFT spectrogram, such as non-negativity and time-frequency shift invariance, the resulting reassigned STFT spectrogram substantially improves the time-frequency resolution. In 1995, *Auger* and *Flandrin* revisited the reassignment method and extended this scheme into general cases, such as Cohen's class [62].

The basic idea of reassignment can be described as follows. Instead of simply smoothing the Wigner-Ville distribution (7.73), the reassigned STFT spectrogram reallocates every $STFT(t,\omega)$ into a new point $(\hat{t}, \hat{\omega})$, the center of gravity of the signal energy in the area centered on (t,ω). If we let

$$I_{t,\omega}(x, y) = WVD_{\gamma}(x, y)WVD_{s}(t-x, \omega-y) \tag{8.54}$$

in (7.73) be a local energy distribution (in the vicinity of (t,ω)), then the center of gravity of the area centered on (t,ω) is

$$\hat{t}(t, \omega) = \frac{\int_{-\infty}^{\infty} x \int_{-\infty}^{\infty} I_{t,\omega}(x, y)dydx}{\int_{-\infty}^{\infty}\int_{-\infty}^{\infty} I_{t,\omega}(x, y)dydx} = \frac{\int_{-\infty}^{\infty} x \int_{-\infty}^{\infty} I_{t,\omega}(x, y)dydx}{|STFT_s(t, \omega)|^2} \tag{8.55}$$

and

$$\hat{\omega}(t, \omega) = \frac{\int_{-\infty}^{\infty} y \int_{-\infty}^{\infty} I_{t,\omega}(x, y)dxdy}{\int_{-\infty}^{\infty}\int_{-\infty}^{\infty} I_{t,\omega}(x, y)dydx} = \frac{\int_{-\infty}^{\infty} y \int_{-\infty}^{\infty} I_{t,\omega}(x, y)dxdy}{|STFT_s(t, \omega)|^2} \tag{8.56}$$

The resulting reassigned STFT spectrogram (RSP) has a form

$$RSP_s(t, \omega) = \int_{-\infty}^{\infty}\int_{-\infty}^{\infty} |STFT_s(t', \omega')|^2 \delta(t - \hat{t}(t', \omega'))\delta(\omega - \hat{\omega}(t', \omega'))dt'd\omega' \tag{8.57}$$

which is non-negative.

One of the most interesting properties of the new point $(\hat{t}, \hat{\omega})$ is that, if we let the phase of the $STFT_S(t,\omega)$ be $\varphi_S(t,\omega)$, i.e.,

$$\varphi_s(t, \omega) = \arg\{STFT_s(t, \omega)\} \tag{8.58}$$

then

$$\hat{t}(t, \omega) = -\frac{\partial}{\partial\omega}\varphi_s(t, \omega) \qquad and \qquad \hat{\omega}(t, \omega) = \omega + \frac{\partial}{\partial t}\varphi_s(t, \omega) \tag{8.59}$$

Therefore, the new point $(\hat{t}, \hat{\omega})$ recovers the phase information that is completely ignored in the STFT spectrogram, even though the phase usually carries important information of the signal.

It can also be proved [62] that the reassigned STFT spectrogram is perfectly localized on linear chirp signals and impulses, i.e.,

$$s(t) = Ae^{-j(\omega_0 t + \beta t^2/2)} \Rightarrow \hat{\omega} = \omega_0 + \beta t \tag{8.60}$$

and

$$s(t) = A\delta(t - t_0) \Rightarrow \hat{t} = t_0 \tag{8.61}$$

However, neither Eqs. (8.55) and (8.56) nor Eq. (8.59) lead to an efficient implementation. For discrete time signals the derivatives must be replaced by the first differences, and the phase of the STFT must be unwrapped. This probably explains why the reassignment algorithm almost fell into oblivion in spite of its promising results.

It was Auger and Flandrin [62] who discovered that the centers of gravity $(\hat{t}, \hat{\omega})$, in fact, can be simply computed by three STFTs with different analysis window functions, i.e.,

$$\hat{t}(t, \omega) = t - Re\left\{\frac{STFT(t\gamma(t);t, \omega)STFT^*(\gamma(t);t, \omega)}{|STFT(\gamma(t);t, \omega)|^2}\right\} \tag{8.62}$$

and

$$\hat{\omega}(t, \omega) = \omega + Im\left\{\frac{STFT\left(\frac{d}{dt}\gamma(t);t, \omega\right)STFT^*(\gamma(t);t, \omega)}{|STFT(\gamma(t);t, \omega)|^2}\right\} \tag{8.63}$$

where the first variable in $STFT(t,\omega;\gamma(t))$ denotes the corresponding analysis window function.

Figure 8-22 plots time-dependent power spectra of a bat sound computed by the reassigned STFT spectrogram (f) as well as several other methods introduced in this book. Compared to the regular STFT spectrogram (a), the reassigned STFT spectrogram (f) has much better time-frequency resolution without a drastic increase of computational complexity. Unlike most other methods, such as the Wigner-Ville distribution (b), the Choi-Williams distribution (c), the cone-shape distribution (d), the signal-dependent representation (e), and the time-frequency distribution series (g), which can go to negative, the reassigned STFT spectrogram (f) is always greater than or equal to zero. It is also time and frequency shift invariant.

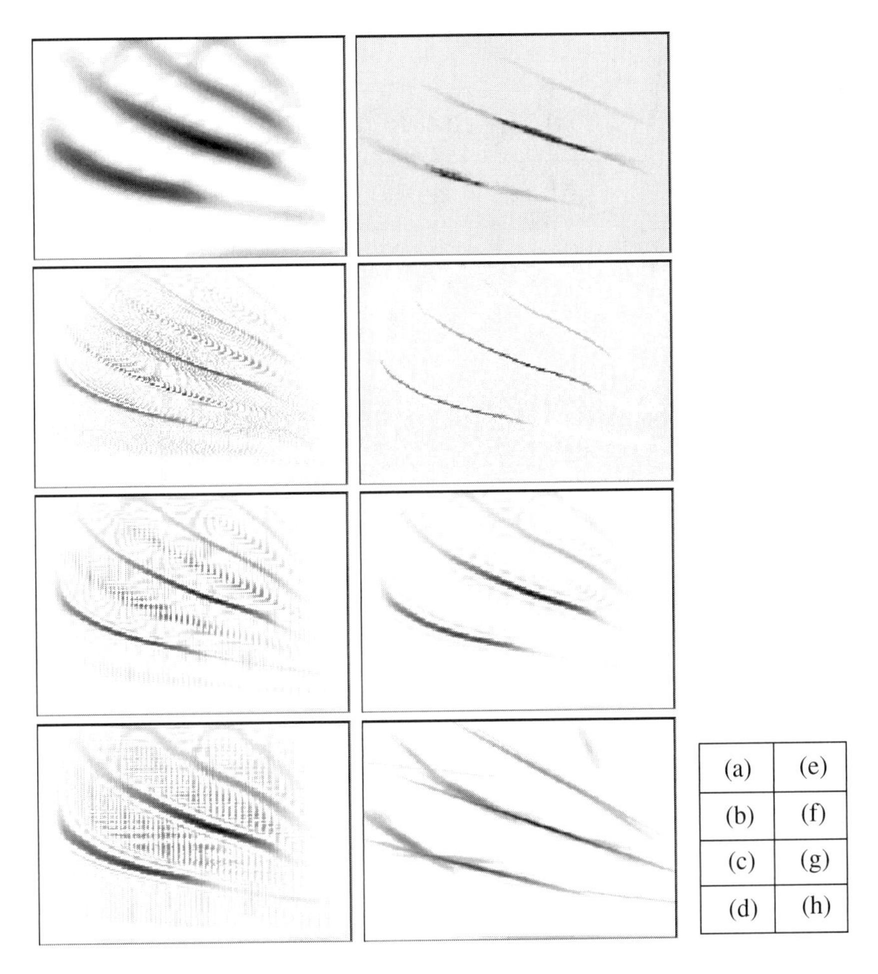

Figure 8-22 Time-frequency distributions of a bat sound: (a) STFT spectrogram, (b) Wigner-Ville distribution, (c) Choi-Williams distribution, (d) cone-shape distribution, (e) signal-dependent time-frequency representations, (f) reassigned STFT spectrogram, (g) fourth order time-frequency distribution series, (h) adaptive Gaussian chirplet decomposition

The time-frequency distribution series (g) and adaptive Gaussian chirplet based spectrogram are two time-frequency representations that do not belong to Cohen's class. We will introduce them in Chapter 9 and Chapter 10, respectively.

Finally, by replacing the Wigner-Ville distribution of the analysis window function in (8.55) and (8.56) by the kernel function $\Phi(t,\omega)$, one can generalize the STFT reassignment into the general Cohen's class. In this case, the local center of gravity is computed by

$$\hat{t}(t, \omega) = \frac{\int_{-\infty}^{\infty} x \int_{-\infty}^{\infty} \phi(x, y) WVD_s(t-x, \omega-y) dy dx}{\int_{-\infty}^{\infty} \int_{-\infty}^{\infty} \phi(x, y) WVD_s(t-x, \omega-y) dy dx} \tag{8.64}$$

and

$$\hat{\omega}(t, \omega) = \frac{\int_{-\infty}^{\infty} y \int_{-\infty}^{\infty} \phi(x, y) WVD_s(t-x, \omega-y) dx dy}{\int_{-\infty}^{\infty} \int_{-\infty}^{\infty} \phi(x, y) WVD_s(t-x, \omega-y) dy dx} \tag{8.65}$$

The reader can find a more detailed treatment about the reassignment method in [16], [62], and [63].

CHAPTER 9

Decomposition of the Wigner-Ville Distribution

*I*n the preceding chapters we introduced an important member of time-frequency representations, the Wigner-Ville distribution. Unlike the STFT spectrogram, the Wigner-Ville distribution possesses many useful properties for time-frequency analysis. The main deficiency of the Wigner-Ville distribution is cross-term interference. At any time instant, if there is more than one frequency tone, then the Wigner-Ville distribution may become messed up because of the presence of undesired terms. It is the cross-term interference that has greatly limited the applications of the Wigner-Ville distribution. Hence, reducing the cross-term interference with a minimum degradation of the desired properties is a major activity in the field of time-frequency analysis.

As discussed in Section 7.3, the cross-terms are highly oscillating and localized, and always midway between the pair of corresponding auto-terms. In the meantime, useful properties, such as the time marginal (7.20), the frequency marginal (7.21), and the mean instantaneous frequency property (7.23), are all obtained by *averaging* the Wigner-Ville distribution. For example, the time marginal condition (7.20) is computed by integrating (*averaging*) the Wigner-Ville distribution over frequency. Conversely, the frequency marginal (7.21) is computed by integrating (*averaging*) the Wigner-Ville distribution over time. These observations suggest that if the Wigner-Ville distribution can be decomposed as the sum of two-dimensional (time and frequency) localized harmonic functions, such as modulated 2D Gaussian functions, then we may be able to use low-order oscillating terms to delineate the time-dependent spectrum with reduced interference. High-order oscillations, creating the cross-terms, have relatively small averages and thereby have negligible influence on the useful properties. The signal energy and useful properties are mainly determined by a few leading low-order harmonic terms. The above conjecture has been successfully verified as being true. The resulting representation is named the time-frequency distribution series. By using a truncated time-frequency distribution series, we can effectively balance cross-term interference with the desired properties.

This presentation starts off with the discussion of the Gabor expansion for the decomposition of the Wigner-Ville distribution. Then, we develop a new time-frequency representation, the time-frequency distribution series. Finally, we apply the similar decomposition technique to the estimation of the mean instantaneous frequency and instantaneous bandwidth. It shows that the resulting estimator balances the variance, particularly in noisy environments, occurring in the Wigner-Ville distribution based estimator and the bias introduced by the STFT based estimator.

In principle, the Wigner-Ville distribution can also be decomposed by expansions other than the Gabor expansion, such as the wavelet series ([92] and [187]). In addition to the Wigner-Ville distribution, the decomposition scheme introduced in this chapter can also be applied to other time-dependent spectra, such as the Choi-Williams distribution. However, the two-dimensional elementary functions $WVD_{h,h'}(t,\omega)$ in the Gabor expansion and Wigner-Ville distribution based decomposition introduced in this chapter are symmetric and optimally concentrated in the joint time-frequency domain. They have simple analytical forms, and the resulting time-frequency distribution series can be efficiently computed by regular separable two-dimensional interpolation filters.

The concept of the decomposition of the Wigner-Ville distribution through the Gabor expansion was first proposed by *Qian* and *Morris* [193] in the early 90s. Shortly after that, *Qian* and *Chen* further developed it into the time-frequency distribution [198]. Since then, the time-frequency distribution has been applied in many areas. One successful application is ISAR (Inverse Synthetic Aperture Radar) image processing [88]. In addition to the Gabor expansion based decomposition, *Pasquier* et al. [187] and *Israel Cohen* et al. [92] also investigated decompositions through the wavelet and wavelet packet, respectively. More recently, *Baranuik* et al. [71] have applied a similar concept for the Wigner-Ville distribution based mean instantaneous frequency and instantaneous bandwidth estimates. The resulting estimate can not only reduce computation of the spectrogram based estimates, but can also balance the variance appearing in the Wigner-Ville distribution based method and the bias emerging in the STFT spectrogram based method.

9.1 Decomposition of the Wigner-Ville Distribution

Compared to other schemes, the Wigner-Ville distribution better characterizes the changes of a signal's frequency content. For example, the conditional mean frequency of the Wigner-Ville distribution is equal to the signal's mean instantaneous frequency, which is traditionally used as a figure of merit for time-frequency representations. The STFT spectrogram is simple and intuitive, but it does not have such an important property (see Example 3.1).

The main deficiency of the Wigner-Ville distribution is the cross-term interference. As discussed in the previous chapter, the Wigner-Ville distribution often creates highly oscillatory terms in places where they are not expected. It is interesting to observe, however, that the cross-term interference usually is highly oscillatory and it is localized. It always occurs midway between a pair of auto-terms. On the other hand, the useful properties of the Wigner-Ville distribution, such as the time marginal (7.7), the frequency marginal (7.7), and the conditional mean frequency property (7.23), depend on the *average* of the Wigner-Ville distribution. These observations motivated researchers to decompose the Wigner-Ville distribution as the sum of two-dimensional harmonics [193]. The procedure is described as follows.

First apply the 1D Gabor expansion to the signal $s(t)$, i.e.,

$$s(t) = \sum_{m=-\infty}^{\infty} \sum_{n=-\infty}^{\infty} c_{m,n} h_{m,n}(t) \tag{9.1}$$

where

$$h_{m,n}(t) = \sqrt[4]{\frac{\alpha}{\pi}} \exp\left\{-\frac{\alpha}{2}(t-mT)^2 + jn\Omega t\right\} \tag{9.2}$$

The Gabor coefficients $c_{m,n}$ are computed by the sampled short-time Fourier transform,

$$c_{m,n} = \int_{-\infty}^{\infty} s(t)\gamma^*_{m,n}(t)dt = \int_{-\infty}^{\infty} s(t)\gamma^*_{opt}(t-mT)e^{-jn\Omega t}dt = STFT(mT, n\Omega) \tag{9.3}$$

where $\gamma_{opt}(t)$ denotes the dual function whose shape is optimally close to the Gaussian elementary function $h(t)$. This ensures that the Gabor coefficients $c_{m,n}$ indeed reflect the behavior of the signal in the vicinity of $[mT - \Delta_t, mT + \Delta_t] \times [n\Omega - \Delta_\omega, n\Omega + \Delta_\omega]$ if the function $h(t)$ is centered at (0,0).

Taking the Wigner-Ville distribution of (9.1) yields,

$$WVD_s(t, \omega) = \sum_{m,n} \sum_{m',n'} c_{m,n} c^*_{m',n'} WVD_{h,h'}(t, \omega) \tag{9.4}$$

where

$$\boxed{\begin{aligned} WVD_{h,h'}(t, \omega) &= 2\exp\left\{j(n-n')\Omega(m+m')\frac{T}{2}\right\} \\ &\times \exp\left\{-\alpha\left(t - \frac{m+m'}{2}T\right)^2 + j(n-n')\Omega\left(t - \frac{m+m'}{2}T\right)\right\} \\ &\times \exp\left\{-\alpha^{-1}\left(\omega - \frac{n+n'}{2}\Omega\right)^2 - j(m-m')T\left(\omega - \frac{n+n'}{2}\Omega\right)\right\} \end{aligned}} \tag{9.5}$$

which is a separable 2D Gaussian function.

Eq. (9.4) shows that the Wigner-Ville distribution can be considered as a superposition of an infinite number of cross Wigner-Ville distributions $WVD_{h,h'}(t,\omega)$. Since the Wigner-Ville distribution is energy conserving, the cross Wigner-Ville distributions $WVD_{h,h'}(t,\omega)$ can also be considered as an *energy atom*, a fundamental element of the signal's energy in the joint time-frequency domain.

Eq. (9.5) shows that the 2D envelope of the energy atom $WVD_{h,h'}(t,\omega)$ is concentrated and symmetrical with respect to the point

$$\left(\frac{m+m'}{2}T, \frac{n+n'}{2}\Omega\right) \tag{9.6}$$

which is midway between the two Gaussian elementary functions, $h_{m,n}(t)$ and $h_{m',n'}(t)$. Moreover, the energy atom $WVD_{h,h'}(t,\omega)$ in Eq. (9.5) is oscillating. The rate of oscillation along the time axis is

$$(n-n')\Omega \tag{9.7}$$

which is the difference between $h_{m,n}(t)$ and $h_{m',n'}(t)$ in the frequency domain. The rate of oscillation along the frequency axis is

$$(m-m')T \tag{9.8}$$

which is the difference between $h_{m,n}(t)$ and $h_{m',n'}(t)$ in the time domain. The further $h_{m,n}(t)$ and $h_{m',n'}(t)$ are apart, the more the oscillations. Figure 9-1 illustrates the real part of $WVD_{h,h'}(t,\omega)$ in Eq. (9.5).

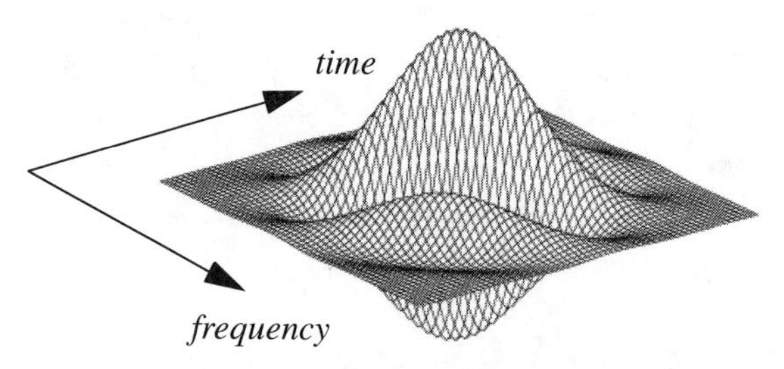

Figure 9-1 The energy atom is concentrated, oscillating, and symmetrical. Its average (or energy) is inversely proportional to its rate of oscillation.

Figure 9-2 illustrates the locations of the energy atoms $WVD_{h,h'}(t,\omega)$. Note that all the pairs of $h_{m,n}(t)$ and $h_{m',n'}(t)$, in which $m_0 = (m+m')/2$ and $n_0 = (n+n')/2$, will create an energy atom $WVD_{h,h'}(t,\omega)$ in the vicinity of $(m_0T, n_0\Omega)$. Hence, each ellipse in Figure 9-2 is the superposition of an infinite number of energy atoms $WVD_{h,h'}(t,\omega)$. Since the pair of complex conjugate energy atoms $WVD_{h,h'}(t,\omega)$ and $WVD_{h',h}(t,\omega)$ occur in the same area, the sum of the energy atoms at each ellipse is always real. The energy contained in a complex conjugate energy pair is

$$\int_{-\infty}^{\infty}\int_{-\infty}^{\infty}\{WVD_{h,h'}(t,\omega)+WVD_{h',h}(t,\omega)\}dtd\omega \tag{9.9}$$

$$= 2e^{-\frac{\alpha}{2}[(m-m')T]^2-\frac{1}{2\alpha}[(n-n')\Omega]^2}\cos\left\{(n-n')\Omega(m+m')\frac{T}{2}\right\}$$

whose magnitude is inversely proportional to the rate of oscillation

$$(m - m')T \qquad and \qquad (n - n')\Omega \tag{9.10}$$

or the distance between the two elementary Gaussian functions, $h_{m,n}(t)$ and $h_{m',n'}(t)$. The higher the frequency of oscillation, the less the energy contained. The energy in the pair of complex conjugate energy atoms exponentially diminishes as the distance between $h_{m,n}(t)$ and $h_{m',n'}(t)$ increases.

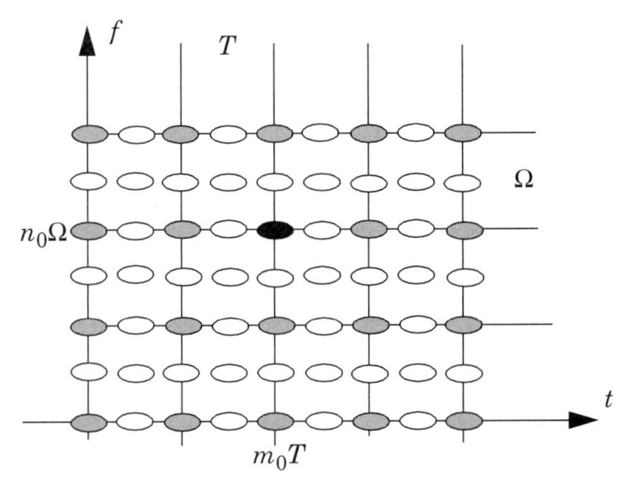

Figure 9-2 Each ellipse represents the superposition of an infinite number of elementary energy atoms $WVD_{h,h'}(t,\omega)$. All $h_{m,n}(t)$ and $h_{m',n'}(t)$, in which $(m + m')/2 = m_0$ and $(n + n')/2 = n_0$, will create a term centered at $(m_0T, n_0\Omega)$.

The concept of the cross-term in fact is ambiguous. A given signal can be broken up in an infinite number of ways. Different decomposition schemes will lead to different cross-terms. The cross-terms are not unique. However, the decomposition scheme introduced in this section is simple, which enables us to better understand the nature of the cross-terms. As a matter of fact, the cross-terms are not always *ghosts*. When the cross-term is created by a pair of elementary functions that are close to each other, it has a significant influence on the useful properties and thereby cannot be simply gotten rid of. It is the higher harmonics that interfere with the meaningful pattern of time-dependent spectra. The useful properties of the Wigner-Ville distribution are mainly determined by the few lowest harmonics. Eq. (9.9) shows the relationship between the signal's energy and the rate of oscillation, or the distance between the elementary Gaussian functions. In a similar manner, the reader can also reach the same conclusions for the marginal condition and the conditional mean frequency properties.

9.2 Time-Frequency Distribution Series

As discussed in the preceding section, the contribution of each individual energy atom to the desired properties is inversely proportional to its rate of oscillation. Further, the rate of oscillation is determined by the distance between the two elementary Gaussian functions. Based on these observations, we reorganize Eq. (9.4) as

$$\boxed{TFDS_D(t,\omega) = \sum_{|m-m'|+|n-n'| \le D} c_{m,n} c^*_{m',n'} WVD_{h,h'}(t,\omega)} \tag{9.11}$$

in which the energy atoms are grouped in terms of the *Manhattan distance* between $h_{m,n}(t)$ and $h_{m',n'}(t)$. As shown in Figure 9-3, unlike the normal Euclidean distance, the Manhattan distance defines the distance between two points as the vertical distance plus the horizontal distance. Since the significance of each individual energy atom $WVD_{h,h'}(t,\omega)$ depends on the rate of its oscillation, and the rate of oscillation is equal to $|n-n'|\Omega$ and $|m-m'|T$ (see Eq. (9.5)), the Manhattan distance better describes the nature of the energy atom than does the Euclidean distance.

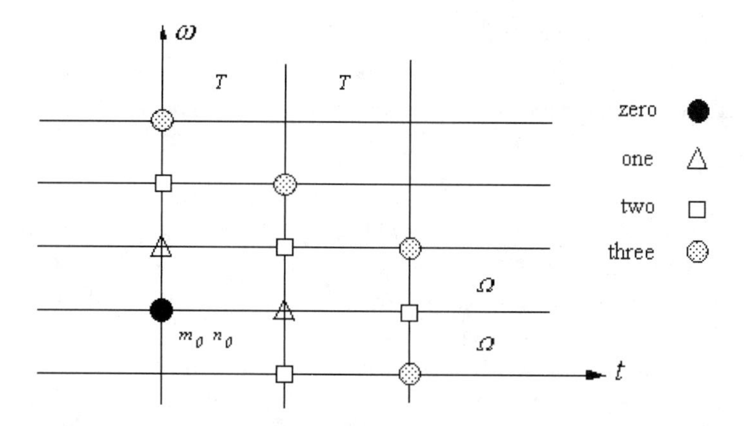

Figure 9-3 Because the rate of oscillation of the energy atom is proportional to $|n-n'|\Omega$ and $|m-m'|T$, the Manhattan distance better characterizes the energy atom's oscillation than does the Euclidean distance.

Eq. (9.11) is named as a *time-frequency distribution series* (TFDS) or the Gabor expansion based spectrogram, the *Gabor spectrogram,* to distinguish it from the STFT spectrogram. The parameter D in Eq. (9.11) denotes the order of the time-frequency distribution series. $TFDS_D(t,\omega)$ contains all $WVD_{h,h'}(t,\omega)$ in which $|m-m'|+|n-n'| \le D$. In general, the TFDS in (9.11) is not time- and frequency-shift invariant unless the time-shift is a multiple of the time sampling step T and the frequency-shift is a multiple of the frequency sampling step Ω.

Substituting (9.5) into (9.11) yields

$$TFDS_D(t, \omega) = 2 \sum_{|m-m'|+|n-n'| \le D} c_{m,n} c^*_{m',n'} \exp\left\{ j(n-n')\Omega(m+m')\frac{T}{2} \right\}$$
$$\times \exp\left\{ -\alpha\left(t - \frac{m+m'}{2}T\right)^2 + j(n-n')\Omega\left(t - \frac{m+m'}{2}T\right) \right\}$$
$$\times \exp\left\{ -\alpha^{-1}\left(\omega - \frac{n+n'}{2}\Omega\right)^2 - j(m-m')T\left(\omega - \frac{n+n'}{2}\Omega\right) \right\} \quad (9.12)$$

When $D = 0$,

$$TFDS_0(t, \omega) = 2\sum_{m,n} |c_{m,n}|^2 \exp\left\{ -\alpha(t - mT)^2 - \frac{1}{\alpha}(\omega - n\Omega)^2 \right\} \quad (9.13)$$

which can be computed by a regular separable 2D linear interpolation filter, i.e.,

$$TFDS_0(t, \omega) = \sum_n \left\{ \sum_m |c_{m,n}|^2 g(t - mT) \right\} f(\omega - n\Omega) \quad (9.14)$$

where $g(t)$ and $f(\omega)$ have forms

$$g(t) = e^{-\alpha t^2} \quad and \quad f(\omega) = e^{-\omega^2/\alpha} \quad (9.15)$$

Obviously, the zero order time-frequency distribution series $TFDS_0(t,\omega)$ in Eq. (9.13) is non-negative. As the order D of the time-frequency distribution series increases, more and more highly oscillating terms are included. As D goes to infinity, $TFDS_\infty(t,\omega)$ in Eq. (9.11) converges to the Wigner-Ville distribution.

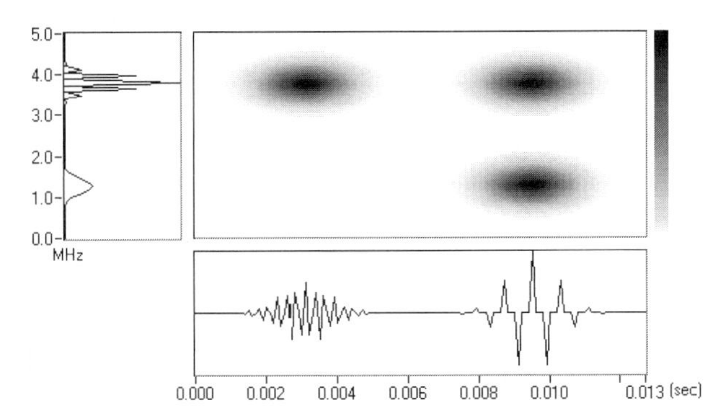

Figure 9-4 *$TFDS_0(t,\omega)$* is non-negative. It does not have cross-terms interference.

Figures 9-4 to 9-6 illustrate the time-frequency distribution series with different orders. When $D = 0$ in Figure 9-4, there is no undesirable term, but the resolution is rather poor (closer to that of the STFT spectrogram). For $D = 3$ in Figure 9-5, the resolution is closer to that of the

Wigner-Ville distribution, but there is no undesirable term. Moreover, it can be shown that $TFDS_3(t,\omega)$ well preserves the useful properties possessed by the Wigner-Ville distribution. As the order D gets larger, the unwanted term becomes more and more visible and the time-frequency distribution series converges to the Wigner-Ville distribution in Figure 9-6. In general, the higher the order, the better the resolution. On the other hand, the higher the order, the more severe the interference. A good compromise usually is from order two to order four.

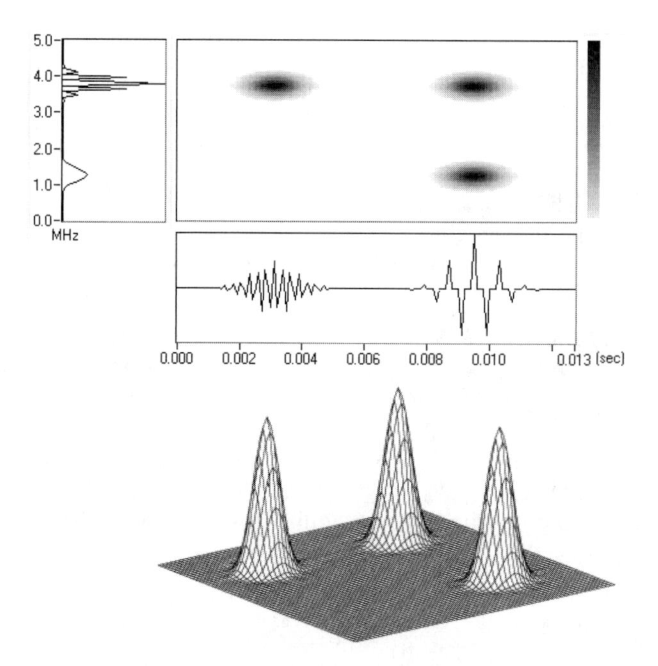

Figure 9-5 The resolution of $TFDS_3(t,\omega)$ is close to the Wigner-Ville distribution in Figure 9-7, but there are no undesired cross-terms. It can be shown that $TFDS_3(t,\omega)$ very well preserves the useful properties possessed by the Wigner-Ville distribution.

As shown above, by adjusting the order D, we can effectively balance the cross-term interference, the useful properties, and the resolution. Traditionally, the cross-terms are always considered as undesirable and bad. Thereby, they should simply be removed. However, the time-frequency distribution series in Eq. (9.11) and the corresponding plots, Figures 9-4 to 9-6, demonstrate that the cross-term is not always unwanted. The significance of each individual cross Wigner-Ville distribution $WVD_{h,h'}(t,\omega)$ depends on the Manhattan distance between the two corresponding auto-terms, $h_{m,n}(t)$ and $h_{m',n'}(t)$. When $WVD_{h,h'}(t,\omega)$ corresponds to a pair of $h_{m,n}(t)$ and $h_{m',n'}(t)$ that are close to each other, they play important roles for the useful properties and, therefore, cannot be neglected, even though they are the "cross-terms." The terms that need to be removed are only those $WVD_{h,h'}(t,\omega)$ in which the two Gaussian elementary functions $h_{m,n}(t)$ and $h_{m',n'}(t)$ are far apart; they have negligible influence on the useful properties but create severe cross-term interference.

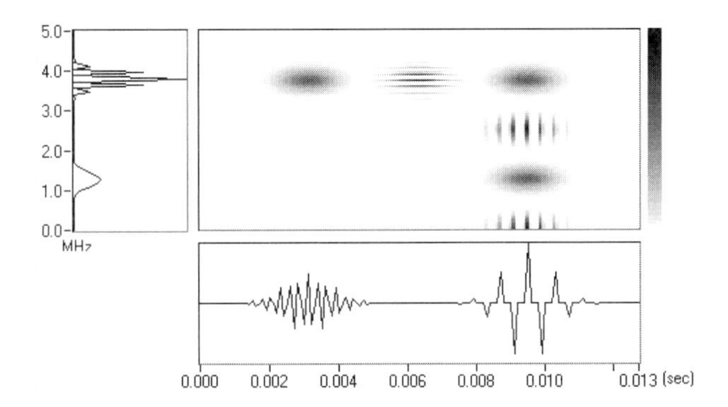

Figure 9-6 $TFDS_{10}(t,\omega)$ converges to the Wigner-Ville distribution. A high-order time-frequency distribution series improves the resolution, but also introduces interference. A good compromise has been found to be from order two to order four.

The idea of the time-frequency distribution series is rather straightforward. Over the years, the time-frequency distribution series presented in this section has been applied to a wide range of applications involving non-stationary signals — from radar and sonar to biomedicine and geophysics. In many applications, the time-frequency distribution series has been found to be a good alternative to the STFT spectrogram and Wigner-Ville distribution.

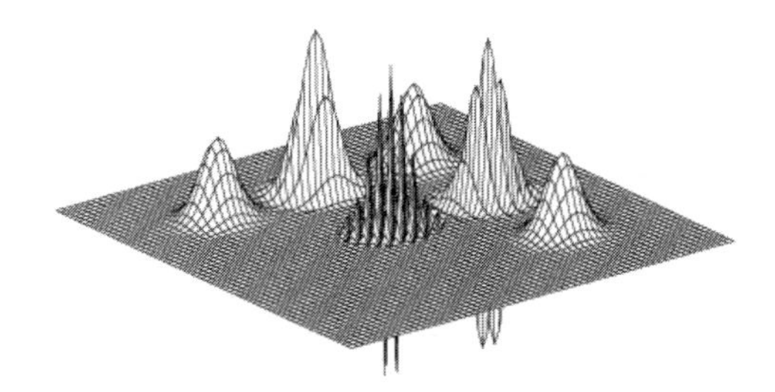

Figure 9-7 The Wigner-Ville distribution possesses severe cross-term interference.

The time-frequency distribution series can be effectively computed by separable 2D linear interpolation filters. Eqs. (9.14) and (9.15) show how to apply separable 2D linear interpolation to compute the zero order Gabor spectrogram. For more general cases, let

$$m_t = m + m' \qquad n_t = n + n' \qquad m_d = m - m' \qquad n_d = n - n' \tag{9.16}$$

Due to the fact that

$$WVD_{h,\,h'}(t,\,\omega) + WVD_{h',\,h}(t,\,\omega) = 2Re\{WVD_{h,\,h'}(t,\,\omega)\} \tag{9.17}$$

$TFDS_D(t,\omega)$ in Eq. (9.12) can be partitioned into

$$TFDS_D(t, \omega) = 2\Gamma_{0,0}(t, \omega) + 4Re\left\{\sum_{n_d=1}^{D} \Gamma_{0, n_d}(t, \omega)\right\} + 4Re\left\{\sum_{m_d=1}^{D} \sum_{n_d=m_d-D}^{D-n_d} \Gamma_{m_d, n_d}(t, \omega)\right\} \tag{9.18}$$

where Γ denotes separable 2D linear interpolation filtering, i.e.,

$$\Gamma_{m_d n_d}(t, \omega) = \sum_{m_t}\sum_{n_t} A_{m_d n_d}(m_t, n_t) g_{n_d}\left(t - m_t\frac{T}{2}\right) f_{m_d}\left(\omega - n_t\frac{\Omega}{2}\right) \tag{9.19}$$

where the weight A is equal to

$$A_{m_d n_d}(m_t, n_t) = c_{\frac{m_t+m_d}{2}, \frac{n_t+n_d}{2}} c^*_{\frac{m_t-m_d}{2}, \frac{n_t-n_d}{2}} \exp\left\{jn_d\Omega m_t\frac{T}{2}\right\} \tag{9.20}$$

which reflects the correlation between the Gabor coefficients. The filters $g(t)$ and $f(\omega)$ are defined as

$$g_{n_d}(t) = \exp\{-\alpha t^2 + jn_d\Omega t\} \tag{9.21}$$

and

$$f_{m_d}(\omega) = \exp\left\{-\frac{1}{\alpha}\omega^2 - jm_d T\omega\right\} \tag{9.22}$$

Apparently, both filters $g(t)$ and $f(\omega)$ are bandpass filters with center frequencies $n_d\Omega$ and m_dT, respectively. While the filter $g(t)$ determines linear interpolation along the time index, $f(\omega)$ determines linear interpolation along the frequency index. When $m_d = n_d = 0$, Eqs. (9.21) and (9.22) reduce to Eq. (9.15), and the weight A in Eq. (9.20) is simply equal to $|c_{m,n}|^2$. In this case, only the correlation between the two identical components is considered. As the order D increases, increasingly less-correlated components are included. As the order D goes to infinity, the time-frequency distribution series converges to the Wigner-Ville distribution.

Note that Γ in Eq. (9.19) represents standard linear interpolation filtering. It can be effectively computed in the Fourier transform domain, particularly when the lengths of the signal and the filter $g(t)$ are close. Moreover, because the Fourier transforms of both $g(t)$ and $f(\omega)$ have simple analytical forms, there is no need to compute the Fourier transforms of $g(t)$ and $f(\omega)$.

The presentation so far has been limited to continuous-time signals. The discrete time-frequency distribution series can be obtained by simply replacing t and ω in Eq. (9.18) by $i\Delta_t$ and $2\pi k/\Delta_t$, respectively. That is,

$$TFDS_D[i, k] = TFDS_D(t, \omega)\Big|_{t = i\Delta_t,\ \omega = \frac{2\pi k}{\Delta_t}} \tag{9.23}$$

where Δ_t denotes the sampling frequency. $TFDS_D[i,k]$ can be considered as a sampled version of $TFDS_D(t,\omega)$. Unlike the Wigner-Ville distribution, as well as many other members of Cohen's class discussed in Chapter 8, there is no aliasing problem for the discrete time-frequency distribution series $TFDS_D[i,k]$.

When the number of discrete-time data samples is finite, the corresponding interpolation filters in Eq. (9.19) become

$$g_{n_d}[i] = \exp\left\{-\alpha i^2 + jn_d\Omega i - m_d\frac{T}{2}\right\} \qquad f_{m_d}[k] = \exp\left\{-\frac{1}{\alpha}\left(\frac{2\pi k}{L}\right)^2 - jm_dT\frac{2\pi k}{L} - n_d\frac{\Omega}{2}\right\} \tag{9.24}$$

where L denotes the length of the dual functions $h[k]$ and $\gamma[k]$, which usually is also equal to the total number of frequency bins.

9.3 Selection of Dual Functions

As long as the functions $h(t)$ and $\gamma(t)$ satisfy the dual relationship discussed in Chapter 3, the time-frequency distribution series always converges to the Wigner-Ville distribution. In other words, the selection of the dual functions $h(t)$ and $\gamma(t)$ is independent of the convergence of the time-frequency distribution series. However, the dual functions $h(t)$ and $\gamma(t)$ have great impact on the low-order time-frequency distribution series. This is because, as discussed in the preceding section, the Gabor coefficients essentially are nothing more than the sampled STFT, and the zeroth order time-frequency distribution series $TFDS_0(t,\omega)$ is close to the STFT spectrogram. Including high order terms can improve the time-frequency resolution, but in the meantime high order terms will also bring undesirable cross-term interference. Hence, we normally keep the order of the time-frequency distribution series from two to four. In this case, as numerical simulations indicate, the selection of the dual functions is rather critical, though it is not as sensitive as for the short-time Fourier transform. In general, we would like

- $h(t)$ and $\gamma(t)$ to have similar time/frequency centers and time/frequency resolutions.
- Both $h(t)$ and $\gamma(t)$ to be localized in the joint time-frequency domain.

These concepts may not be obvious at first glance. In order to get a better understanding, let's look at a couple of examples. First, let's examine the zeroth order time-frequency distribution series, e.g.,

$$TFDS_0(t,\omega) = 2\sum_{m,n}|c_{m,n}|^2\exp\left\{-\alpha(t-mT)^2 - \frac{1}{\alpha}(\omega - n\Omega)^2\right\} \tag{9.25}$$

Notice that the exponential parts are the Wigner-Ville distributions of $h_{m,n}(t)$, which are concentrated at $(mT,n\Omega)$. The weight $c_{m,n}$ is the projection of the signal $s(t)$ on $\gamma_{m,n}(t)$, that is, $c_{m,n} = \langle s,\gamma_{m,n}\rangle$. If the time and frequency centers of $h(t)$ and $\gamma(t)$ are substantially different, for instance, the time/frequency center of $h(t)$ is (0,0) and that of $\gamma(t)$ is (t_0,ω_0), then $|c_{m,n}|^2$ will reflect the signal's behavior in the vicinity of $(mT+t_0,n\Omega+\omega_0)$ rather than $(mT,n\Omega)$, as one might anticipate. Hence, $h(t)$ and $\gamma(t)$ must have similar time/frequency centers and time/frequency resolutions.

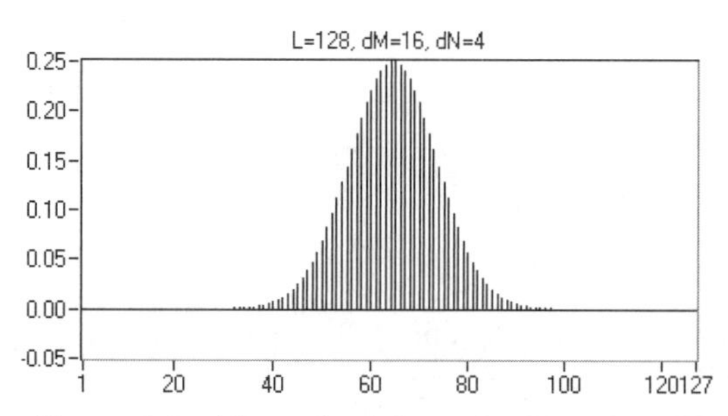

Figure 9-8 Normalized Gaussian Function $h[i]$

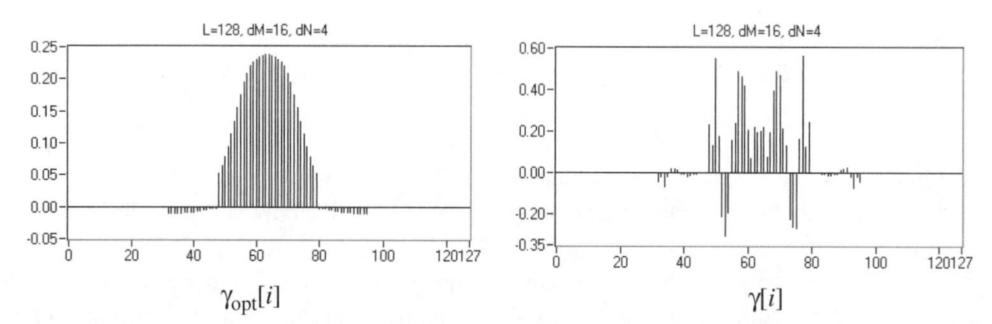

$\gamma_{opt}[i]$ $\gamma[i]$

Figure 9-9 Although both dual functions lead to perfect reconstruction, their shapes are dramatically different. $\gamma_{opt}[i]$ is optimally close to $h[i]$. Therefore, it has similar time and frequency centers to those of $h[i]$. $\gamma_{opt}[i]$ is optimally concentrated in the joint time-frequency domain, whereas the dual function on the right-side is neither concentrated in time nor in frequency.

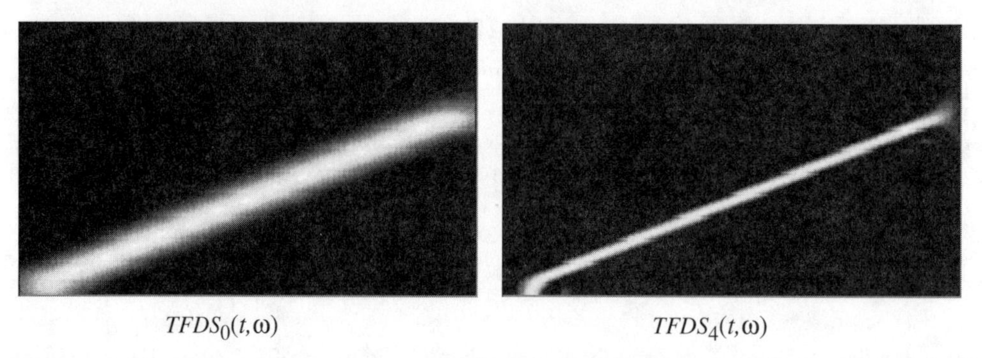

$TFDS_0(t, \omega)$ $TFDS_4(t, \omega)$

Figure 9-10 When dual functions form the orthogonal-like Gabor expansion, we can obtain the desired representation with the low-order time-frequency distribution series.

Figures 9-8 and 9-9 depict the discrete Gabor elementary function $h[i]$ and the corresponding dual functions $\gamma[i]$. Although both dual functions in Figure 9-9 lead to perfect reconstruction, their shapes are dramatically different. $\gamma_{opt}[i]$ is optimally close to $h[i]$. In other words, it has similar time and frequency centers as those of $h[i]$. Because $h[i]$ is a Gaussian-type function that is optimally concentrated in the joint time-frequency domain, $\gamma_{opt}[i]$ is also localized. On the other hand, the dual function depicted on the right side of Figure 9-9 is neither concentrated in time nor in frequency.[1]

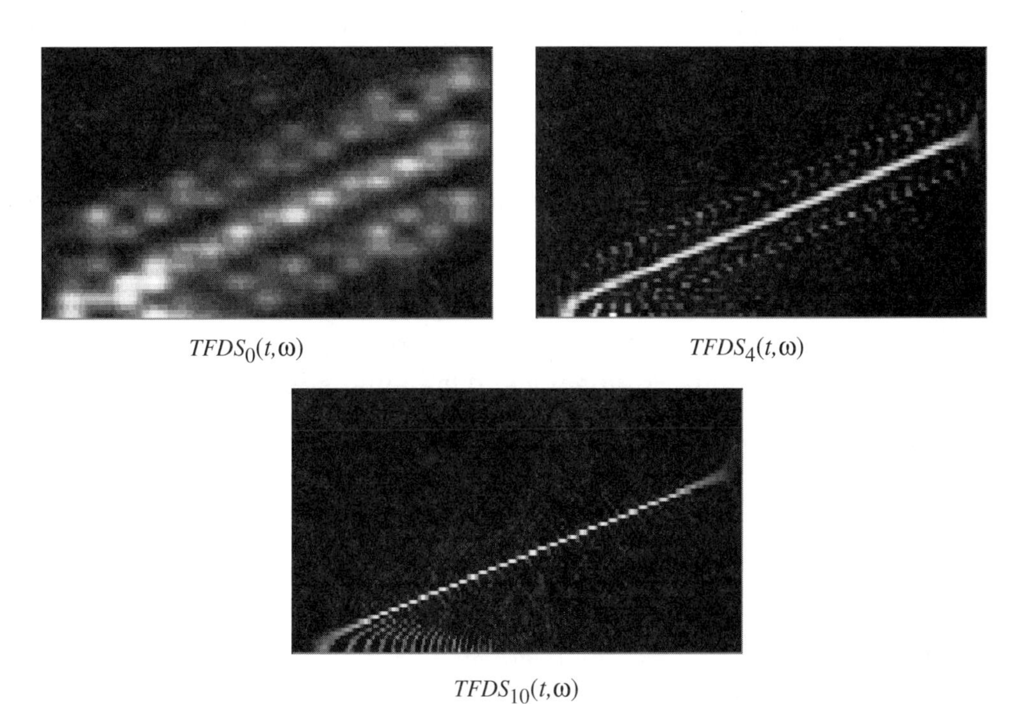

Figure 9-11 With improperly selected dual functions, we need substantially high-order time-frequency distribution series for meaningful time-frequency resolution at the cost of cross-term interference.

Figures 9-10 and 9-11 illustrate the time-frequency distribution series computed by the two different dual functions plotted in Figure 9-9 for the linear chirp signal whose frequency linearly increases with respect to time. In Figure 9-10, because $\gamma_{opt}[i]$ has the same time and frequency center as $h[i]$ and is localized, the low-order time-frequency distribution series well delineates the linear chirp signal with negligible interference. Figure 9-11 plots the results of he time-frequency distribution series for the same chirp signal but with dual function $\gamma[i]$ (plotted as shown on the right side of Figure 9-9), whose shape considerably differs from $h[i]$. Although as the order increases the time-frequency distribution series eventually converges to the Wigner-Ville distribution, its low-order version does not reflect the time-dependent spectrum as well as that computed by using $\gamma_{opt}[i]$. The meaningful linear chirp pattern does not appear until the order increases to ten. In the meantime, however, the resulting time-frequency distribution series has already included substantial interference. Therefore, to achieve the desired time-frequency

1. The algorithm for computing $\gamma[k]$ that is intentionally different from $h[k]$ is introduced in Appendix.

representation, $\gamma[i]$ not only has to be localized, but also should be close to $h[i]$. The solution is the orthogonal-like Gabor expansion that is introduced in Section 3.4.

Finally, the variance α that controls the time and frequency resolution of the Gabor elementary function also affects the performance of the lower order time-frequency distribution series, though its impact is much smaller than that in the STFT spectrogram. A good choice depends on the application at hand. The principle of selection, however, is similar to the selection of the window function for the short-time Fourier transform. If time resolution is important, for instance, when we try to catch a short duration pulse, then a larger value of α is favored. Otherwise, we should use a smaller value of α, which leads to better frequency resolution. Usually, we start with order zero and vary the variance α to obtain the best resolution. Then, we gradually increase the order until we reach the best compromise between the time-frequency resolution and cross-term interference.

9.4 Mean Instantaneous Frequency and Instantaneous Bandwidth

In this section, we shall further study the time-frequency distribution series, in terms of the conditional mean frequency and instantaneous bandwidth. While the conditional mean frequency $\langle\omega\rangle_t$ is given in Eq. (7.23), the instantaneous bandwidth is usually defined by

$$\Delta_\omega^2(t) = \frac{\int_{-\infty}^{\infty} (\omega - \langle\omega\rangle_t)^2 P(t, \omega) d\omega}{\int_{-\infty}^{\infty} P(t, \omega) d\omega} \tag{9.26}$$

which characterizes the energy spread of the signal. Intuitively, a good time-dependent spectrum $P(t,\omega)$ should be such that its conditional mean frequency is equal to the mean instantaneous frequency of the signal, the first derivative of a signal's phase. The instantaneous bandwidth should be as small as possible.

Figures 9-12 illustrates the fourth order time-frequency distribution series $TFDS_4(t,\omega)$ for a synthetic Gaussian chirplet. The black line represents the corresponding conditional mean frequency. The numerical simulation indicates that the difference between the first derivative of a signal's phase and the conditional mean frequency computed from the fourth order time-frequency distribution series is almost negligible. Figures 9-13 shows the result of the STFT spectrogram for the same signal. As introduced in Example 3.1, the conditional mean frequency computed from the STFT spectrogram in general is not equal to the first derivative of the phase of the signal. In this example, however, due to the narrow window applied the conditional mean frequency computed from the STFT spectrogram is very close to that computed from the time-frequency distribution series. However, the energy spread of the STFT spectrogram is much wider than that in the fourth order time-frequency distribution series in Figures 9-12. The same observation can also be obtained from the RF signal in Figures 9-14 and 9-15. As numerical simulations indicate, the conditional mean frequency of the time-frequency distribution series not only well approximates the signal's mean instantaneous frequency, but also has much smaller instantaneous bandwidth than that computed from its STFT spectrogram counterpart.

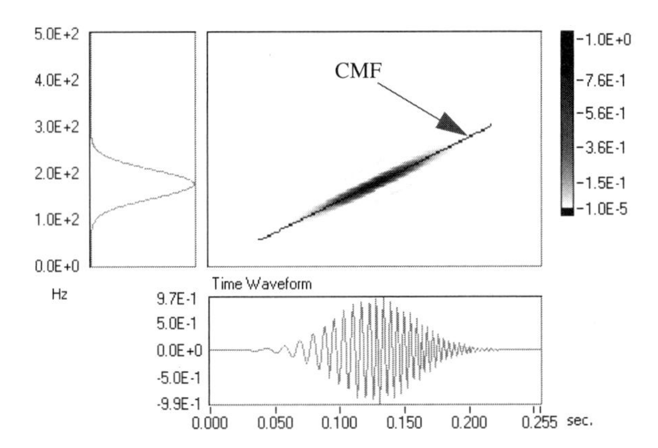

Figure 9-12 *$TFDS_4(t,\omega)$* and the corresponding conditional mean frequency (CMF) for a Gaussian chirplet

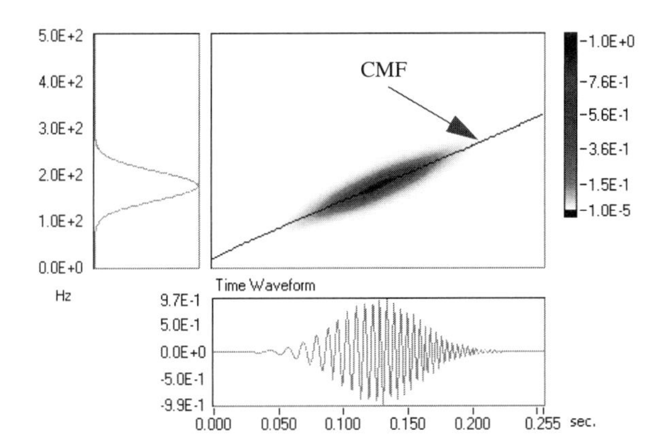

Figure 9-13 The STFT spectrogram (with 0.032 seconds Hanning window) and the corresponding conditional mean frequency (CMF) for the Gaussian chirplet. Due to the narrow window applied, the difference of the conditional mean frequencies computed by the STFT spectrogram and the fourth order time-frequency distribution series (see Figure 9-12) is almost negligible. However, the energy spread of the STFT spectrogram is much wider than that in $TFDS_4(t,\omega)$.

The mean instantaneous frequency and instantaneous bandwidth have been found useful quantities for characterizing the nature of the signal's time-varying frequencies. The mean instantaneous frequency is defined as the rate of change of the phase at a given time. This is equivalent to the first moment in frequency of the Wigner-Ville distribution of the signal normalized by its instantaneous energy. Unfortunately, when we estimate the mean instantaneous frequency either directly from the phase of the signal or from the Wigner-Ville distribution, we generate estimates with high variance, particularly in noisy environments. Estimates based on the first moment in frequency of other time-frequency representations, such as the STFT spectrogram, reduce the variance at the expense of some bias. The task can become even more challenging when the signal is real.

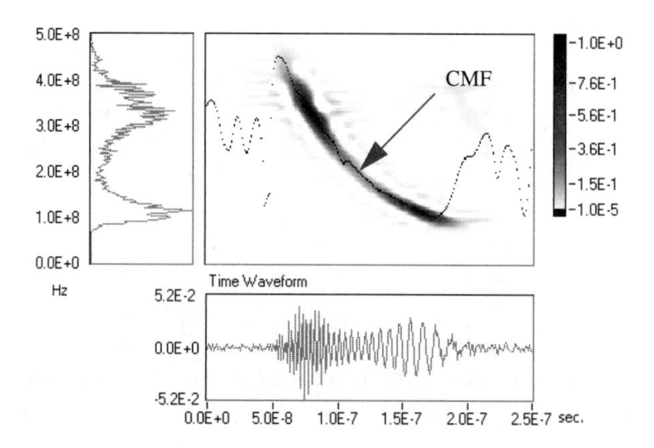

Figure 9-14 $TFDS_4(t,\omega)$ and the corresponding conditional mean frequency (CMF) for the RF signal

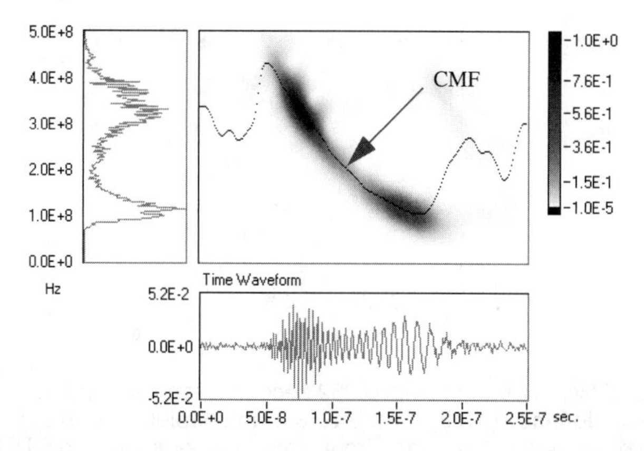

Figure 9-15 STFT spectrogram (with 32 nanoseconds Hanning window) and the corresponding conditional mean frequency (CMF) for the RF signal. Due to the narrow window applied, the conditional mean frequency computed by the STFT spectrogram is close to that computed by the fourth order time-frequency distribution (see Figure 9-14). However, the energy spread of the STFT spectrogram is much wider than that in $TFDS_4(t,\omega)$.

If we define the rate of change of the phase of the corresponding analytical signal at a given time as the mean instantaneous frequency, then we first need to compute the analytic function. As shown in Section 7.5, this can bring about tremendous distortion. A signal and its analytic function can have substantially different behavior in the frequency domain, particularly in the low frequency band. Moreover, the analytic signal is supposed to have only non-negative frequencies. However, there is no guarantee that the derivative of its phase is always bigger than or equal to zero. Hence, for real-valued signals, the first moment of the time-dependent power spectrum is a more desirable way of estimating the signal's instantaneous frequency and bandwidth.

For the Wigner-Ville distribution, they are

$$\langle \omega \rangle_t = \frac{\int_0^{\infty} \omega WVD_s(t, \omega) d\omega}{\int_0^{\infty} WVD_s(t, \omega) d\omega} \tag{9.27}$$

and

$$\Delta_{\omega}^2(t) = \frac{\int_0^{\infty} (\omega - \langle \omega \rangle_t)^2 WVD_s(t, \omega) d\omega}{\int_0^{\infty} WVD_s(t, \omega) d\omega} \tag{9.28}$$

Note that the time-dependent power spectrum $P(t,\omega)$ of a real-valued signal is symmetrical in frequency. Therefore, the range of integration is from zero to infinity rather than from minus infinity to plus infinity. Otherwise, the conditional mean frequency $<\omega>_t$ will be uniformly equal to zero.

Inserting the time-frequency distribution series Eq. (9.11) into Eqs. (9.27) and (9.28) yields [71]

$$\langle \omega \rangle_t = \frac{\sum_{|m-m'|+|n-n'| \le D} c_{m,n} c^*_{m',n'} \int_0^{\infty} \omega WVD_{h,h'}(t, \omega) d\omega}{\sum_{|m-m'|+|n-n'| \le D} c_{m,n} c^*_{m',n'} \int_0^{\infty} WVD_{h,h'}(t, \omega) d\omega} \tag{9.29}$$

and

$$\Delta_{\omega}^2(t) = \frac{\sum_{|m-m'|+|n-n'| \le D} c_{m,n} c^*_{m',n'} \int_0^{\infty} (\omega - \langle \omega \rangle_t)^2 WVD_{h,h'}(t, \omega) d\omega}{\sum_{|m-m'|+|n-n'| \le D} c_{m,n} c^*_{m',n'} \int_0^{\infty} WVD_{h,h'}(t, \omega) d\omega} \tag{9.30}$$

The main advantage of using Eqs. (9.29) and (9.30) to compute the mean instantaneous frequency and instantaneous bandwidth is the ability of balancing the estimate's variance introduced by the Wigner-Ville distribution based method and the bias generated by the STFT spectrogram based method. Intuitively, when the order $D = 0$, the result of Eq. (9.29) contains some bias, similar to that obtained by the STFT spectrogram. As D increases, the result of Eq. (9.29) gets closer and closer to that computed by the Wigner-Ville distribution. A good choice for the order D has been found to be from 2 to 4.

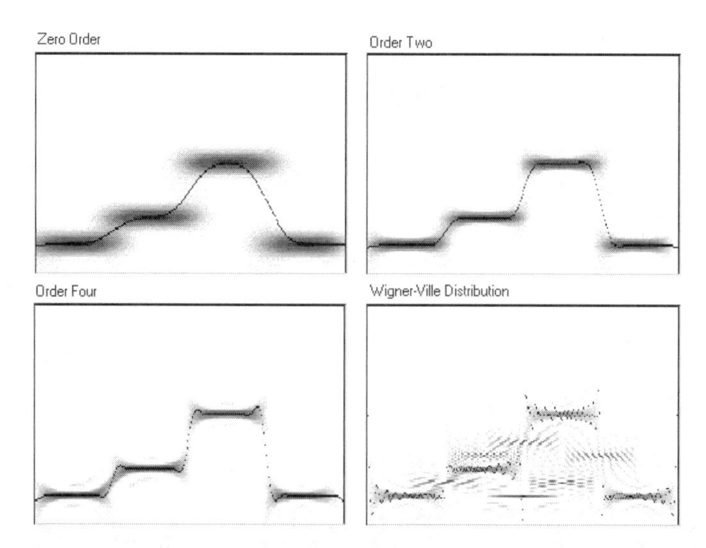

Figure 9-16 Mean instantaneous frequencies of a frequency hopper signal computed by Eq. (9.29) and the Wigner-Ville distribution. While the mean instantaneous frequency for $D = 0$ (top left) contains substantial bias, the one computed by the Wigner-Ville distribution (bottom right) is too noisy to identify. The estimated mean instantaneous frequency becomes closer and closer to the true value as the order increases. A good compromise between the variance of the estimator and the bias of the estimator appears when the order is equal to four (bottom left).

Figure 9-16 depicts the mean instantaneous frequencies (black lines) of a frequency hopper signal (the corresponding time waveform and power spectrum are plotted in Figure 7-7) computed by Eq. (9.29) and the Wigner-Ville distribution. While the mean instantaneous frequency for $D = 0$ (top left) contains substantial bias, the one computed by the Wigner-Ville distribution (bottom right) is too noisy to identify. The estimated mean instantaneous frequency becomes closer and closer to the true value as the order increases. The difference between the true value and the estimated value is basically negligible when the order is equal to four (bottom left).

9.5 Application for Earthquake Engineering

Because it provides a relatively simple and effective way to balance time-frequency resolution and cross-term interference, the time-frequency distribution series has been used in many areas. One successful story is an application in the area of earthquake engineering for the detection of soil liquefaction conducted at the University of Tokyo [156].

Soil liquefaction is an earthquake-related phenomenon that takes place in saturated soils with a soft sub-surface layer. The cause of liquefaction is the rise of the water pore pressure under undrained conditions during shaking of the ground. The increase of the pore pressure reduces the soil shear resistance to almost zero, causing the soil to behave as a liquid. Consequently, the energy of horizontal vibrations (seismic shear waves from deep within the ground) transferred by the soil will be substantially reduced, particularly in the high frequency band. Soil liquefaction has been recognized as the main reason for the collapse of earth dams and slopes, and for the failure of foundations, superstructures, and lifelines, such as gas and electrical power supplies (see Figure 1-15).

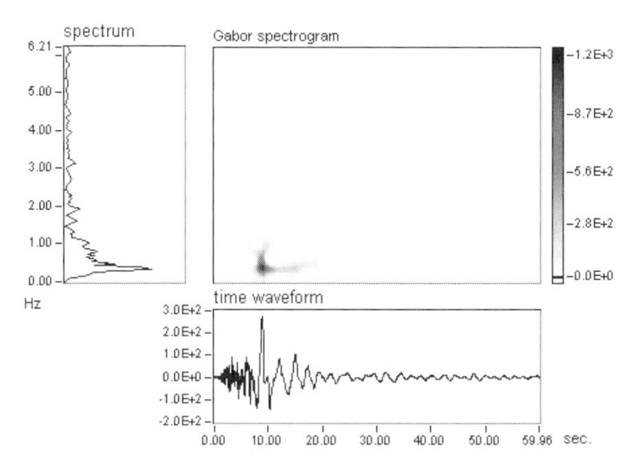

Figure 9-17 $TFDS_2(t,\omega)$ of the East-West component of the ground acceleration record at Higashi-Kobe Bridge from the 1995 Hyogoken-Nanbu earthquake. Extensive liquefaction took place at the site (Data from Strong Motion Array Observation Record Database CD-ROM Vol.5, Association for Earthquake Disaster Prevention, Tokyo, Japan, 1997).

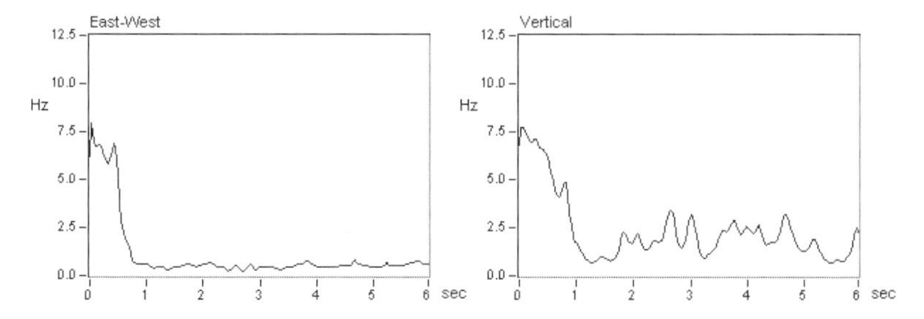

Figure 9-18 Mean instantaneous frequencies at the site where extensive liquefaction took place. The mean instantaneous frequency for the East-West component, computed from Figure 9-17, is obviously lower than that for the non-liquefaction case (see Figure 9-20).

Recent earthquakes in the US and Japan have shown the need of advanced earthquake disaster mitigation management in order to prevent or minimize damage, especially in urban areas. The basic idea is to monitor the seismic motion and perform actions based on the information obtained during an earthquake. The main components for advanced earthquake disaster mitigation management are the early warning and preliminary damage assessment systems, the backbone of which is a ground-motion monitoring network consisting of three-axial accelerometers. Such systems are used to stop the high-speed bullet-trains in Japan before the arrival of the destructive seismic waves, to shut off the gas supply in the areas where the ground shaking is too severe, and to estimate the damages and casualties immediately after an earthquake when the exact data have not yet been collected. Early localization of the liquefied areas is also of interest, particularly for the lifelines, and efforts have been made to determine the occurrence of liquefaction shortly after an event.

Proposals for detectors of soil liquefaction include downhole piesometers and sensors for measuring the rise of the water level in a hollow pipe inserted into the ground. Besides the consideration of the transducer reliability, however, such approaches require costly placing work and are of low durability.

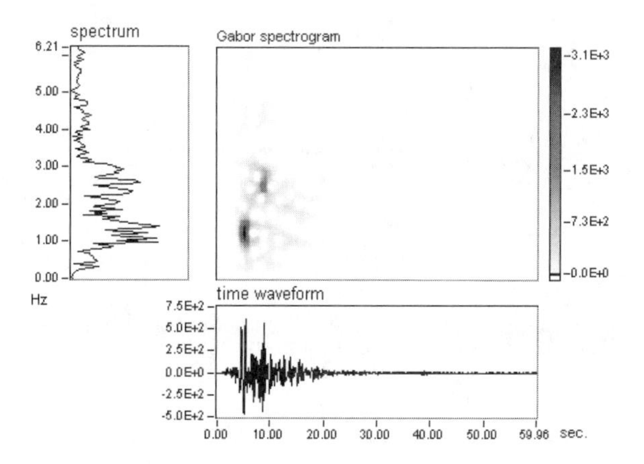

Figure 9-19 *$TFDS_2(t,\omega)$* of the East-West component of the ground acceleration record at JMA Kobe station from the 1995 Hyogoken-Nanbu earthquake. No liquefaction was reported at the site (Data from JMA Kobe Observatory, Japan Weather Association CD-ROM).

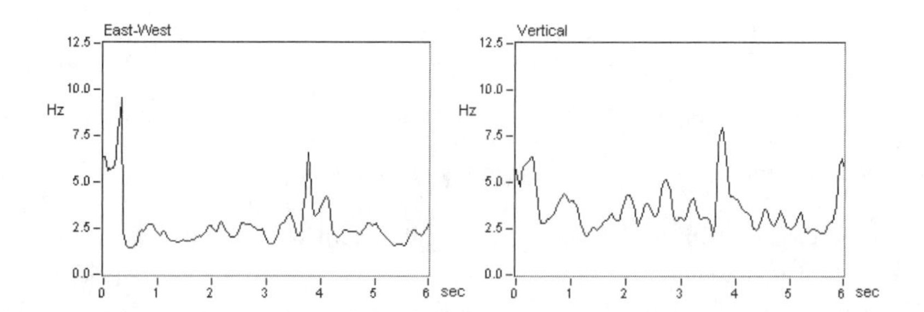

Figure 9-20 Mean instantaneous frequencies at the site where no liquefaction was reported. The mean instantaneous frequency for the East-West component is computed from Figure 9-19.

An alternative is to use the seismic records, collected from the ground-motion monitoring networks, for detection of liquefaction. Since the 1964 Niigata earthquake, a number of seismic records from liquefied-soil sites have been obtained and studied. The records show that the horizontal ground acceleration alters uniquely after the onset of liquefaction – its frequency abruptly drops off within the 0.5 - 1 Hz range and its amplitude decreases – while the vertical acceleration is rather stable. This alternation of the horizontal acceleration is triggered by the decrease of the soil shear modulus, as a consequence of the increase in the water pore-pressure. Methods directly based on the seismic records for liquefaction detection have been developed. Recently, researchers from the International Center for Disaster-Mitigation Engineering at the University of Tokyo have employed the instantaneous spectrum of the seismic-signal to quantify the alternation of the horizontal ground acceleration.

Figure 9-17 illustrates the second order time-frequency distribution series of a ground acceleration record when extensive liquefaction took place. Figure 9-18 depicts the mean instantaneous frequencies computed by Eq. (9.29), which quantitatively characterizes the frequency changes. The occurrence of liquefaction is judged by the relative difference of the mean instantaneous frequencies of the horizontal acceleration and the vertical acceleration. Figure 9-19 plots the second order time-frequency distribution series of a ground acceleration record where no liquefaction was reported. Figure 9-20 depicts the resulting mean instantaneous frequencies computed by Eq. (9.29). It shows that the relative difference in the mean instantaneous frequency of the horizontal and vertical acceleration is much larger than that when liquefaction occurs (see Figure 9-18). Based on the relative difference in the mean instantaneous frequency of the horizontal and vertical acceleration, researchers from the University of Tokyo are now able to remotely detect the occurrence of soil liquefaction.

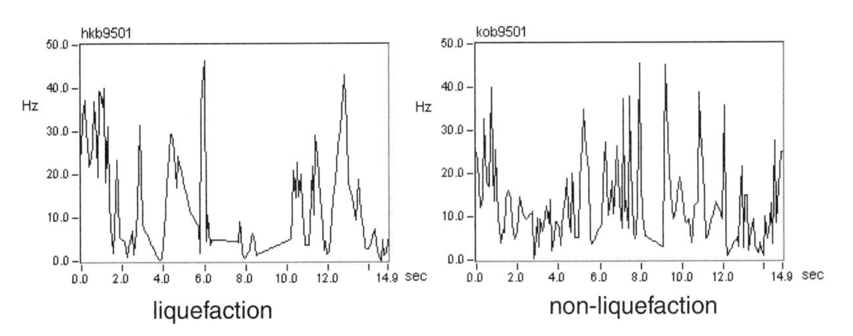

Figure 9-21 Due to the presence of random noise, the mean instantaneous frequency computed from the Wigner-Ville distribution based estimate changes dramatically. No useful information can be extracted.

As a comparison, Figure 9-21 illustrates the mean instantaneous frequencies computed from the Wigner-Ville distribution. While the plot on the left is for a liquefaction signal, the plot on the right corresponds to non-liquefaction. Compared to their counterparts computed by Eq. (9.29) (the left plots in Figures 9-18 and 9-20), the Wigner-Ville distribution based estimates look very noisy. Due to the presence of random noise, the mean instantaneous frequency computed from the Wigner-Ville distribution changes dramatically. No useful information can easily be extracted. Theoretically, the conditional mean frequency of the Wigner-Ville distribution is equal to the signal's mean instantaneous frequency. In real applications, the Wigner-Ville distribution based estimate is rarely used because of its high variance, particularly when noise is involved.

CHAPTER 10

Adaptive Gabor Expansion and Matching Pursuit

*I*n previous chapters, we introduced the Gabor transform (which can be computed by the sampled STFT), the wavelet transform, and several bilinear transforms. A primary motivation of these different schemes is to improve the time-frequency resolution with the least amount of cross-term interference. In the case of the Gabor transform, to achieve a better time resolution we chose a short duration window. But a shorter window reduces the frequency resolution. A longer window will improve the frequency resolution but reduce the time resolution. Window selection is a nontrivial problem in the Gabor transform and is often at the expense of sacrificing one property or the other.

On the other hand, the Wigner-Ville distribution does not suffer from the window problem and possesses the best time-frequency resolution. However, the Wigner-Ville distribution has the problem of cross-term interference, which greatly limits its applications. The time-frequency distribution series (TFDS), the Gabor expansion based spectrogram, decomposes the Wigner-Ville distribution. Selectively removing the high harmonic terms leads to a high time-frequency resolution representation with limited interference. But for the situation where both very short duration and long duration signals are present, the effectiveness of the TFDS is also limited.

Figures 10-1 and 10-2 show the sum of two impulses, one in the time domain and the other in the frequency domain (a sinusoidal function in the time domain). Figure 10-1 (a) and Figure 10-1 (b) show the STFT spectrogram (square of the STFT) computed by a 16-point Hanning window and a 64-point Hanning window, respectively. The shorter window, Figure 10-1 (a), reveals the impulse as indicated by the vertical line in the middle of the picture. The wide horizontal stripe indicates very poor frequency resolution. The longer window, Figure 10-1 (b), clearly improves the frequency resolution at the expense of the time resolution. Thus, by using the STFT it is difficult to achieve good time and frequency resolutions simultaneously.

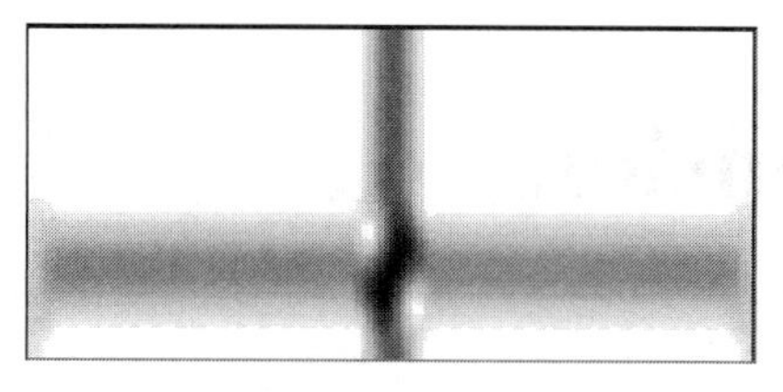

(a) 16-point Hanning window

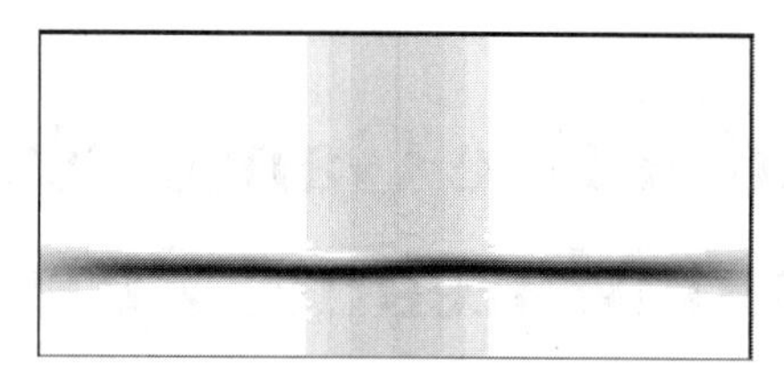

(b) 64-point Hanning window

Figure 10-1 STFT spectrogram with a short window in (a) reveals the impulse as indicated by the vertical line in the middle of the picture. The wide horizontal stripe indicates very poor frequency resolution. The longer window in (b) clearly improves the frequency resolution at the expense of the time resolution. By using the STFT, it is difficult to achieve good time and frequency resolutions simultaneously.

Figure 10-2 shows the fourth-order TFDS, in which the wideband elementary function is used to catch a very short duration time-domain pulse. Compared with the STFT spectrogram, the TFDS holds up quite well in both the time and the frequency resolutions, though the low-order TFDS is also subject to the selection of elementary functions.

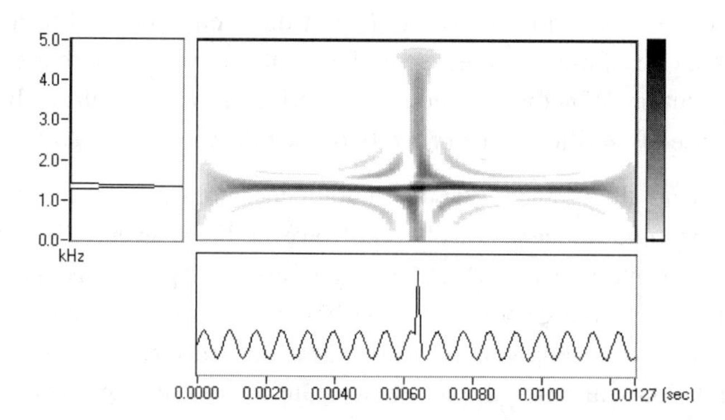

Figure 10-2 Compared with the STFT spectrogram, the $TFDS_4(t,\omega)$ holds up quite well in both the time and the frequency resolutions, though the low-order TFDS is also subject of the selection of elementary functions.

The desire for good resolution in both time and frequency domains occurs in many real applications. Figure 10-3 illustrates a simple radar target of a strip containing an open cavity. A fundamental interest in the radar community is the nature of reflected electromagnetic signals. Based on the scattering physics of this structure, we anticipate there to be three scattering centers, corresponding to the left and right edges of the strip, and the cavity exterior. Therefore, the reflected signal must contain three wideband time pulses that are caused by these scattering centers. Besides these three scattering centers, we also expect to see some narrowband signals that correspond to the energy coupled into and re-radiated from the cavity. These portions of the signal contain information on the resonant frequencies of the cavity. A fundamental task in radar image processing is to distinguish between these two types of signals, the wideband scattering centers which can be used to form spatially meaningful radar images and the narrowband signals which convey target resonance information.

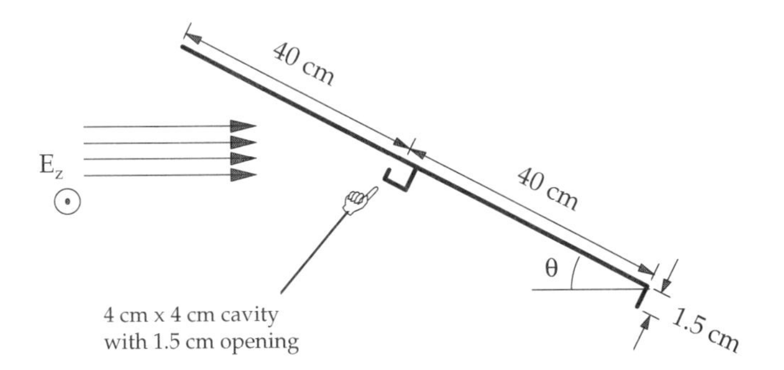

Figure 10-3 Based on the scattering physics of this conducting strip with a small cavity and fin, we anticipate there to be three scattering centers, corresponding to the left and right edges of the strip, and the cavity exterior. Therefore, the reflected signal must contain three wideband time pulses that are caused by these scattering centers. Besides these three scattering centers, we also expect to see some narrowband signals that correspond to the energy coupled into and re-radiated from the cavity. These portions of the signal contain information on the resonant frequencies of the cavity (Diagram courtesy of Hao Ling, Department of Electrical and Computer Engineering, University of Texas, Austin).

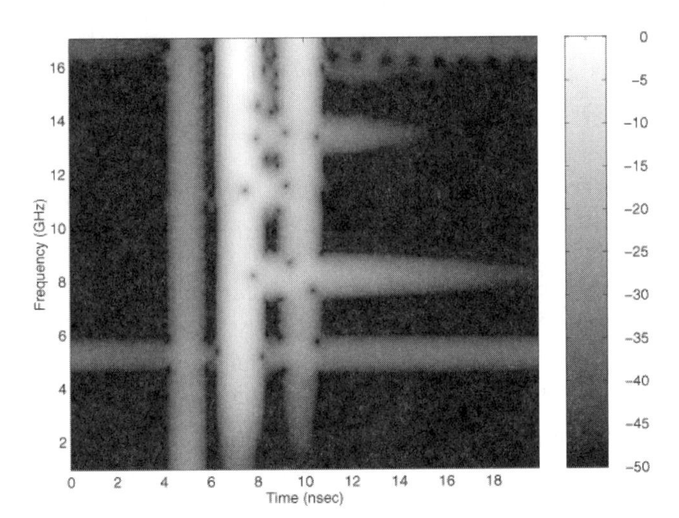

Figure 10-4 STFT spectrogram of reflected signals in Figure 10-3. The frequency resolution comes at the cost of losing time resolution, and vice versa. Because the quality of radar imagery and the accuracy of target identification algorithms are closely related to the sharpness of both time and frequency resolutions, the STFT spectrogram has been found to be inadequate in radar imaging applications (Image courtesy of Hao Ling, Department of Electrical and Computer Engineering, University of Texas, Austin).

Figure 10-4 displays an STFT spectrogram of the reflected electromagnetic signal. As expected, three scattering centers can be identified as vertical lines (wideband signals) in the STFT spectrogram. In addition to these scattering centers, we also see horizontal lines (narrowband signals), which correspond to the energy coupled into the cavity and re-radiated at its corresponding resonances. Although we gained some insights with the STFT spectrogram, the frequency resolution came at the cost of losing time resolution, and vice versa. Because the quality of radar imagery and the accuracy of target identification algorithms are closely related to the sharpness of both time and frequency resolutions, the STFT spectrogram has been found to be inadequate in radar imaging applications.

Motivated by adaptive windows, Qian and Chen developed the adaptive Gabor expansion in the early 90s ([192] and [197]). Unlike the classical Gabor expansion introduced in Chapter 3 in which the time and the frequency resolutions are fixed, for the adaptive Gabor expansion the time and frequency resolutions of the elementary function can be adjusted for best matching. About the same time period, *Stéphane Mallat* and *Zhifeng Zhang* proposed the so-called matching pursuit (MP) method [172]. It turns out that the adaptive Gabor expansion is a special but most important member of the MP class, which can be thought of as a combination of time-frequency and time-scale representation.

The matching pursuit is very similar to the projection pursuit principle discussed in statistics by *Huber* [132], whose strong convergence when the dictionary is complete was proved by *Jones* [150]. In addition to time-frequency analysis, the MP in fact addresses a common signal processing application — the decomposition of signals whose fundamental models do not form bases, such as the set of Gaussian functions with adjustable variances. In such applications, we may employ redundant frames. But, for a redundant frame, one component may have projections on every elementary vector.

In Section 10.1, we introduce the basics of MP. In most cases, we only present the result without justification. The reader can find a rigorous mathematical treatment in Mallat [27]. Section 10.2 is dedicated to the adaptive Gabor expansion — one of the most important members of the MP class. Generally speaking, the accuracy of the MP methods hinges on the size of the dictionary. The larger the dictionary, the better the accuracy. On the other hand, the larger the dictionary, the more the computation involved. Generally speaking, the MP algorithm is slow. However, the computation speed may be substantially improved if the analytical form for the inner product of a pair of elementary functions exists, such as in the case of the adaptive Gabor expansion. For the adaptive Gabor transform, we first use a coarse dictionary to obtain a rough estimate. Then, based on the rough estimate, we apply a unique curve fitting algorithm to perform a detailed matching. As introduced in Section 10.3, the resulting refinement process is extremely efficient; no iterations are involved. In principle, we can apply general optimization algorithms, such as the *Newton-Raphson* method, for refinement. However, for such multidimensional optimization applications, the generic digital optimization algorithms often have convergence problems. Compared to other optimization approaches, the algorithm introduced in Section 10.3 is fast, robust, and accurate. Moreover, the basic idea behind the fast algorithm presented in this chapter is not only valid for three-tuple adaptive Gabor expansion, but can also be directly applied to more general cases. As an example, we introduce a fast refinement algorithm for four-tuple Gaussian chirplet decomposition in Section 10.5.

10.1 Matching Pursuit

Many signals can be modeled as the superposition of a single prototype function, i.e.,

$$s(t) = \sum_k A_k h_k(t) \tag{10.1}$$

where $h_k(t)$ denote the elementary functions that are completely determined by a set of parameters. A_k represents the weight of each individual function $h_k(t)$. If the set of functions $\{h_k\}_{k \in Z}$ forms an orthogonal basis, then we can easily find the weight or coefficient A_k by the regular inner product, i.e.,

$$A_k = \langle s, h_k \rangle \tag{10.2}$$

In this case, the resulting representation truly describes the nature of the signal. However, if $h_k(t)$ do not form an orthogonal basis, a faithful decomposition of the signal $s(t)$ in (10.1) can become very challenging.

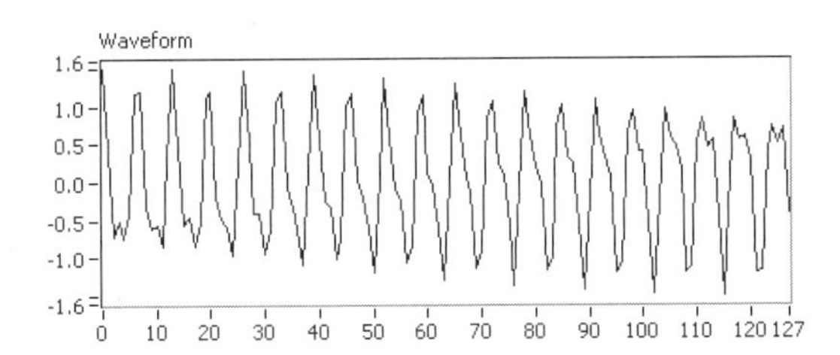

Figure 10-5 Sum (real part) of two complex sinusoidal functions

Figure 10-5 plots the sum of two complex sinusoidal functions,

$$s[k] = \exp\left\{j19.5\frac{2\pi}{128}k\right\} + 0.5\exp\left\{j39.5\frac{2\pi}{128}k\right\} \qquad 0 \le k < 128 \tag{10.3}$$

At first look, we may think that it should be solved by using the DFT. The top plot of Figure 10-6 illustrates the resulting coefficients computed by a 128-point FFT. Note that the frequencies of both terms in Eq. (10.3) are not multiples of $2\pi/128$. In other words, neither of the terms can find an exact match from the set of elementary functions

$$\{e^{j\Delta\omega nk}\}_{n = 0, 1, \ldots 127} \qquad \Delta\omega = \frac{2\pi}{128} \tag{10.4}$$

Consequently, the true weights at $\omega = 39\pi/128$ and $79\pi/128$ "leak" into their neighbors. Although the set of elementary functions in Eq. (10.4) forms an orthogonal basis, it does not lead to a representation that reflects the real structure of the signal. Engineers usually give this phenomenon a fancy term, *energy leakage*.

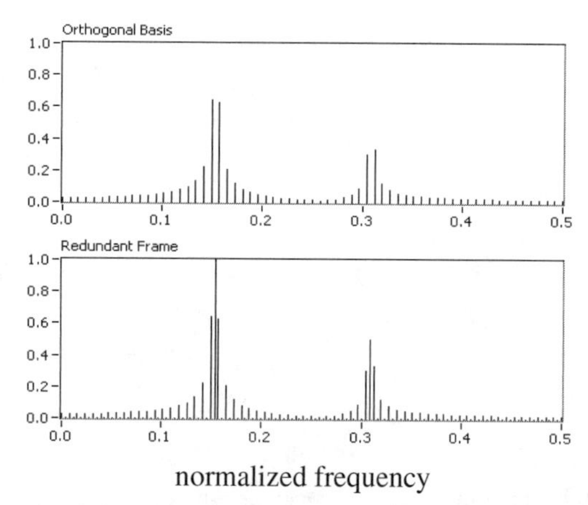

Figure 10-6 The orthogonal decomposition (top) does not yield correct weights. By using a redundant frame (bottom), we do obtain the right weights, but we also obtain many other non-zero components that actually do not exist.

To find an exact match for the samples $s[k]$ in (10.3), we can double the number of frequency bins, forming a new set of elementary functions,

$$\{e^{j\Delta\omega nk}\}_{n = 0, 1, \dots 255} \qquad \Delta\omega = \frac{2\pi}{256} \tag{10.5}$$

Since there are only 128 data samples available, the set with 256 elements in Eq. (10.5) is linearly dependent. Therefore, it is redundant. The bottom plot in Figure 10-6 illustrates the coefficients computed by a 256-point DFT. It shows true weights at $\omega = 39\pi/128$ and $79\pi/128$. However, due to the non-linearity of the set of elementary functions in (10.5) there are also many non-zero terms that actually do not exist in the original signal $s[k]$ in (10.3)!

To better represent signals that are superpositions of non-orthogonal elementary functions, in what follows we will introduce the *Matching Pursuit* (MP) algorithm. The procedure of the MP approach can roughly be described by Figure 10-7.

First, start with $k = 0$ and $s_0(t) = s(t)$, which is the original signal. Then, find the $h_0(t)$ among the set of the desired elementary functions that is most similar to $s_0(t)$, in the sense of

$$|A_k|^2 = \max_{h_k} |\langle s_k(t), h_k(t)\rangle|^2 \tag{10.6}$$

for $k = 0$. The next step is to compute the residual $s_1(t)$ by

$$s_{k+1}(t) = s_k(t) - A_k h_k(t) = s_k(t) - v_k(t) \tag{10.7}$$

where

$$v_k(t) = A_k h_k(t) \tag{10.8}$$

which denotes the component parallel to the optimal elementary function $h_k(t)$. Without loss of generality, let $h_k(t)$ have unit energy. That is,

$$\|h_k(t)\|^2 = 1 \tag{10.9}$$

Then, the energy contained in the residual is

$$\|s_{k+1}(t)\|^2 = \|s_k(t)\|^2 - |A_k|^2 \tag{10.10}$$

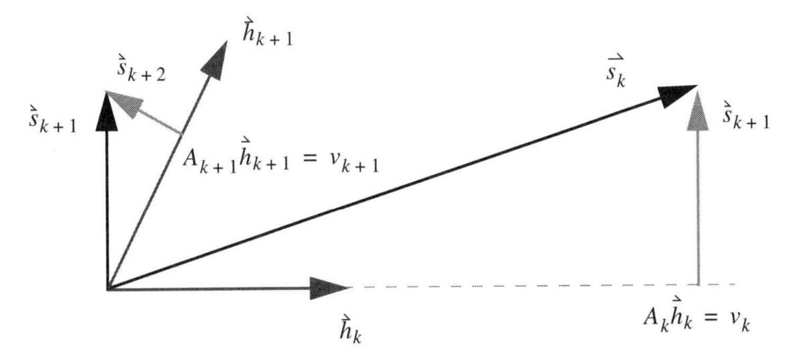

Figure 10-7 The elementary function h_k is selected to best match the signal being analyzed.

Repeat (10.6) to find $h_1(t)$ that best matches $s_1(t)$, and so on. At each step, we find one elementary function, $h_k(t)$, that has the best match with $s_k(t)$. Table 10-1 depicts the operations up to the kth step.

Table 10-1 Matching Pursuit

Residual	Projection	Energy in Residual
$s_0(t) = s(t)$		$\|s_0(t)\|^2$
$s_1(t) = s_0(t)\text{-}A_0h_0(t) = s_0(t)\text{-}v_0(t)$	$A_0 = <s_0(t), h_0(t)>$	$\|s_1(t)\|^2 = \|s_0(t)\|^2 - \|A_0\|^2$
$s_2(t) = s_1(t)\text{-}A_1h_1(t) = s_1(t)\text{-}v_1(t)$	$A_1 = <s_1(t), h_1(t)>$	$\|s_2(t)\|^2 = \|s_1(t)\|^2 - \|A_1\|^2$
...		
$s_k(t) = s_{k+1}(t)\text{-}A_{k+1}h_{k+1}(t) = s_{k+1}(t)\text{-}v_{k+1}(t)$	$A_{k+1} = <s_{k+1}(t), h_{k+1}(t)>$	$\|s_k(t)\|^2 = \|s_{k+1}(t)\|^2 - \|A_{k+1}\|^2$
...		

Intuitively, the residual will vanish if we continue to carry on the decompositions described by (10.6) and (10.7). To see this, let θ_k be the angle between $s_k(t)$ and $h_k(t)$, then

$$\cos\theta_k = \frac{\langle s_k, h_k\rangle}{\|s_k\|} = \frac{A_k}{\|s_k\|} \tag{10.11}$$

Applying the relationship (10.11) to (10.10) yields

$$\|s_k(t)\|^2 = \|s_{k+1}(t)\|^2(\sin\theta_{k+1})^2 \tag{10.12}$$

Similarly, $\|s_{k+1}(t)\|^2$ can be written in terms of $s_{k+2}(t)$. Continuing this process, we finally have

$$\|s_k(t)\|^2 = \|s_0(t)\|^2\prod_{i=0}^{k-1}(\sin\theta_i)^2 \le \|s_0(t)\|^2(\sin\theta_{max})^{2k} \tag{10.13}$$

where

$$|\sin\theta_{max}| = \max_{\theta_k}|\sin\theta_k| \tag{10.14}$$

Assume that at each step k, we are always able to find the optimal $h_k(t)$ that is not perpendicular to $s_k(t)$, that is,

$$\cos\theta_k \ne 0 \qquad or \qquad |\sin\theta_k| < 1 \qquad \forall\, k \tag{10.15}$$

Then,

$$|\sin\theta_{max}| < 1 \tag{10.16}$$

Substituting (10.16) into (10.13) yields

$$\|s_k(t)\|^2 \le \|s_0(t)\|^2(\sin\theta_{max})^{2k} \to 0 \qquad for\ k \to \infty \tag{10.17}$$

which says that the energy of the residual signal exponentially decays and converges to zero. The reader can find a more rigorous treatment about the convergent properties from [27] and [38].

It can be shown that the residual $\|s_k(t)\|^2$ is always bounded. For continuous time signals $s(t)$, the residual will be reduced to zero as the number of iterations k goes to infinity. For a finite number of discrete time samples $s[n]$, where $0 \le n < N$, we are able to find at most N linearly independent functions $h_k[n]$ for $0 \le k < N$. After N iterations, the residual may not be equal to zero. If one needs a zero residual decomposition, then we can use the MP along with the back projection algorithm [172]. It is interesting to note that the convergence property of the MP is independent of the selection of the elementary function $h_k[n]$. Irrespective of whether or not the elementary function $h_k[n]$ achieves a global optimum, the residual $\|s_k(t)\|^2$ is always bounded. The matching pursuit is very similar to the projection pursuit principle discussed in statistics by Huber [132], whose strong convergence when the dictionary is complete was proved by Jones [150].

Obviously, for a different signal its set of adaptive elementary functions $\{h_k\}$ must also be different. Each set of adaptive elementary functions $\{h_k\}$ only works for one particular signal, not for arbitrary signals in L^2. Unlike the regular Gabor expansion (as well as for wavelets), the set $\{h_k\}$ constructed above is not complete in L^2, even if the residual converges to zero. Mallat named the set $\{h_k\}_{k \in Z}$ as a dictionary. The main task of the adaptive signal expansion is to find a set of elementary functions $\{h_k\}$ that most resembles the signal's time-frequency structures and, in the meantime, satisfy the formulae (10.1) and (10.2). When the number of elements of the dictionary is finite, then the MP algorithm can be described as follows.

a. *Initialization*: Compute $\{\langle s_0, h_m\rangle\}_{m \in \Gamma}$, where $\{h_m\}_{m \in \Gamma}$ denotes a predetermined dictionary with finite number of elements Γ.
b. *Best Match*: Select the best match h_0 among $\{\langle s_0, h_m\rangle\}_{m \in \Gamma}$, such that $|<s_0,h_0>|^2$ is maximum.
c. *Update*: Compute the inner product of Eq. (10.7) and $\{h_m\}_{m \in \Gamma}$, i.e.,

$$\langle s_{k+1}, h_m\rangle = \langle s_k, h_m\rangle - A_k\langle h_k, h_m\rangle \tag{10.18}$$

The term $<s_k,h_m>$ is known from the previous process. We only need to compute the inner product $<h_k,h_m>$ to update $\langle s_{k+1}, h_m\rangle_{m \in \Gamma}$.

Note that the above algorithm is independent of the dictionary selected. The accuracy (or resolution) of the estimates is determined by the size of the dictionary. On the other hand, the larger the dictionary, the more the computation required. In general, the speed of the MP algorithm is very slow. However, if the analytical form of the inner product $<h_n,h_m>$ exists, we may derive more efficient and accurate algorithms to compute the optimal elementary function, as introduced in Section 10.3.

Let's rewrite (10.10) as

$$\|s_k(t)\|^2 = \|s_{k+1}(t)\|^2 + |A_k|^2 \tag{10.19}$$

which shows that the signal energy residual at the kth stage could be determined by the signal residual at the $k+1^{th}$ stage plus A_k. Continuing to carry out this process, we have

$$\|s(t)\|^2 = \sum_{k=0}^{\infty} |A_k|^2 \tag{10.20}$$

which is the energy conservation equation and is similar to Parseval's relation in the Fourier transform.

Apply the Wigner-Ville distribution to both sides of (10.1) and arrange the terms in two groups

$$WVD_s(t, \omega) = \sum_k A_k^2 WVD_{h_k}(t, \omega) + \sum_{k \neq q} A_k A_q^* WVD_{h_k h_q}(t, \omega) \tag{10.21}$$

The first group represents the auto-terms and the second group represents the cross-terms. Since the function $h_k(t)$ has unit energy and the Wigner-Ville distribution is energy conserving, i.e.,

$$\frac{1}{2\pi}\int_{-\infty}^{\infty}\int_{-\infty}^{\infty} WVD_{h_k}(t, \omega)dtd\omega = \|h_k(t)\|^2 = 1 \tag{10.22}$$

we have

$$\|s(t)\|^2 = \frac{1}{2\pi}\int_{-\infty}^{\infty}\int_{-\infty}^{\infty} WVD_s(t, \omega)dtd\omega \tag{10.23}$$

$$= \sum_k |A_k|^2 \frac{1}{2\pi}\int_{-\infty}^{\infty}\int_{-\infty}^{\infty} WVD_{h_k}(t, \omega)dtd\omega + \frac{1}{2\pi}\int_{-\infty}^{\infty}\int_{-\infty}^{\infty}\sum_{k \neq q} A_k A_q{}^* WVD_{h_k h_q} dtd\omega$$

$$= \sum_k |A_k|^2 + \frac{1}{2\pi}\int_{-\infty}^{\infty}\int_{-\infty}^{\infty}\sum_{k \neq q} A_k A_q{}^* WVD_{h_k h_q} dtd\omega$$

Because of the relation described in (10.20), it is obvious that

$$\frac{1}{2\pi}\int_{-\infty}^{\infty}\int_{-\infty}^{\infty}\sum_{k \neq q} A_k A_q{}^* WVD_{h_k h_q} dtd\omega = 0 \tag{10.24}$$

which implies that the second term in (10.23) contains zero energy. This gives us the reason to define a new time-dependent spectrum as

$$AS(t, \omega) = \sum_k |A_k|^2 WVD_{h_k}(t, \omega) \tag{10.25}$$

Because it is an adaptive representation-based time-dependent spectrum, we call it an *adaptive spectrogram* (AS). Clearly, the adaptive spectrogram does not contain the cross-term interference, as occurred in the Wigner-Ville distribution, and it also satisfies the energy conservation condition

$$\|s(t)\|^2 = \frac{1}{2\pi}\int_{-\infty}^{\infty}\int_{-\infty}^{\infty} AS_s(t, \omega)dtd\omega \tag{10.26}$$

The MP method introduced in this section is a very powerful tool to decompose signals whose fundamental models do not form a basis. In general, the set of optimal elementary functions $\{h_k\}$ is not complete, but they are at least linearly independent.

Unfortunately, the MP computed by the algorithm introduced in this section is usually far from being useful for real world applications because of the massive computations involved. The number of computations mainly depends on the number of waveforms in the dictionary. The higher the accuracy, the more the number of waveforms that need to be included in the dictionary. Consequently, high-resolution analysis can take days, even on today's powerful computers. One exception is the adaptive Gabor expansion, a most useful member in the MP family in which there is a nice analytical form of the inner product $<h_m,h_n>$. As we will see shortly, the adaptive Gabor expansion can be computed very efficiently.

10.2 Adaptive Gabor Expansion

In Chapter 3, we discussed the Gabor expansion and Gabor transform (which can be computed by the short-time Fourier transform). For the classical Gabor expansion, the time and frequency resolutions of the elementary functions are fixed. In some applications (such as those discussed at the beginning of this chapter) it is desired that the time and the frequency resolutions of the elementary functions are adjustable so that we can better describe the features of the signal that are of interest. One selection of such elementary functions is

$$h_k(t) = \sqrt[4]{\frac{\alpha_k}{\pi}}\exp\left\{-\frac{\alpha_k}{2}(t-t_k)^2 + j\omega_k(t-t_k)\right\} \qquad \alpha_k > 0 \qquad t_k, \omega_k \in R \tag{10.27}$$

where (t_k, ω_k) is the time-frequency center of the elementary function. The Gaussian functions used in the regular Gabor expansions are located at fixed time and fixed frequency grid points $(mT, n\Omega)$, whereas the centers of the elementary functions in (10.27) are not fixed and can be located anywhere. The parameter α_k is the inverse of the variance of the Gaussian function at (t_k, ω_k), which is similar to the scale factor in the wavelet transform. Adjusting the parameters (t_k, ω_k) will change the time and frequency centers of the elementary function; adjusting the variance will increase or decrease the duration of the elementary function. The overall effect of adjusting the variance and the time-frequency centers will allow us to better match the local time-frequency feature of the signal $s(t)$.

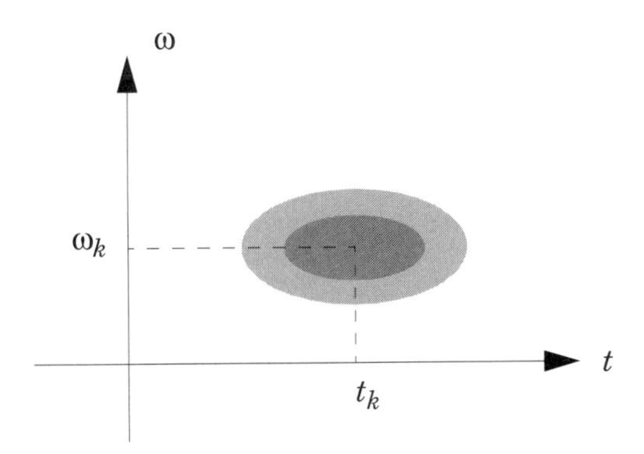

Figure 10-8 The Wigner-Ville distribution of adaptive Gaussian functions is concentrated and symmetrical in the joint time-frequency domain.

Figure 10-8 plots the time-frequency density function of the adaptive Gaussian functions, where

$$WVD_{h_k}(t, \omega) = 2\exp\left\{-\alpha_k(t-t_k)^2 - \frac{1}{\alpha_k}(\omega-\omega_k)^2\right\} \tag{10.28}$$

which is computed by the Wigner-Ville distribution. The time-frequency density function of the adaptive Gaussian functions is an ellipse and is centered at (t_k, ω_k). As shown in Chapter 7, the energy concentration of the Gaussian-type functions is optimal. By using Gaussian functions of different variances and different time-frequency centers, we can well characterize the local behavior of the signal $s(t)$ that is being analyzed. If the signal at a particular time instant exhibits an abrupt change, then a Gaussian function with a very small variance may be used to match the abrupt change. If the signal exhibits a stable frequency for a long time, a Gaussian function with a large variance value may be used.

Substituting (10.27) into (10.2) yields

$$s(t) = \sum_k A_k h_k(t) = \sum_k A_k \sqrt[4]{\frac{\alpha_k}{\pi}} \exp\left\{-\frac{\alpha_k}{2}(t-t_k)^2 + j\omega_k(t-t_k)\right\} \tag{10.29}$$

which can be thought of as an *adaptive Gabor expansion* (AGE). Unlike the Gabor expansion, where the analysis and synthesis functions in general are not identical (they are dual functions), the adaptive representation has the same analysis and synthesis functions. Therefore, once we have obtained the optimal synthesis function $h_k(t)$, we can readily compute the adaptive coefficients A_k through the regular inner product operation, i.e.,

$$A_k = \int_{-\infty}^{\infty} s_k(t) h^*_k(t) dt = \sqrt[4]{\frac{\alpha_k}{\pi}} \int_{-\infty}^{\infty} s_k(t) \exp\left\{-\frac{\alpha_k}{2}(t-t_k)^2 + j\omega_k(t-t_k)\right\} dt \tag{10.30}$$

which guarantees that A_k indeed reflects the local behavior of the signal.

Substituting (10.28) into (10.25), we obtain the adaptive Gabor expansion-based spectrogram,

$$AS(t, \omega) = 2\sum_k |A_k|^2 \exp\left\{-\alpha_k(t-t_k)^2 - \frac{1}{\alpha_k}(\omega-\omega_k)^2\right\} \tag{10.31}$$

As is evident from the above equation, the adaptive spectrogram $AS(t,\omega)$ in (10.31) is non-negative.

To illustrate the usefulness and power of the adaptive Gabor expansion and adaptive spectrogram, let's look at the signal

$$s(t) = \delta(t-t_1) + \delta(t-t_2) + e^{j\omega_1 t} + e^{j\omega_2 t} \tag{10.32}$$

which is made up of two time impulses and two frequency impulses. The corresponding STFT spectrogram, scalogram, and adaptive spectrogram are illustrated in Figure 10-9.

Because the wavelet uses wideband windows at higher frequency bands and narrowband windows at lower frequency bands, the wavelet transform offers excellent time resolution but poor frequency resolution at the higher frequency bands, and vice versa. Therefore, the wavelet is ideal for constant Q analysis, such as the octave analysis commonly used in machine vibration analysis.

The STFT spectrogram offers a uniform time and frequency resolution at any frequency band. Once the windowing function is selected, so is the frequency resolution. Compared with the wavelet transform, the STFT spectrogram cannot offer very high time resolution at a higher frequency band without sacrificing frequency resolution. Similarly, the STFT spectrogram cannot offer very good frequency resolution at low frequency without sacrificing time resolution. This was the original motivation for *Qian* and *Chen* to investigate the adaptive spectrogram ([192] and [197]).

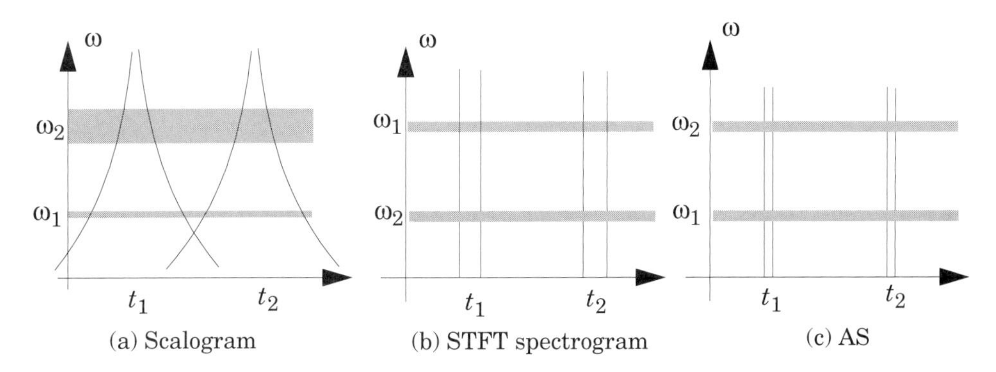

Figure 10-9 Unlike the wavelet that provides good time resolution at higher frequency and good frequency resolution at lower frequency, the adaptive spectrogram offers good frequency resolution at any frequency band.

The adaptive spectrogram is suitable for many signals. By selecting a different variance, one can obtain good frequency resolution at any frequency band. The same is true for the time resolution. Moreover, the implementation of the adaptive Gabor expansion can be much more efficient and accurate than the algorithm developed for the general MP.

10.3 Fast Refinement

As shown in Section 10.1, the accuracy of the MP method directly hinges on the size of the dictionary. The larger the dictionary, the better the accuracy. On the other hand, the larger the dictionary, the more the computation involved. One solution is to use a small size dictionary to obtain a coarse estimate. Based on the coarse estimate, one can then perform a more detailed search. The refinement can work extremely well if the analytical form of the inner product of the elementary functions exists. In this case, such as for the adaptive Gaussian function or the Gaussian chirplet, we can apply the curve fitting technique to find the optimal vector $h_{opt}(t)$. Unlike general optimization algorithms, such as the *Newton-Raphson* method, this method does not suffer convergence problems. Its resolution is virtually unlimited. Compared to other digital optimization approaches, the method presented in this section is not only fast and stable, but also is more accurate.

At the kth stage, assume that the rough estimate $h_{k,0}$ with a set of parameters $(\alpha_{k,0}, t_{k,0}, \omega_{k,0})$ is close enough to the optimal function. Then, we have

$$\begin{aligned}|\langle s_k(t), h_{k,0}(t)\rangle| &= |\langle v_k(t) + s_{k+1}(t), h_{k,0}(t)\rangle| \\ &= |\langle v_k(t), h_{k,0}(t)\rangle + \aleph| \approx |\langle v_k(t), h_{k,0}(t)\rangle| \\ &= P(\alpha_k, t_k, \omega_k; \alpha_{k,0}, t_{k,0}, \omega_{k,0})\end{aligned} \tag{10.33}$$

As shown in Eq. (10.7), the residual s_{k+1} is orthonormal to v_k. $\aleph$ denotes the correlation between $h_{p,0}$ and s_{k+1}. When $h_{k,0}$ is close enough to v_k (parallel), $\aleph$ can be negligible. In this case, $P(\alpha_k, t_k, \omega_k;\ \alpha_{k,0}, t_{k,0}, \omega_{k,0})$ can be approximated by $|\langle s_k, h_{k,0}\rangle|$. While the set of parameters $(\alpha_k, t_k, \omega_k)$ represents unknown parameters at the kth stage, the set $(\alpha_{k,m}, t_{k,m}, \omega_{k,m})$ denotes the mth testing point. Consequently, the multi-dimensional optimization problem becomes the solution of a group of simultaneous equations in four variables (including the amplitude A_k), i.e.,

$$|\langle s_k(t), h_{k,m}(t)\rangle| = |\langle s_k, h_{\alpha_{k,m} t_{k,m} \omega_{k,m}}\rangle| = P(\alpha_k, t_k, \omega_k; \alpha_{k,m}, t_{k,m}, \omega_{k,m}) \tag{10.34}$$

where $P(\alpha_k, t_k, \omega_k;\ \alpha_{k,m}, t_{k,m}, \omega_{k,m})$ indicates a functional relationship between the estimated parameters $(\alpha_k, t_k, \omega_k)$ and the testing variables $(\alpha_{k,m}, t_{k,m}, \omega_{k,m})$, i.e.,

$$\begin{aligned}P(\alpha_k, t_k, \omega_k; \alpha_{k,m}, t_{k,m}, \omega_{k,m}) &= \left|\int_{-\infty}^{\infty} v_k(t) h^*{}_{k,m}(t)\, dt\right| \\ &= \left|A_k \int_{-\infty}^{\infty} h_k(t) h^*{}_{k,m}(t)\, dt\right| \\ &= \frac{|A_k|}{\sqrt[4]{\frac{1}{4\alpha_k \alpha_{k,m}}(\alpha_k + \alpha_{k,m})^2}} e^{-\frac{\alpha_k \alpha_{k,m}}{2(\alpha_k + \alpha_{k,m})}(t_k - t_{k,m})^2 - \frac{1}{2(\alpha_k + \alpha_{k,m})}(\omega_k - \omega_{k,m})^2} \\ &= a(t_{k,m}) e^{-b(\omega_k - \omega_{k,m})^2}\end{aligned} \tag{10.35}$$

which leads to

$$a(t_{k,n}) = \frac{|A_k|}{\sqrt[4]{\frac{1}{4\alpha_k \alpha_{k,0}}(\alpha_k + \alpha_{k,0})^2}} e^{-\frac{\alpha_k \alpha_{k,0}}{2(\alpha_k + \alpha_{k,0})}(t_k - t_{k,n})^2} \qquad n = 0, 1 \tag{10.36}$$

and

$$b = \frac{1}{2(\alpha_k + \alpha_{k,0})} \tag{10.37}$$

Note that $a(t_{k,n})$, b, and ω_k in Eq. (10.35) are not functions of the testing frequencies $\omega_{k,m}$. Hence, in principle, we should be able to solve $(a(t_{k,n}), b, \omega_k)$ by a set of equations (10.35) with different testing frequencies $\omega_{k,m}$ (while the other testing variables $\alpha_{k,0}$ and $t_{k,n}$ are unchanged).

Because both sides of Eq. (10.35) are greater than zero, we can have

$$\ln\frac{|\langle s_k, h_{\alpha_{k,o}, t_{k,n}, \omega_{k,i}}\rangle|}{|\langle s_k, h_{\alpha_{k,o}, t_{k,n}, \omega_{k,j}}\rangle|} = b[(\omega_k - \omega_{k,j})^2 - (\omega_k - \omega_{k,i})^2] \tag{10.38}$$

$$= 2b\omega_k(\omega_{k,i} - \omega_{k,j}) - b(\omega_{k,i}^2 - \omega_{k,j}^2)$$

By inserting the following three testing points into Eq. (10.38),

$$\begin{array}{c}(\alpha_{k,0}, t_{k,0}, \omega_{k,-1}, \beta_{k,0})\\ (\alpha_{k,0}, t_{k,0}, \omega_{k,0}, \beta_{k,0})\\ (\alpha_{k,0}, t_{k,0}, \omega_{k,1}, \beta_{k,0})\end{array} \tag{10.39}$$

we obtain the following linear system

$$\begin{bmatrix}(\omega_{k,0} - \omega_{k,-1}) & (\omega_{k,-1}^2 - \omega_{k,0}^2)\\ (\omega_{k,0} - \omega_{k,1}) & (\omega_{k,1}^2 - \omega_{k,0}^2)\end{bmatrix}\begin{bmatrix}x_\omega\\ b\end{bmatrix} = \begin{bmatrix}\ln\frac{|\langle s_k, h_{\alpha_{k,0}, t_{k,0}, \omega_{k,0}}\rangle|}{|\langle s_k, h_{\alpha_{k,0}, t_{k,0}, \omega_{k,-1}}\rangle|}\\ \ln\frac{|\langle s_k, h_{\alpha_{k,0}, t_{k,0}, \omega_{k,0}}\rangle|}{|\langle s_k, h_{\alpha_{k,0}, t_{k,0}, \omega_{k,1}}\rangle|}\end{bmatrix} \tag{10.40}$$

where

$$x_\omega = 2b\omega_k \qquad b > 0 \qquad \omega_k \in R \tag{10.41}$$

From Eq. (10.37), we get

$$\alpha_k = \frac{1}{2b} - \alpha_{k,0} \tag{10.42}$$

Once b and ω_k have been calculated, we can also compute $a(t_{k,0})$ through Eq. (10.35), i.e.,

$$\ln a(t_{k,0}) = \ln|\langle s_k, h_{\alpha_{k,0}, t_{k,0}, \omega_{k,m}}\rangle| + b(\omega_k - \omega_{k,m})^2 \tag{10.43}$$

Similarly, by applying

$$\begin{array}{c}(\alpha_{k,0}, t_{k,1}, \omega_{k,-1}, \beta_{k,0})\\ (\alpha_{k,0}, t_{k,1}, \omega_{k,0}, \beta_{k,0})\\ (\alpha_{k,0}, t_{k,1}, \omega_{k,1}, \beta_{k,0})\end{array} \tag{10.44}$$

we can determine $a(t_{k,1})$. Having known α_k, $a(t_{k,0})$, and $a(t_{k,0})$, we can further determine the time parameter t_k from[1]

1. Note that each set of testing points, Eq. (10.39) and Eq. (10.43), yields one a_k and one ω_k. In principle, the quantity a_k and one ω_k computed from different sets of testing points should be identical, but actually they may not be so due to the presence of noise. If this is the case, we usually take the average.

$$t_k = \frac{\alpha_k^{-1} + \alpha_{k,0}^{-1}}{t_{k,0} - t_{k,1}} \ln\frac{a(t_{k,0})}{a(t_{k,1})} + \frac{t_{k,0} + t_{k,1}}{2} \tag{10.45}$$

The set of parameters $(\alpha_{k,0}, t_{k,0}, \omega_{k,0})$ denotes initial values that can be estimated by the algorithm presented in Section 10.1 with a very small dictionary. The rest of the testing points can be determined by

$$t_{k,n} = t_{k,0} + n\Delta_t \qquad n = 0, 1 \tag{10.46}$$

and

$$\omega_{k,n} = \omega_{k,0} + n\Delta_\omega \qquad n = -1, 0, 1 \tag{10.47}$$

The time increment Δ_t is inversely proportional to the initial variance $\alpha_{k,0}$. The bigger the variance $\alpha_{k,0}$, the smaller the time increment Δ_t, and vice versa. On the other hand, the frequency increment Δ_ω is proportional to the variance $\alpha_{k,0}$. That is, the bigger the variance $\alpha_{k,0}$, the bigger the frequency increment Δ_ω, and vice versa.

Once the best match $h_k(t)$ has been found, we can compute the residual from (10.7). Then, we apply the same process to compute the next elementary function $h_{k+1}(t)$. The process continues until $\|s_{k+1}(t)\| < \varepsilon$, where ε is a predetermined error threshold.

Note that the resulting parameter $(\alpha_k, t_k, \omega_k)$ is not always valid. For instance, the variance α_k might go negative. In this case, we can permute the testing points until valid values are achieved. Moreover, $h_k(t)$ computed by the procedure introduced in this section may not be globally optimal. However, due to the fact that the convergence of the residual $\|s_{k+1}(t)\|$, as stated in Section 10.1, is independent of the selection of the elementary functions $h_k(t)$, the residual $\|s_{k+1}(t)\|$ always converges. In fact, as long as $(\alpha_k, t_k, \omega_k)$ leads to

$$\|s_{k+1}(t)\| < \|s_k(t)\| \qquad k \geq 0 \tag{10.48}$$

the decomposition scheme always works. Of course, the speed of the convergence will be faster if $h_k(t)$ indeed achieves global optimality.

10.4 Applications of the Adaptive Gabor Expansion

In what follows, we will briefly discuss the applications and limitations of the adaptive Gabor expansion. The presentation starts with the problem posed at the beginning of this chapter.

Figure 10-10 illustrates the STFT spectrogram, time-frequency distribution series, and adaptive spectrogram for the signal introduced earlier. In this example the time-frequency resolution of the adaptive spectrogram is obviously superior to all other schemes.

Figure 10-11 and Figure 10-12 plot the adaptive spectrogram, WVD, STFT spectrogram, and $TFDS_4(t,\omega)$ of a frequency hopper signal. As expected, the WVD suffers from the cross-term interference problem. Both the STFT and the fourth-order $TFDS_4(t,\omega)$ give a good and easy-to-understand spectrogram, but they fail to compare with the adaptive spectrogram for the same signal. The adaptive spectrogram not only shows good time-frequency resolution, but also does not suffer from cross-term interference. In addition, it is non-negative.

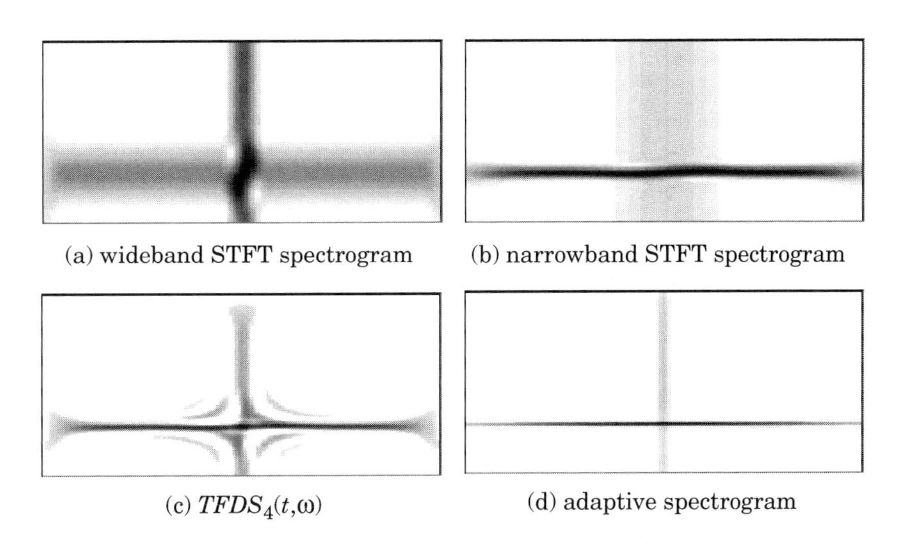

Figure 10-10 The adaptive spectrogram possesses the best time-frequency resolution for a signal consisting of two types of impulses, the time domain and frequency domain (sinusoidal time waveform) impulses.

Figures 10-13 and 10-14 depict an application of the adaptive Gabor expansion for inverse synthetic aperture radar (ISAR) image processing.[2] This example is based on numerically simulated data for a perfectly conducting plate containing a long duct open at one end, as shown in Figure 10-13. Figure 10-14 (a) shows its ISAR image at 40° from edge on. The scattering data were simulated numerically using a method of moments electromagnetics code. We notice in the image that, in addition to the three scattering centers corresponding to the two edges of the plate and the mouth of the duct, there is also a very strong return outside of the target. This return corresponds to the energy that gets coupled into the duct, travels down the length of the duct, hits the end and reflects back to the duct mouth, and is finally re-radiated towards the radar. Figure 10-14 (b) shows the enhanced ISAR image of Figure 10-14 (a), obtained by applying the adaptive Gabor expansion and keeping only the small-variance Gaussians. We see that, as expected, only the scattering-center part of the original signal remains in the image. Figure 10-14 (c) shows the frequency-aspect display of the high-variance Gaussians. We see three equally spaced vertical lines. They correspond to the cut-off frequencies of the waveguide modes excited in the duct, which should occur at 3.75, 7.5, and 11.25 GHz based on the transverse dimension of the duct. Indeed, we see that they occur close to these frequencies and are almost aspect independent.

2. The reader can find detailed discussions in [38], [89], and [218].

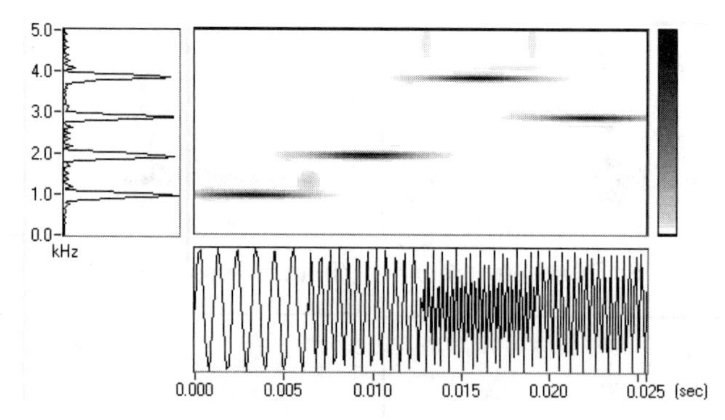

Figure 10-11 The adaptive spectrogram of a frequency hopper signal possesses the best time-frequency resolution.

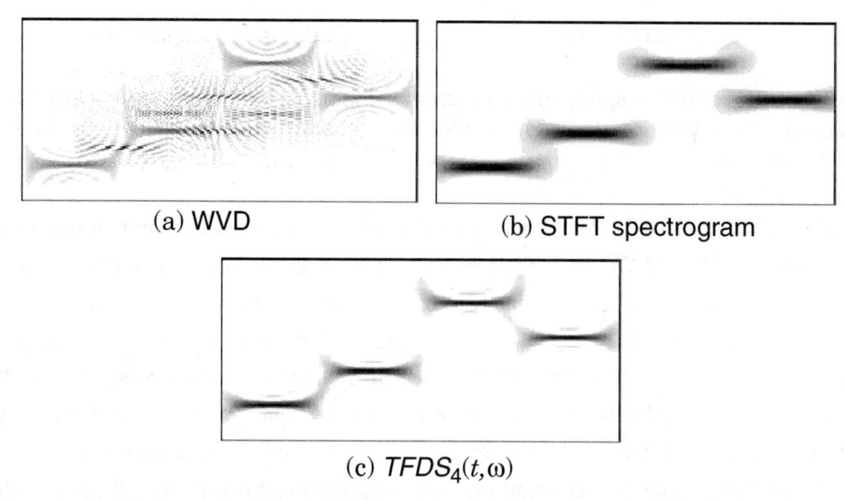

(a) WVD

(b) STFT spectrogram

(c) $TFDS_4(t,\omega)$

Figure 10-12 Frequency hopper signal represented by the Wigner-Ville distribution, the STFT spectrogram, and the fourth order time-frequency distribution series.

The adaptive Gabor expansion has also been found useful in analyzing an electroencephalogram (EEG), which is a record of the electrical activity of the brain made from electrodes placed on the scalp. The signal is extremely complex and reveals both rhythmic and transient features.

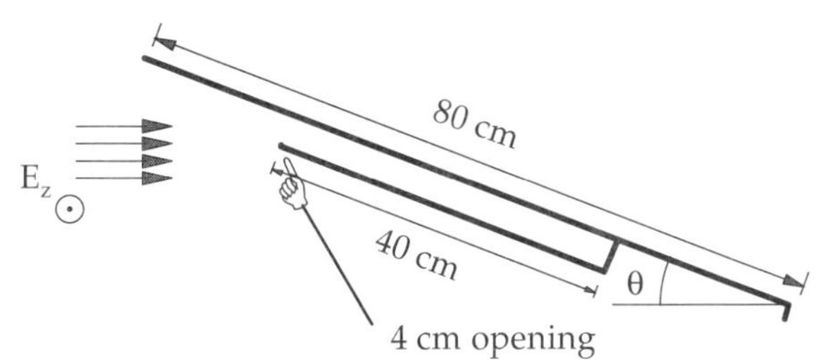

Figure 10-13 Conducting strip with a long cavity (Diagram courtesy of Hao Ling, Department of Electrical and Computer Engineering, University of Texas, Austin).

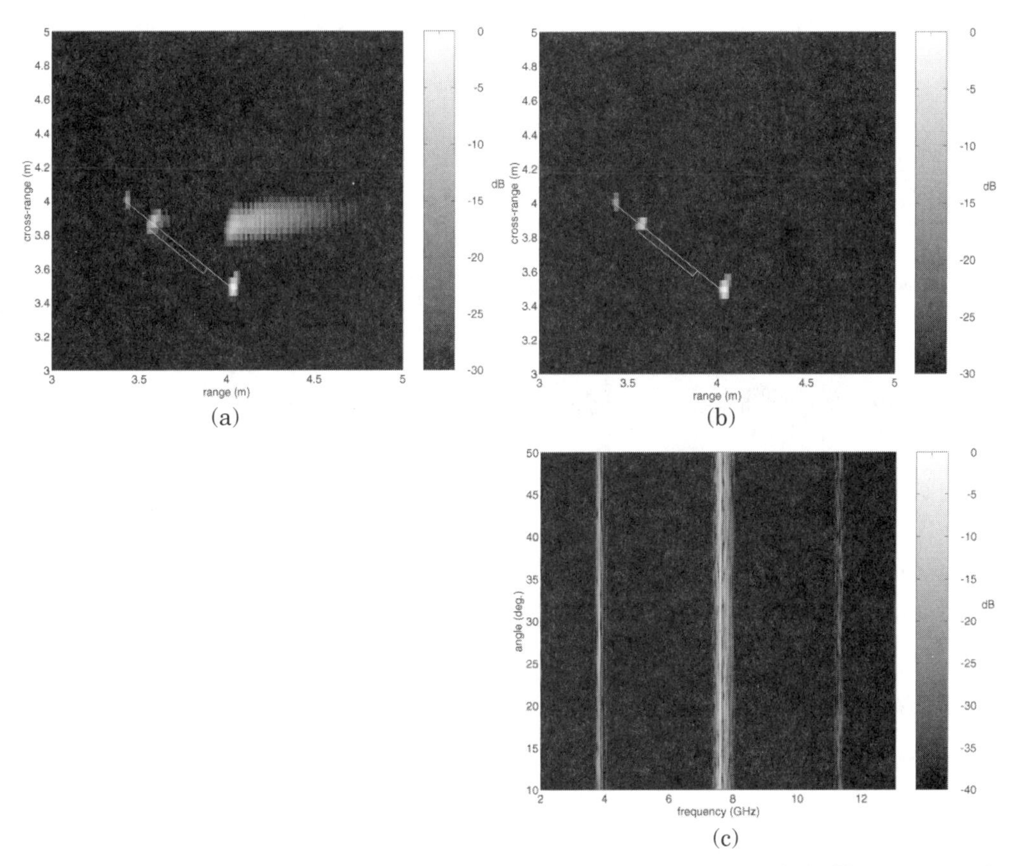

Figure 10-14 (a) Standard ISAR image of conducting strip with a long cavity in Figure 10-13, obtained for f = 2-13 GHz, $\theta = 25°\text{-}55°$. Enhanced ISAR image (b) and frequency aspect display (c) obtained by applying AGR to the ISAR image in (a) (Image courtesy of Hao Ling, Department of Electrical and Computer Engineering, University of Texas, Austin).

Dr. *Durka* of the Laboratory of Medical Physics, Warsaw University, said "The computer revolution in EEG analysis still consists mainly of the fact that what was on paper now is watched at the monitor screen, mainly because the new signal processing methods were incompatible with knowledge of visual EEG analysis. Matching Pursuit (more specifically, the adaptive Gabor expansion) corresponds directly to the traditional way of analyzing EEG, namely visual analysis. For over 40 years of clinical practice, structures present in EEG being described in terms like 'cycles per second' or 'time width,' are directly interpretable in terms of time-frequency parameters. Matching Pursuit is one of the first methods that allow for direct utilization of a huge amount of knowledge collected over years of clinical EEG" [22].

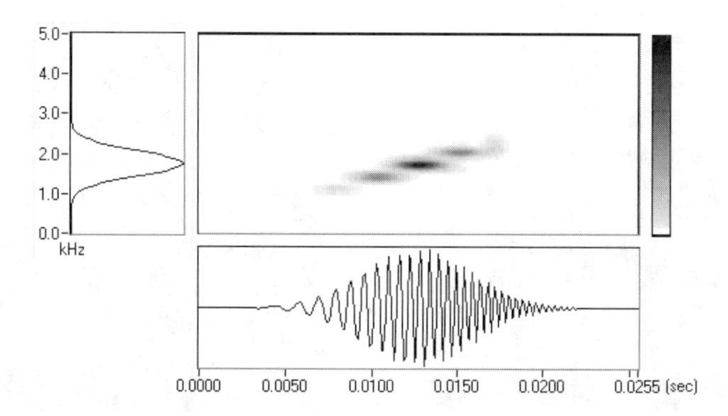

Figure 10-15 The adaptive Gabor expansion based spectrogram is not suitable for representing the linear chirp signal.

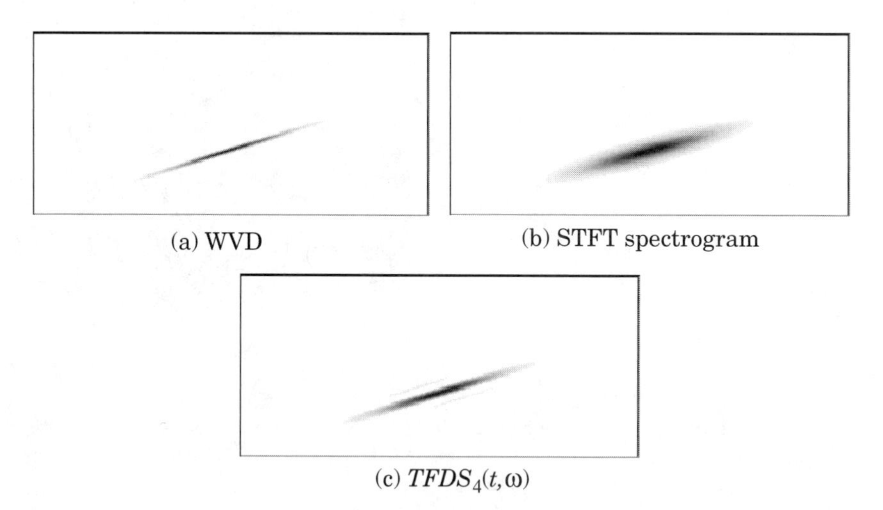

Figure 10-16 A linear chirp signal represented by the Wigner-Ville distribution, STFT spectrogram, and the fourth order time-frequency distribution series.

The definition given in a classical reference [203] states the following: "The presence of sleep spindle should not be defined unless it is at least 0.5 seconds duration, i.e., one should be able to count six or seven distinct waves within the half-second period... The term should be used only to describe activity between 12 and 14 cps (cycles per second)." Although intended for standardization of visual detection, the above definition is almost directly translatable into the time-frequency parameters of adaptive Gaussian functions in Eq. (10.27). The shape of sleep spindles is so similar to Gaussian functions that we can assume a one-to-one relationship between sleep spindles and adaptive Gaussian functions fitted to an EEG. Compared to other time-frequency representation schemes, the adaptive Gabor expansion offers extra advantages: high resolution, local adaptability to transient structures, and compatibility of its time-frequency parameters with definitions of EEG structures used in traditional visual analysis ([101], [102], and [178]).

Figures 10-15 and 10-16 plot the adaptive spectrogram, WVD, STFT spectrogram, and $TFDS_4(t,\omega)$ of a linear chirp signal. In this case, the adaptive Gabor expansion based spectrogram does not offer good time-frequency resolution due to the limitation of the number of degrees of freedom in the elementary functions used.

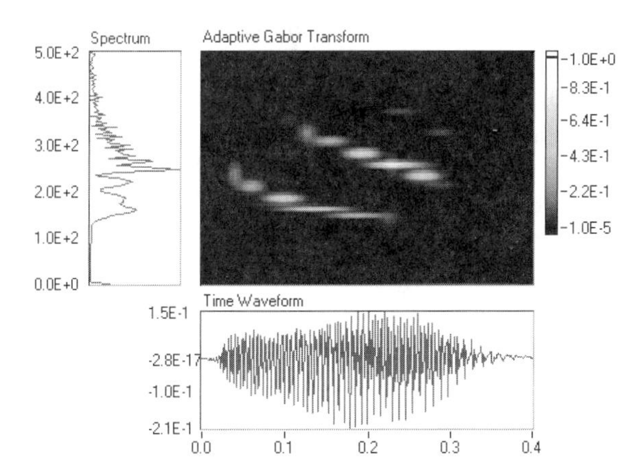

Figure 10-17 Compared with the other methods shown in Figure 8-22, the performance of the adaptive Gabor transform of the bat sound is rather poor. Generally speaking, a parametric algorithm, such as the adaptive Gabor expansion, works extremely well only for the signal that fits the model.

Figure 10-17 illustrates the adaptive spectrogram of a bat sound computed by the adaptive Gabor transform. Compared with the other methods shown in Figure 8-22, the performance of the adaptive Gabor transform is rather poor. Generally speaking, a parametric algorithm, such as the adaptive Gabor expansion, works extremely well only for the signal that fits the model.

10.5 Adaptive Gaussian Chirplet Decomposition

Figures 10-11 and 10-15 are two extreme examples that show the best and the worst scenarios for the adaptive spectrogram. They point out general guidelines in determining how the adaptive spectrogram will perform. In general, the adaptive spectrogram using the Gaussian functions will perform well if the signal of interest is a combination of short duration pulses with quasi-stationary signals such as the frequency hopper. It does not do well for chirp-type signals. For chirp type signals, we need to use the Gaussian chirplet-based MP algorithm.

The *Gaussian chirplet* is defined as

$$h_k(t) = \sqrt[4]{\frac{\alpha_k}{\pi}} \exp\left\{-\frac{\alpha_k}{2}(t-t_k)^2 + j\left(\omega_k(t-t_k) + \frac{\beta_k}{2}(t-t_k)^2\right)\right\} \tag{10.49}$$

$$\alpha_k > 0 \qquad t_k, \omega_k, \beta_k \in R$$

which has a short and smooth envelope — a Gaussian envelope. (t_k,ω_k) indicates the time and frequency center of the linear chirp function. The variance α_k controls the width of the chirp function. The parameter β_k determines the rate of change of frequency. Figure 10-18 illustrates the time-frequency distribution of $h_k(t)$ in Eq. (10.49). It shows that not only can we adjust the variance and time-frequency center, but we can also regulate the orientation of $h_k(t)$ in the time-frequency domain by varying the parameter β_p. Hence, the Gaussian chirplet fits more signals than the adaptive Gaussian function in Eq. (10.29). The chirp is one of the most important functions in nature. Many natural phenomena, for instance, signals encountered in radar systems [89], the impulsive signal that is dispersed by the ionosphere [199], and seismic signals [223], can all be modeled as chirp type functions.

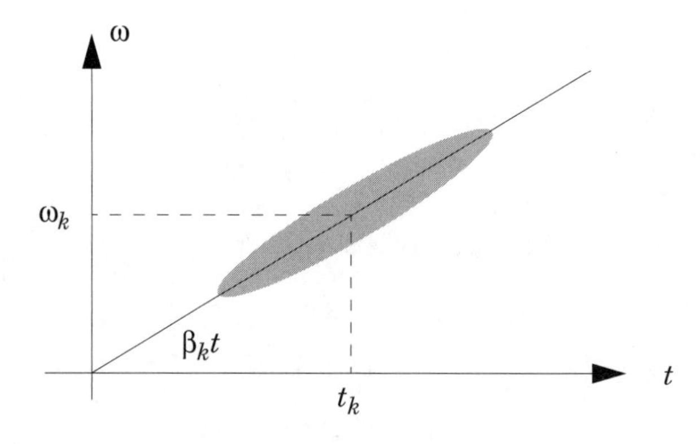

Figure 10-18 The Wigner-Ville distribution of the Gaussian chirplet is concentrated in the joint time-frequency domain. Its dispersion is proportional to $1/\alpha_k$ in the time domain and α_k in the frequency domain.

For the Gaussian chirplet decomposition, the corresponding adaptive spectrogram is

$$AS(t, \omega) = 2\sum_{k}|B_k|^2 \exp\left\{-\alpha_k(t-t_k)^2 - \frac{1}{\alpha_k}(\omega-\omega_k-\beta_k t)^2\right\} \tag{10.50}$$

which is non-negative.

Although the structure of the Gaussian chirplet is more complicated than that of the adaptive Gaussian function, we can apply the same principle as that introduced in Section 10.3 to compute an optimal function.

For Gaussian chirplet, Eq. (10.35) becomes

$$P(\alpha_k, t_k, \omega_k, \beta_k; \alpha_{k,m}, t_{k,m}, \omega_{k,m}, \beta_{k,m}) = \frac{|A_k|}{\sqrt[4]{\frac{1}{4\alpha_{k,m}\alpha_k}[(\alpha_k+\alpha_{k,m})^2+(\beta_k-\beta_{k,m})^2]}} \tag{10.51}$$

$$\times \exp\left\{-\frac{\alpha_{k,m}^2\alpha_k + \alpha_{k,m}\alpha_k^2 + \alpha_{k,m}\beta_k^2 + \alpha_k\beta_{k,m}^2}{2[(\alpha_p+\alpha_{p,m})^2+(\beta_p-\beta_{p,m})^2]}(t_k-t_{k,m})^2\right\}$$

$$\times \exp\left\{\frac{\alpha_{k,m}\beta_k + \alpha_k\beta_{k,m}}{(\alpha_k+\alpha_{k,m})^2+(\beta_k-\beta_{k,m})^2}(t_k-t_{k,m})(\omega_k-\omega_{k,m})\right\}$$

$$\times \exp\left\{-\frac{\alpha_k+\alpha_{k,m}}{2[(\alpha_k+\alpha_{k,m})^2+(\beta_k-\beta_{k,m})^2]}(\omega_k-\omega_{k,m})^2\right\}$$

which can be rewritten as

$$P(\alpha_k, t_k, \omega_k, \beta_k; \alpha_{k,0}, t_{k,n}, \omega_{k,m}, \beta_{k,0}) = \left|\langle s_k, h_{\alpha_{k,o}, t_{k,n}, \omega_{k,m}, \beta_{k,0}}\rangle\right| = a(t_{k,n})e^{-b[\varpi(t_{k,n})-\omega_{k,m}]^2} \tag{10.52}$$

where

$$a(t_{k,n}) = \frac{|A_k|}{\sqrt[4]{\frac{1}{4\alpha_{k,0}\alpha_k}[(\alpha_k+\alpha_{k,0})^2+(\beta_k-\beta_{k,0})^2]}}\exp\left\{-\frac{(t_k-t_{k,n})^2}{2(\alpha_k^{-1}+\alpha_{k,0}^{-1})}\right\} \tag{10.53}$$

$$\varpi(t_{k,n}) = \omega_k - \frac{\alpha_{k,0}\beta_k + \alpha_k\beta_{k,0}}{\alpha_k+\alpha_{k,0}}(t_k-t_{k,n}) \tag{10.54}$$

and

$$b = \frac{\alpha_k+\alpha_{k,0}}{2[(\alpha_k+\alpha_{k,0})^2+(\beta_k-\beta_{k,0})^2]} \tag{10.55}$$

Obviously, they are general cases of (10.35), (10.36), and (10.37). When $\beta_k = \beta_{k,0} = 0$, they reduce to Eqs. (10.35), (10.36), and (10.37).

Because both sides of Eq. (10.52) are greater than zero, we can have

$$\ln\frac{|\langle s_k, h_{\alpha_{k,o}, t_{k,n}, \omega_{k,i}, \beta_{k,0}}\rangle|}{|\langle s_k, h_{\alpha_{k,o}, t_{k,n}, \omega_{k,j}, \beta_{k,0}}\rangle|} = b[(\varpi(t_{k,n}) - \omega_{k,j})^2 - (\varpi(t_{k,n}) - \omega_{k,i})^2] \tag{10.56}$$

$$= 2b\varpi(t_{k,n})(\omega_{k,i} - \omega_{k,j}) - b(\omega_{k,i}^2 - \omega_{k,j}^2)$$

By inserting the following three testing points into Eq. (10.38),

$$\begin{aligned}&(\alpha_{k,0}, t_{k,0}, \omega_{k,-1}, \beta_{k,0})\\&(\alpha_{k,0}, t_{k,0}, \omega_{k,0}, \beta_{k,0})\\&(\alpha_{k,0}, t_{k,0}, \omega_{k,1}, \beta_{k,0})\end{aligned} \tag{10.57}$$

we obtain the following linear system

$$\begin{bmatrix}(\omega_{k,0} - \omega_{k,-1}) & (\omega_{k,-1}^2 - \omega_{k,0}^2)\\(\omega_{k,0} - \omega_{k,1}) & (\omega_{k,1}^2 - \omega_{k,0}^2)\end{bmatrix}\begin{bmatrix}x_\omega\\b\end{bmatrix} = \begin{bmatrix}\ln\dfrac{|\langle s_k, h_{\alpha_{k,0}, t_{k,0}, \omega_{k,0}, \beta_{k,0}}\rangle|}{|\langle s_k, h_{\alpha_{k,0}, t_{k,0}, \omega_{k,-1}, \beta_{k,0}}\rangle|}\\\ln\dfrac{|\langle s_k, h_{\alpha_{k,0}, t_{k,0}, \omega_{k,0}, \beta_{k,0}}\rangle|}{|\langle s_k, h_{\alpha_{k,0}, t_{k,0}, \omega_{k,1}, \beta_{k,0}}\rangle|}\end{bmatrix} \tag{10.58}$$

where

$$x_\omega = 2b\varpi(t_{k,0}) \qquad b > 0 \qquad \varpi(t_{k,0}) \in R \tag{10.59}$$

Once b and $\varpi(t_{k,0})$ have been calculated, we can further compute $a(t_{k,0})$ through (10.52), i.e.,

$$\ln a(t_{k,0}) = \ln|\langle s_k, h_{\alpha_{k,0}, t_{k,0}, \omega_{k,m}, \beta_{k,0}}\rangle| + b(\varpi(t_{k,0}) - \omega_{k,m})^2 \tag{10.60}$$

Similarly, by applying

$$\begin{aligned}&(\alpha_{k,0}, t_{k,1}, \omega_{k,-1}, \beta_{k,0})\\&(\alpha_{k,0}, t_{k,1}, \omega_{k,0}, \beta_{k,0})\\&(\alpha_{k,0}, t_{k,1}, \omega_{k,1}, \beta_{k,0})\end{aligned} \tag{10.61}$$

we can obtain another set of intermediate variables $\varpi(t_{k,1})$ and $a(t_{k,1})$.

Let

$$r = \frac{\alpha_k\beta_{k,0} + \alpha_{k,0}\beta_k}{\alpha_k + \alpha_{k,0}} \tag{10.62}$$

and substituting it into (10.54) yields

$$\varpi(t_{k,n}) = \omega_k - r(t_k - t_{k,n}) \tag{10.63}$$

Eq. (10.62) shows that if the initial chirp rate $\beta_{k,0}$ is close to the true value β_k, regardless of α_k and $\alpha_{k,0}$, then r is a good approximation of the chirp rate β_k. It can be computed from

$$r = \frac{\varpi(t_{k,0}) - \varpi(t_{k,1})}{t_{k,0} - t_{k,1}} \tag{10.64}$$

Then, the parameters α_k and β_k can be determined from the following equations,

$$\begin{cases} b = \dfrac{\alpha_k + \alpha_{k,0}}{2[(\alpha_k + \alpha_{k,0})^2 + (\beta_k - \beta_{k,0})^2]} \\ r = \dfrac{\alpha_k \beta_{k,0} + \alpha_{k,0}\beta_k}{\alpha_k + \alpha_{k,0}} \end{cases} \tag{10.65}$$

With elementary mathematical manipulations, we can obtain

$$\left\{2b\left[\left(\frac{\alpha_{k,0}}{r - \beta_{k,0}}\right)^2 + 1\right](\beta_k - \beta_{k,0}) - \frac{\alpha_{k,0}}{r - \beta_{k,0}}\right\}(\beta_k - \beta_{k,0}) = 0 \tag{10.66}$$

whose solution is

$$\beta_k = \begin{cases} \beta_{k,0} + \dfrac{\alpha_{k,0}}{2b(r - \beta_{k,0})\left[\left(\dfrac{\alpha_{k,0}}{r - \beta_{k,0}}\right)^2 + 1\right]} \\ \beta_{k,0} \end{cases} \tag{10.67}$$

If $\beta_k = \beta_{k,0}$, based on (10.62), then it must be true that $r = \beta_{k,0}$. In this case, Eq. (10.55) leads to

$$\alpha_k = \frac{1}{2b} - \alpha_{k,0} \tag{10.68}$$

Otherwise,

$$\alpha_k = \frac{\beta_k - r}{r - \beta_{k,0}}\alpha_{k,0} \qquad \alpha_k > 0 \tag{10.69}$$

Note that each set of testing points, Eq. (10.39) and Eq. (10.44), yields one b. In principle, the quantity b computed from different sets of testing points should be identical, but actually it may be not due to the presence of noise. If this is the case, we usually take the average.

Knowing α_k and β_k, we can calculate the time parameter t_k by $a(t_{k,0})$ and $a(t_{k,1})$ in (10.53), i.e.,

$$t_k = \frac{\alpha_k^{-1} + \alpha_{k,0}^{-1}}{t_{k,0} - t_{k,1}} \ln\frac{a(t_{k,0})}{a(t_{k,1})} + \frac{t_{k,0} + t_{k,1}}{2} \tag{10.70}$$

Once we obtain t_k, the frequency parameter ω_k can be calculated from Eq. (10.63).

In short, the fast Gaussian chirplet refinement algorithm can be summarized as follows.

1. At stage k, assume that we have a good estimate $(\alpha_{k,0}, t_{k,0}, \omega_{k,0}, \beta_{k,0})$. Then, use the following three testing points

$$\begin{aligned}&(\alpha_{k,0}, t_{k,0}, \omega_{k,-1}, \beta_{k,0})\\&(\alpha_{k,0}, t_{k,0}, \omega_{k,0}, \beta_{k,0})\\&(\alpha_{k,0}, t_{k,0}, \omega_{k,1}, \beta_{k,0})\end{aligned} \tag{10.71}$$

to solve the linear system

$$\begin{bmatrix}(\omega_{k,0}-\omega_{k,-1}) & (\omega_{k,-1}^2-\omega_{k,0}^2)\\(\omega_{k,0}-\omega_{k,1}) & (\omega_{k,1}^2-\omega_{k,0}^2)\end{bmatrix}\begin{bmatrix}x_\omega\\b\end{bmatrix} = \begin{bmatrix}\ln\dfrac{|\langle s_k, h_{\alpha_{k,0}, t_{k,0}, \omega_{k,0}, \beta_{k,0}}\rangle|}{|\langle s_k, h_{\alpha_{k,0}, t_{k,0}, \omega_{k,-1}, \beta_{k,0}}\rangle|}\\ \ln\dfrac{|\langle s_k, h_{\alpha_{k,0}, t_{k,0}, \omega_{k,0}, \beta_{k,0}}\rangle|}{|\langle s_k, h_{\alpha_{k,0}, t_{k,0}, \omega_{k,1}, \beta_{k,0}}\rangle|}\end{bmatrix} \tag{10.72}$$

where

$$x_\omega = 2b\varpi(t_{k,0}) \qquad b>0 \qquad \varpi(t_{k,0}) \in R \tag{10.73}$$

and the testing frequencies are computed by

$$\omega_{k,m} = \omega_{k,0} + m\Delta_\omega \qquad m = -1, 0, 1 \tag{10.74}$$

2. Compute

$$\ln a(t_{k,0}) = \ln|\langle s_k, h_{\alpha_{k,0}, t_{k,0}, \omega_{k,0}, \beta_{k,0}}\rangle| + b[\varpi(t_{k,0}) - \omega_{k,0}]^2 \tag{10.75}$$

3. Repeat steps 1 and 2 with a new set of testing points

$$\begin{aligned}&(\alpha_{k,0}, t_{k,1}, \omega_{k,-1}, \beta_{k,0})\\&(\alpha_{k,0}, t_{k,1}, \omega_{k,0}, \beta_{k,0})\\&(\alpha_{k,0}, t_{k,1}, \omega_{k,1}, \beta_{k,0})\end{aligned} \tag{10.76}$$

to compute $\varpi(t_{k,1})$ and $\ln a(t_{k,1})$, respectively.

4. Compute another intermediate variable

$$r = \frac{\varpi(t_{k,0}) - \varpi(t_{k,1})}{t_{k,0} - t_{k,1}} \tag{10.77}$$

5. If $r = \beta_{k,0}$, then

$$\begin{aligned}\beta_k &= \beta_{k,0} \qquad \beta_k \in R\\ \alpha_k &= \frac{1}{2b} - \alpha_{k,0} \qquad \alpha_k > 0\end{aligned} \tag{10.78}$$

6. Otherwise,

$$\beta_k = \beta_{k,0} + \frac{\alpha_{k,0}}{2b(r-\beta_{k,0})\left[\left(\frac{\alpha_{k,0}}{r-\beta_{k,0}}\right)^2 + 1\right]} \qquad \beta_k \in R \tag{10.79}$$

$$\alpha_k = \frac{\beta_k - r}{r-\beta_{k,0}}\alpha_{k,0} \qquad \alpha_k > 0$$

7. Finally,

$$t_k = \frac{\alpha_k^{-1} + \alpha_{k,0}^{-1}}{t_{k,0} - t_{k,1}} \ln\frac{a(t_{k,0})}{a(t_{k,1})} + \frac{t_{k,0} + t_{k,1}}{2} \qquad t \in R \tag{10.80}$$

$$\omega_k = \varpi(t_{k,0}) + r(t_k - t_{k,0}) \qquad \omega_k \in R$$

Due to the complexity of the target function, classical numerical approaches, such as the *Newton-Raphson* method, virtually cannot be used for such a multi-dimensional optimization problem. Even for simple synthetic data samples, the computation time required by these iterative methods is too long for real-world applications. The approach presented in this section is not only fast, but also is very accurate.

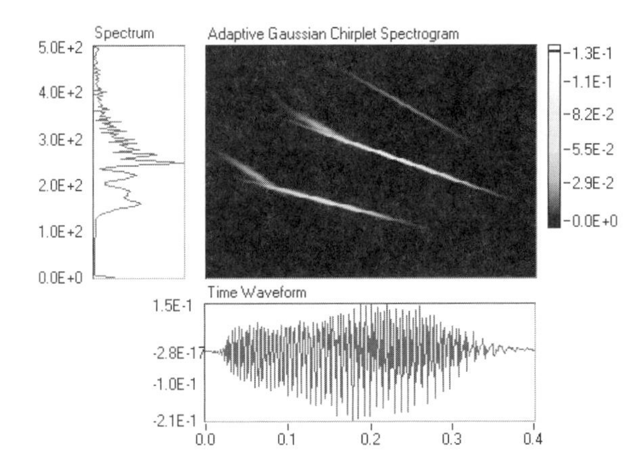

Figure 10-19 Compared with Figure 10-17, the Gaussian chirplet matches the bat sound much better in terms of fewer terms used. This is because the Gaussian chirplet fits the bat sound better than does the regular Gaussian function. The regular Gaussian function is a special case of the Gaussian chirplet.

Figure 10-19 illustrates an adaptive spectrogram of a bat sound computed by the adaptive Gaussian chirplet algorithm, Eqs. (10.71) to (10.79). Compared with one based on the adaptive Gabor transform (see Figure 10-17), the Gaussian chirplet matches the bat sound much better in terms of fewer terms used. In this example, to achieve the same amount of the residual $\|s_k(t)\|$, the adaptive Gabor expansion needs ten times more terms than that needed by its adaptive Gaussian chirplet counterpart. This is because the Gaussian chirplet fits the bat sound better than does the regular Gaussian function. The regular Gaussian function is a special case of the Gaussian chirplet.

Note that the refinement algorithms presented in this chapter are not unique. For example, we can also formulate (10.52) as

$$P(\alpha_k, t_k, \omega_k, \beta_k; \alpha_{k,0}, t_{k,n}, \omega_{k,m}, \beta_{k,0}) = |\langle s_k, h_{\alpha_{k,o}, t_{k,m}, \omega_{k,n}, \beta_{k,0}} \rangle| = a(\omega_{k,n}) e^{-b[\tilde{t}(\omega_{k,n}) - t_{k,m}]^2} \tag{10.81}$$

The corresponding testing points then become

$$\begin{matrix} (\alpha_{k,0}, t_{k,-1}, \omega_{k,0}, \beta_{k,0}) & & (\alpha_{k,0}, t_{k,-1}, \omega_{k,1}, \beta_{k,0}) \\ (\alpha_{k,0}, t_{k,0}, \omega_{k,0}, \beta_{k,0}) & \text{and} & (\alpha_{k,0}, t_{k,0}, \omega_{k,1}, \beta_{k,0}) \\ (\alpha_{k,0}, t_{k,1}, \omega_{k,0}, \beta_{k,0}) & & (\alpha_{k,0}, t_{k,1}, \omega_{k,1}, \beta_{k,0}) \end{matrix} \tag{10.82}$$

By applying the same technique as that introduced earlier, the reader should readily obtain an alternative fast refinement algorithm, based on Eqs. (10.81) and (10.82), to compute the optimal Gaussian chirplet.

APPENDIX

Optimal Dual Functions

For a given function $h[k]$, the corresponding dual function $\gamma[k]$ may not be unique. In Section 3.4, we discussed the solution of $\gamma[k]$ that is optimally close to $h[k]$. The resulting Gabor expansion is called the orthogonal-like Gabor expansion. In some applications [171], the Gabor transform and Gabor expansion can also be formulated as perfect reconstruction filter banks. In those applications, the analysis and synthesis functions may need to have different properties; one may intentionally pursue $\gamma[k]$ that differs from $h[k]$.[1] In what follows, we will investigate a general optimal algorithm that allows $\gamma[k]$ to be optimally close to an arbitrarily desired function $d[k]$ for a given $h[k]$.

We formulate the problem as follows. For a given function $h[k]$ and the sampling pattern (determined by the time sampling interval ΔM and the number of frequency channels N), find a dual function $\gamma[k]$ that is most similar to a desired function $d[k]$, in the sense of the least mean square error (LMSE), i.e.,

$$\xi = \min_{\gamma : H\gamma^* = \vec{\mu}} \left\| \frac{\vec{\gamma}}{\|\vec{\gamma}\|} - \vec{d} \right\|^2 \tag{A.1}$$

where $d[k]$ is a unit energy function. As discussed in Section 3.4, when $d[k] = h[k]$, (A.1) will be identical to (3.52), which leads to the orthogonal-like Gabor expansion.

In general, (A.1) is a least square problem with an equality constraint given by

$$H\vec{\gamma}^* = \vec{\mu} \tag{A.2}$$

1. As mentioned in previous sections, $h[k]$ and $\gamma[k]$ are dual functions; they are exchangeable. We can use $\gamma[k]$ as an analysis filter and $h[k]$ as a synthesis filter, and vice versa.

where H is a p-by-L matrix and $\vec{\mu}$ is a vector with p-element. Both of them are defined in Section 3.3. Without loss of generality, let's assume that H has a full row rank. (Otherwise, we always can employ SVD to alleviate the rank deficiency problem.)

By QR decomposition [19], we have

$$H_{LP}^T = Q_{LL}\begin{bmatrix} R_{pp} \\ 0 \end{bmatrix} \tag{A.3}$$

where Q is orthonormal and R is upper triangular. Substituting (A.3) into (A.2) obtains

$$\begin{bmatrix} R^T & 0 \end{bmatrix} Q^T \vec{\gamma} = \begin{bmatrix} R^T & 0 \end{bmatrix} \begin{bmatrix} \vec{x} \\ \vec{y} \end{bmatrix} = \vec{\mu} \tag{A.4}$$

where

$$Q^T \vec{\gamma} = \begin{bmatrix} \vec{x} \\ \vec{y} \end{bmatrix} \tag{A.5}$$

From (A.4),

$$\vec{x} = (R^T)^{-1} \vec{\mu} \tag{A.6}$$

Because $QQ^T = I$, left multiplying Q to both sides (A.5) yields

$$\vec{\gamma} = Q\begin{bmatrix} \vec{x} \\ \vec{y} \end{bmatrix} = \begin{bmatrix} Q_x & Q_y \end{bmatrix} \begin{bmatrix} \vec{x} \\ \vec{y} \end{bmatrix} = Q_x\vec{x} + Q_y\vec{y} \tag{A.7}$$

where

$$x \in R^P \qquad y \in R^{L-p} \tag{A.8}$$

Hence, $\vec{\gamma}$ is the sum of two orthogonal vectors,

$$Q_x\vec{x} + Q_y\vec{y} \tag{A.9}$$

Consequently,

$$\|\vec{\gamma}\|^2 = \|\vec{x}\|^2 + \|\vec{y}\|^2 \tag{A.10}$$

Expanding the error formula (A.1) yields

$$\xi = \min_{\gamma:H\gamma^* = \vec{\mu}} \left\| \frac{\vec{\gamma}}{\|\vec{\gamma}\|} - \vec{d} \right\|^2 = \min_{\gamma:H\gamma^* = \vec{\mu}} 2\left(1 - \frac{Re(\vec{\gamma}^T\vec{d})}{\|\vec{\gamma}\|}\right) \tag{A.11}$$

Then, minimizing ξ with respect to $\vec{\gamma}$ is equivalent to

$$\max_{\gamma:H\gamma^* = \vec{\mu}} \frac{Re(\vec{\gamma}^T\vec{d})}{\|\vec{\gamma}\|} = \max_{\gamma:H\gamma^* = \vec{\mu}} \zeta \tag{A.12}$$

Replacing $\vec{\gamma}$ with (A.7) and $\|\vec{\gamma}\|$ with (A.10), Eq. (A.12) becomes

$$\max_{y \in R^{L-p}} \zeta = \max_{y \in R^{L-p}} \frac{Re(\vec{x}^T Q_x^T \vec{d}) + Re(\vec{y}^T Q_y{}^T \vec{d})}{\sqrt{\|\vec{x}\|^2 + \|\vec{y}\|^2}} \tag{A.13}$$

It is interesting to note that the maximum of ζ occurs when $\vec{y}$ is in the same direction as the vector $Q_y^T \vec{d}$, regardless of the value of $\|\vec{y}\|$. Let

$$\vec{y} = tQ_y^T \vec{d} \tag{A.14}$$

where τ is real and positive. Then the problem of (A.13) is equivalent to maximizing ζ with respect to τ. By replacing $\vec{y}$ with (A.14), (A.13) becomes

$$\max_{t>0} \frac{Re(\vec{x}^T Q_x^T \vec{d}) + \left\|Q_y{}^T \vec{d}\right\|^2 t}{\sqrt{\|\vec{x}\|^2 + \left\|Q_y{}^T \vec{d}\right\|^2 t^2}} = \max_{t>0} \frac{A + Bt}{\sqrt{C + Bt^2}} = \max_{t>0} \zeta \tag{A.15}$$

where

$$A = Re(\vec{x}^T Q_x^T \vec{d}) \qquad B = \left\|Q_y{}^T \vec{d}\right\|^2 \qquad C = \|\vec{x}\|^2 \tag{A.16}$$

Obviously,

$$\lim_{t \to \infty} \zeta = \sqrt{B} = \left\|Q_y^T \vec{d}\right\| \tag{A.17}$$

For $A = 0$,

$$\zeta' = \frac{BC}{\sqrt{(C + Bt^2)^3}} \tag{A.18}$$

In this case, ζ does not have an extremum for the finite τ. In other words, (A.1) does not have a solution if the desired function $\vec{d}$ is perpendicular to $Q_x\vec{x}$.

When $A \neq 0$, the first derivative of ζ, with respect to τ equal to zero, yields a unique solution $\tau = C/A$. Next, we compute the second derivative to see whether $\tau = C/A$ corresponds to the maximum of ζ.

The second derivative of ζ with respect to τ is

$$\zeta'' = \frac{BC}{\sqrt{(C + Bt^2)^5}}(2ABt^2 - 3BC - AC) \tag{A.19}$$

Replacing $\tau = C/A$, (A.19) becomes

$$\zeta'' = -\frac{C}{A}(BC + A^2) \tag{A.20}$$

When $A > 0$, $\zeta(t)$ is convex and attains its maximum at $\tau = C/A > 0$. When $A < 0$, $\zeta(t)$ is concave and increases to $\|Q_y^T \vec{d}\|$ as t goes to infinity; this corresponds to an unbounded $\vec{y}$, which is impractical. Hence, the solution of (A.1) exists only for $A = Re(\vec{x}^T Q_x^T \vec{d}) > 0$. Figure 1 depicts the error curves.

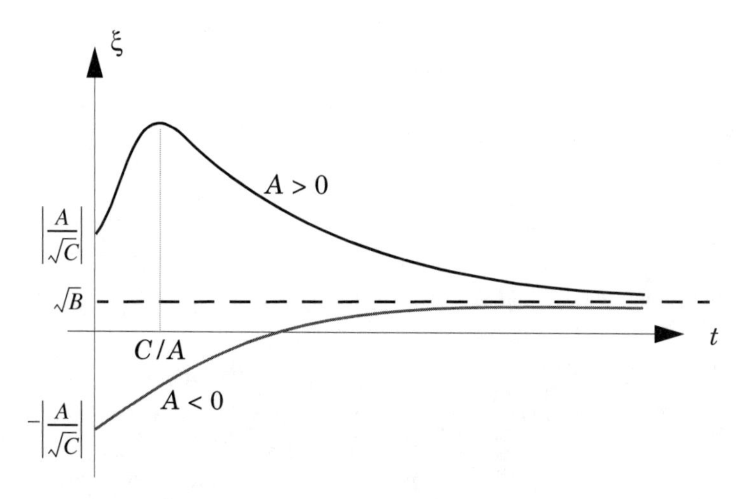

Figure 1 Error curves (ζ attains the maximum only for $A > 0$).

Substituting (A.14) and $\tau = C/A$ into (A.7), we can readily obtain the solution of (A.1) as

$$\vec{\gamma} = Q_x\vec{x} + \frac{\|\vec{x}\|^2}{Re(\vec{x}^T Q_x^T \vec{d})} Q_y Q_y^T \vec{d} \tag{A.21}$$

and the minimum error is

$$\xi = 2\left(1 - \sqrt{\frac{\|Re(\vec{x}^T Q_x^T \vec{d})\|^2}{\|\vec{x}\|^2} + \|Q_y^T \vec{d}\|^2}\right) \tag{A.22}$$

Bibliography

*T*his bibliography consists of three parts; books for time-frequency and wavelet analysis, papers for time-frequency analysis, and papers for wavelet analysis and filter banks. The purpose of this bibliography is to provide the reader with sources for additional and more advanced treatments of topics in time-frequency and wavelet transforms. References are selected mainly on the basis of their relevance to the topics discussed in this book. This is by no means meant to be an exhaustive list but rather it is intended to indicate directions for further study.

Books for time-frequency and wavelet analysis

[1] A. N. Akansu and R. A. Haddad, *Multiresolution Signal Decomposition, Transforms, Sub-bands, and Wavelets*, Academic Press, San Diego, California, 1992.

[2] A. N. Akansu and Mark J. T. Smith, *Subband and Wavelet Transforms, Design and Applications*, Kluwer Academic Publishers, Boston, Massachusetts, 1996.

[3] Akram Aldroubi and Michael Unser, editors, *Wavelets in Medicine and Biology*, CRC Press, Boca Raton, Florida, 1996.

[4] John J. Benedetto and Michael W. Frazier, editors, *Wavelets: Mathematics and Applications*, CRC Press, Boca Raton, Florida, 1993.

[5] B. Boashash, editor, *Time-Frequency Signal Analysis: Methods and Applications*, Wiley Halsted Press, 1992.

[6] C. S. Burrus, R. A. Gopinath, and H. Guo, *Introduction to Wavelets and Wavelets Transforms — A Primer*, Prentice Hall, Upper Saddle River, New Jersey, 1998.

[7] Charles K. Chui, *An Introduction to Wavelets*, Academic Press, New York, 1992. Volume 1 in the series: Wavelet Analysis and its Applications.

[8] Charles K. Chui, editor, *Wavelets: A Tutorial in Theory and Applications*, Academic Press, San Diego, California, 1992. Volume 2 in the series: Wavelet Analysis and its Applications.

[9] Charles K. Chui, Laura Montefusco, and Luigia Puccio, editors, *Wavelets: Theory, Algorithms, and Applications*, Academic Press, San Diego, California, 1994. Volume 5 in the series: Wavelet Analysis and its Applications.

[10] Leon Cohen, *Time-Frequency Analysis*, Prentice Hall, Upper Saddle River, New Jersey, 1995.

[11] J. M. Combes, A. Grossmann, and P. Tchamitchian, editors, *Wavelets, Time-Frequency Methods and Phase Space*, Springer-Verlag, Berlin, 1989.

[12] R. E. Crochiere and L. R. Rabiner, *Multirate Digital Signal Processing*, Prentice Hall, Upper Saddle River, New Jersey, 1983.

[13] I. Daubechies, *Ten Lectures on Wavelets, SIAM*, Philadelphia, Pennsylvania, 1992.

[14] Wolfgang Dahmen, Andrew Durdila, and Peter Oswald, editors, *Multiscale Wavelet Methods for Partial Differential Equations*, Academic Press, San Diego, California, 1997.

[15] H. Feirchtinger and T. Stroms, *Gabor Analysis and Algorithms: Theory and Applications*, Birkhauser, 1998.

[16] P. Flandrin, *Time-Frequency and Time-Scale Analysis*, Academic Press, 1998. Volume 10 in the series: Wavelet Analysis and its Applications. Translated by Joachim Stockler from the French edition.

[17] F. J. Fliege, *Multirate Digital Signal Processing: Multirate Systems, Filter Banks, and Wavelets.* Wiley & Sons, New York, 1994.

[18] Efi Foufoula-Georgiou and Praveen Kumar, editors, *Wavelets in Geophysics*, Academic Press, San Diego, California, 1994.

[19] G. H. Golub and C. F. Van Loan, *Matrix Computations*, Second edition, Johns Hopkins University Press, Baltimore, Maryland, 1989.

[20] Eugenio Hernández and Guido Weiss, *A First Course on Wavelets*, CRC Press, Boca Raton, Florida, 1996.

[21] R. A. Horn and C. R. Johnson, *Matrix Analysis*, Cambridge University Press, 1991.

[22] B. B. Hubbard, *The World According to Wavelets*, Second edition, A. K. Peters, Wellesley, Massachusetts, 1998.

[23] G. Kaiser, *A Friendly Guide to Wavelets*, Birkhäuser, Boston, Massachusetts, 1994

[24] Tom H. Koornwinder, editor, *Wavelets: An Elementary Treatment of Theory and Applications*, World Scientific, Singapore, 1993.

[25] C. Longo and B. Picinbono, editors, *Time and Frequency Representation of Signals and Systems*, Springer-Verlag, New York, 1989.

[26] R. J. Marks II, editor, *Introduction to Shannon Sampling and Interpolation Theory*, Springer-Verlag, New York, 1991.

[27] S. Mallat, *A Wavelet Tour of Signal Processing*, Second edition, Academic Press, 2000.

[28] S. L. Maple, *Digital Spectral Analysis*, Prentice Hall, Upper Saddle River, New Jersey, 1987.

[29] Peter R. Massopust, *Fractal Functions, Fractal Surfaces, and Wavelets*, Academic Press, San Diego, California, 1994.

[30] Yves Meyer, editor, *Wavelets and Applications*, Springer-Verlag, Berlin, 1992. Proceedings of the Marseille Workshop on Wavelets, France, May, 1989.

[31] Yves Meyer, *Wavelets and Operators*, Cambridge, 1992. Translated by D. H. Salinger from the 1990 French edition.

[32] Yves Meyer, *Wavelets, Algorithms and Applications*, SIAM, Philadelphia, Pennsylvania, 1993. Translated by R. D. Ryan based on lectures given for the Spanish Institute, Madrid, February 1991.

[33] J. Von Neumann, *The Geometry of Orthogonal Spaces*, vol. II, Princeton University Press, Princeton, New Jersey, 1950.

[34] A. V. Oppenheim, A. S. Willsky, with S. H. Nawab, *Signals & Systems*, Second edition, Prentice Hall, Upper Saddle River, New Jersey, 1996.

[35] A. V. Oppenheim, R. W. Schafer, and J. R. Buck, *Digital Signal Processing*, Second edition, Prentice Hall, Upper Saddle River, New Jersey, 1998.

[36] A. Papoulis, *Signal Analysis*, McGraw-Hill, New York, 1977.

[37] T. W. Parks and C. S. Burrus, *Digital Filter Design*, John Wiley & Sons, New York, 1987.

[38] Shie Qian and Dapang Chen, *Joint Time-Frequency Analysis*, Prentice Hall, Upper Saddle River, New Jersey, 1996.

[39] K. R. Rao and P. Yip, *Discrete Cosine Transform — Algorithms, Advantages and Applications*, Academic Press, 1990.

[40] R. M. Rao and A. S. Bopardikar, *Wavelet Transforms — Introduction to Theory and Applications*, Addison-Wesley, Reading, Massachusetts, 1998.

[41] R. D. Crochiere and L. R. Rabiner, *Multirate Digital Signal Processing*, Prentice Hall, Upper Saddle River, New Jersey, 1983.

[42] H. L. Resnikoff and R. O. Wells, Jr., *Wavelet Analysis and the Scalable Structure of Information*, Springer-Verlag, New York, 1997.

[43] E. A. Robinson and T. S. Durrani, *Geophysical Signal Processing*, Prentice Hall, Upper Saddle River, New Jersey, 1986.

[44] M. B. Ruskai, G. Beylkin, R. Coifman, I. Daubechies, S. Mallat, Y. Meyer, and L. Raphael, editors, *Wavelets and their Applications*, Jones and Bartlett, Boston, Massachusetts, 1992.

[45] Larry L. Schumaker and Glenn Webb, editors, *Recent Advances in Wavelet Analysis*, Academic Press, San Diego, 1993. Volume 3 in the series: Wavelet Analysis and its Applications.

[46] C. E. Shannon and W. Weaver, *The Mathematical Theory of Communication*, The University of Illinois Press, Illinois, 1964.

[47] G. Strang and T. Q. Nguyen, *Wavelet and Filter Banks*, Wellesley-Cambridge Press, Wellesley, Massachusetts, 1995.

[48] P. P. Vaidyanathan, *Multirate Systems and Filter Banks*, Prentice Hall, Upper Saddle River, New Jersey, 1993.

[49] M. Vetterli and J. Kovacevic, *Wavelets and Subband Coding*, Prentice Hall, Upper Saddle River, New Jersey, 1995.

[50] Mladen Victor Wickerhauser, *Adapted Wavelet Analysis from Theory to Software*, A. K. Peters, Wellesley, Massachusetts, 1994.

[51] Gilbert G. Walter, editor, *Wavelets and Other Orthogonal Systems with Applications*, CRC Press, Boca Raton, Florida, 1994

[52] P. M. Woodward, *Probability and Information Theory with Applications to Radar*, Pergamon Press, Elmsford, New York, 1953.

[53] R. M. Young, *An Introduction to Nonharmonic Fourier Series*, Academic Press, New York, 1980.

[54] R. K. Young, *Wavelet Theory and Its Applications*, Kluwer Academic Publishers, Boston, Massachusetts, 1993.

Papers for Time-Frequency Analysis

[55] M. G. Amin, "Time-varying spectrum estimation of a general class of nonstationary processes," *Proceedings of the IEEE*, vol. 74, pp. 1800-1802, 1986.

[56] M. G. Amin, "Mixed time-based spectrum estimators," *IEEE Trans. Acoustic, Speech, Signal Processing*, vol. 36, pp. 739-750, 1988.

[57] M. G. Amin, "Time-frequency spectrum analysis and estimation for non-stationary random processes," in B. Boashash, editor, *Time-Frequency Signal Analysis: Methods and Applications*, pp. 208-232, 1992.

[58] M. G. Amin, P. Davis, and P. Allen, "An application of the LMS algorithm in smoothing pseudo Wigner distribution," *IEEE Trans. Signal Processing*, vol. 41, pp. 930-934, 1993.

[59] M. G. Amin, "Spectral decomposition of the time-frequency distribution kernels," *IEEE Trans. Signal Processing*, vol. 42, pp. 1156-1166, 1994.

[60] M. G. Amin, "Minimum variance time-frequency distribution kernels for signals in additive noise," *IEEE Trans. Signal Processing*, vol. 44, no. 9, pp. 2352-2356, September 1996.

[61] M. G. Amin and W. J. Williams, "High spectral resolution time-frequency distribution kernels," *IEEE Trans. Signal Processing*, vol. 46, no. 10, pp. 2796-2804, October 1998.

[62] F. Auger and P. Flandrin, "Improving the readability of time-frequency and time-scale representations by the reassignment method," *IEEE Trans. Signal Processing*, vol. 43, no. 5, pp. 1068-1089, May 1995.

[63] F. Auger, P. Flandrin, P. Goncalves, and O. Lemoine, *Tutorial of Time-Frequency Toolbox for Use with Matlab*, http://www-isis.enst.fr/Applications/tftb/iutsn.univ-nantes.fr/auger/tftbftp.html, 1996.

[64] D. A Ausherman, A. Kozma, J. L. Waker, H. M. Jones, and E. C. Poggio, "Developments in radar imaging," *IEEE Trans. Aerospace and Electronic Systems*, vol. 20, no. 4, pp. 363-400, 1984.

[65] L. Auslanders, I. C. Gertner, and R. Tolimieri, "The discrete zak transform application to time-frequency analysis and synthesis of nonstationary signals," *IEEE Trans. Signal Processing*, vol. 39, no. 4, pp. 825-835, April 1991.

[66] R. Balart, "Matrix reformulation of the Gabor transform," *Optical Engineering*, vol. 31, no. 6, pp. 1235-1242, June 1992.

[67] R. Balian, "Un prinsipe d'incertitude fort en theorie du signal ou mecanique quantique," *C. Roy. Acad. Sci. Paris*, ser. 2, vol. 292, 1981.

[68] R. G. Baraniuk and D. L. Jones, "A signal-dependent time-frequency representation: fast algorithm for optimal kernel design," *IEEE Trans. Signal Processing*, vol. 42, pp. 134-146, April 1993.

[69] R. G. Baraniuk and D. L. Jones, "A radially Gaussian, signal-dependent time-frequency representation: optimal kernel design," *Signal Processing*, vol. 32, pp. 263-284, June 1993.

[70] R. G. Baraniuk and D. L. Jones, "A signal-dependent time-frequency representation," *IEEE Trans. Signal Processing*, vol. 41, pp. 1589-1602, January 1994.

[71] R.G. Baraniuk, M. Coates, and P. Steeghs, "Hybrid linear/quadratic time-frequency attributes," *IEEE Trans. Signal Processing*, vol. 49, no. 4, pp. 760-766, April 2001.

[72] M.J. Bastiaans, "Gabor's expansion of a signal into Gaussian elementary signals," *Proceedings of the IEEE*, vol. 68, pp. 538-539, April 1980.

[73] M.J. Bastiaans, "A sampling theorem for the complex spectrogram, and Gabor's expansion of a signal in Gaussian elementary signals," *Optical Engineering*, vol. 20, no. 4, pp. 594-598, July/August 1981.

[74] M.J. Bastiaans, "On the sliding-window representation in digital signal processing," *IEEE Trans. Acoustics, Speech, Signal Processing*, vol. ASSP-33, no. 4, pp. 868-873, August 1985.

[75] M.J. Bastiaans, "Gabor's signal expansion and its relation to sampling of the sliding-window spectrum," in R. J. Marks II, editor, *Introduction to Shannon Sampling and Interpolation Theory*, pp. 1-37, Springer-Verlag, 1992.

[76] M.J. Bastiaans, "Gabor's signal expansion and the Zak transform," *Appl. Opt.*, vol. 33, pp. 5241-5255, 1994.

[77] R. H. Bayley and P. M. Berry, "The electrical field produced by the eccentric current dipole in the nonhomogeneous conductor," *Am. Heart J.*, vol. 63, pp. 808-820, 1962.

[78] A. Belouchrani and M. G. Amin, "Blind source separation based on time-frequency signal representations," *IEEE Trans. Signal Processing*, Vol. 46, no. 11, pp. 2888 -2897, November 1998.

[79] B. Boashash and P. J. Black, "An efficient real-time implementation of the Wigner-Ville distributions," *IEEE Trans. Acoustics, Speech, Signal Processing*, vol. 35, no. 11, pp. 1611-1618, November 1987.

[80] B. Boashash, "Note on the use of the Wigner distribution for time-frequency signal analysis," *IEEE Trans. Acoustics, Speech, Signal Processing*, vol. 36, no. 9, pp. 1518-1521, September 1988.

[81] M. Born and P. Jordan, "Zur quantenmechanik," *Zeit. F. Phys.*, vol. 34, pp. 858-888, 1925.

[82] G. F. Boudreaux-Bartels and T. W. Parks, "Time-varying filtering and signal estimation using Wigner distribution synthesis techniques," *IEEE Trans. Acoustics, Speech, Signal Processing*, vol. 34, pp. 442-451, 1986.

[83] A. C. Bovik, P. Maragos, and T. F. Quatieri, "AM-FM energy detection and separation in noise using multiband energy operators," *IEEE Trans. Signal Processing*, vol. 41, no. 12, pp. 3245-3265, December 1993.

[84] N. G. de Bruijn, "A theory of generalized functions, with applications to Wigner distribution and Weyl correspondence," *Nieuw Arch. Wiskunde* (3), vol. 21, pp. 205-280, 1973.

[85] N. Cartwright, "A non-negative quantum mechanical distribution function," Physica, vol. 83A, pp. 210-212, 1976.

[86] Victor Chen, "Radar ambiguity function, time-varying matched filter, and optimal wavelet correlator," *Optical Engineering*, vol. 33, no. 7, pp. 2212-2217, 1994.

[87] Victor Chen, "Reconstruction of inverse synthetic aperture radar image using adaptive time-frequency wavelet transform," (invited paper), *SPIE Wavelet Applications*, vol. 2491, pp. 373-386, 1995.

[88] Victor Chen and Shie Qian, "Joint time-frequency transform for radar range-doppler imaging," *IEEE Trans. Aerospace and Electronic Systems*, vol. 34, no. 2, pp. 496-499, April 1998.

[89] Victor Chen and Hao Lin, "Joint time-frequency analysis for radar signal and image processing," *IEEE Signal Processing Magazine*, vol. 16, no. 2, pp. 81-93, March 1999.

[90] H. Choi and W. J. Williams, "Improved time-frequency representation of multi-component signals using exponential kernels," *IEEE Trans. Acoustics, Speech, Signal Processing*, vol. 37, no. 6, pp. 862-871, June 1989.

[91] T. A. C. M. Claasen and W. F. G. Mecklenbräuker, "The Wigner Distribution–A tool for time-frequency signal analysis," *Phillips J. Res.*, vol. 35, pp. 217-250, 276-300, and 1067-1072, 1980.

[92] Israel Cohen, Shalom Raz, and David Malah, "Adaptive suppression of Wigner interference-terms using shift-invariant wavelet packet decomposition," submit to publication.

[93] Leon Cohen, "Generalized phase-space distribution functions," *J. Math. Phys.*, vol. 7, pp. 781-806, 1966.

[94] L. Cohen and T. Posch, "Positive time-frequency distribution functions," *IEEE Trans. Acoustics, Speech, Signal Processing*, vol. 33, pp. 31-38, June 1985.

[95] Leon Cohen, "On a fundamental property of the Wigner distribution," *IEEE Trans. Acoustics, Speech, Signal Processing,* vol. 35, pp. 559-561, 1987.

[96] Leon Cohen, "Wigner distribution for finite duration or band limited signals and limited cases," *IEEE Trans. Acoustics, Speech, Signal Processing*, vol. 35, pp. 796-806, 1987.

[97] Leon Cohen, "Time-frequency distribution–a review," *Proceedings of the IEEE*, vol. 77, no. 7, pp. 941-981, July 1989.

[98] L. Cohen and C. Lee, "Instantaneous frequency," in B. Boashash, editor, *Time-Frequency Signal Analysis: Methods and Applications*, pp. 98-117, Wiley Halsted Press, 1992.

[99] J. G. Daugman, "Complete discrete 2-D Gabor transforms by neural networks for image analysis and compression," *IEEE Trans. Acoustics, Speech, Signal Processing*, vol. 36, no. 7, pp. 1169-1179, 1988.

[100] R. J. Duffin and A. C. Schaeffer, "A class of nonharmonic Fourier series," *Trans. Amer. Math. Soc.*, no. 72, pp. 341-366, 1952.

[101] P. J. Durka and K. L. Blinowska, "Analysis of EEG transients by means of matching pursuit," *Ann. Biomed. Eng.*, vol. 23, pp. 608-611, 1995.

[102] P. J. Durka, "Time-frequency analyses of EEG," Ph.D. dissertation, Warsaw Univ., Warsaw, Poland, 1996.

[103] P. D. Einziger, "Gabor expansion of an aperture field in exponential elementary beams," *IEE Electron. Lett.*, vol. 24, pp. 665-666, 1988.

[104] S. Farkash and S. Raz, "Linear systems in Gabor time-frequency space," *IEEE Trans. Signal Processing*, vol. 42, no. 3, pp. 611-617, March 1994.

[105] P. Flandrin, "When is the Wigner-Ville spectrum non-negative?" *Signal Processing III, Theories and Applications*, vol. 1, pp. 239-242, 1986.

[106] P. Flandrin, "On detection-estimation procedures in the time-frequency plane," *Proceedings of IEEE ICASSP-86*, vol. 4, pp. 2331-2334, 1986.

[107] P. Flandrin and F. Hlawatsch, "Signal representation geometry and catastrophes in the time-frequency plane," in T. S. Durrani, et al., editors, *Mathematics in Signal Processing,* pp. 3-14, Clarendon, Oxford, 1987.

[108] P. Flandrin, "Maximum signal energy concentration in a time-frequency domain," *Proceedings of IEEE ICASSP-88*, vol. 4, pp. 2176-2179, 1988.

[109] P. Flandrin, "A time-frequency formulation of optimal detection," *IEEE Trans. Acoustics, Speech, Signal Processing*, vol. 36, no. 9, pp. 1377-1384, September 1988.

[110] P. Flandrin and O. Rioul, "Affine smoothing of the Wigner-Ville distribution," *Proceedings of IEEE ICASSP-90*, pp. 2455-2458, Albuquerque, New Mexico, 1990.

[111] M. J. Freeman and M. E. Dunham, "Trans-ionospheric signal detection by time-scale representation," *UK Symposium on Applications of Time-Frequency and Time-Scale Methods, Proceedings of IEEE SP,* pp. 152-158.

[112] B. Friedlander and B. Porat, "Detection of transient signal by the Gabor representation," *IEEE Trans. Acoustics, Speech, Signal Processing*, vol. 37, no. 2, pp. 169-180, February 1989.

[113] D. Gabor, "Theory of communication," *J. IEE*, vol. 93, no. III, pp. 429-457, London, November 1946.

[114] T. Genossar and M. Porat, "Can one evaluate the Gabor expansion using Gabor's iterative algorithm?" *IEEE Trans. Signal Processing*, vol. 40, no. 8, pp. 1852-1861, August 1992.

[115] A. B. Gershman and M. G. Amin, "Wideband direction-of-arrival estimation of multiple chirp signals using spatial time-frequency distributions," *IEEE Signal Processing Letters*, vol. 7, no. 6, pp. 152-155, June 2000.

[116] A. Gersho, "Characterization of time-varying linear systems," *Proceedings of the IEEE*, p. 238, 1963.

[117] J. P. Gollub and H. L. Swinney, "Onset of turbulence in a rotating fluid," *Physics Review Letters*, vol. 35, pp. 929-930, 1975.

[118] D. W. Griffin and J. S. Lim, "Signal estimation from modified short-time Fourier transform," *IEEE Trans. Acoustics, Speech, Signal Processing*, vol. 32, no. 2, pp. 236-243, April 1984.

[119] B. Harms, "Computing time-frequency distributions," *IEEE Trans. Signal Processing*, vol. 39, no. 3, pp. 727-729, March 1991.

[120] F. Hlawatsch, "Interference terms in the Wigner distribution," in V. Cappellini and A. G. Constantinides, editors, *Digital Signal Processing — 84*, pp. 363-367, Elsevier Science Publishers, Amsterdam, 1984.

[121] F. Hlawatsch and W. Krattenthaler, "Phase matching algorithms for Wigner distribution signal synthesis," *IEEE Trans. Signal Processing*, vol. 39, no. 3, pp. 612-619, March 1991.

[122] F. Hlawatsch, "Duality and classification of bilinear time-frequency signal representations," *IEEE Trans. Signal Processing*, vol. 39, no. 7, pp. 1564-1574, July 1991.

[123] F. Hlawatsch and W. Krattenthaler, "Bilinear signal synthesis," *IEEE Trans. Signal Processing*, vol. 40, no. 2, pp. 352-363, February 1992.

[124] F. Hlawatsch and G. F. Boudreaux-Bartels, "Linear and quadratic time-frequency signal representations," *IEEE Signal Processing Magazine*, vol. 9, pp. 21-67, 1992.

[125] F. Hlawatsch, "The Wigner distribution of a linear signal space," *IEEE Trans. Signal Processing*, vol. 41, no. 3, pp. 1248-1258, March 1993.

[126] F. Hlawatsch, A. H. Costa, and W. Krattenthaler, "Time-frequency signal synthesis with time-frequency extrapolation and don't-care regions," *IEEE Trans. Signal Processing*, vol. 42, no. 9, pp. 2513-2520, September 1994.

[127] F. Hlawatsch and P. Flandrin, "The interference structure of the Wigner distribution and related time-frequency signal representations," in W. F. G. Mecklenbräuker, editor, *The Wigner Distribution — Theory and Applications in Signal Processing*, Elsevier, Amsterdam, 1994.

[128] F. Hlawatsch and W. Kozek, "Time-frequency projection filters and time-frequency signal expansions," *IEEE Trans. Signal Processing*, vol. 42, no. 12, pp. 3321-3334, December 1994.

[129] F. Hlawatsch, T. G. Manickam, R. L. Urbanke, and W. Jones, "Smoothed pseudo-Wigner distribution, Choi-Williams distribution, and cone-kernel representation: ambiguity-domain analysis and experimental comparison," *Signal Processing*, vol. 43, no. 2, pp. 149-168, May 1995.

[130] F. Hlawatsch, G. S. Edelson, and P. Podlucky, "Multipulse maximum-likelihood range/Doppler estimation and the ambiguity function of a linear signal space, "*IEEE Trans. Aerospace and Electronic Systems*, submitted.

[131] F. Hlawatsch and W. Krattenthaler, "Signal synthesis algorithms for bilinear time-frequency signal representations" in W. Mecklenbr, editor, *The Wigner Distribution — Theory and Applications in Signal Processing*, pp. 135-209, Elsevier, Amsterdam, 1996.

[132] P. J. Hubber, "Projection pursuit," *Ann. Statist.*, vol. 13, no. 2, pp. 435-475, 1985.

[133] L. D. Jacobson and H. Wechsler, "The Wigner distribution and its usefulness for 2-D image processing," *Proceedings of 6th Int. Conf. on Pattern Recognition*, Munich, October 19-22, 1982.

[134] L. D. Jacobson and H. Wechsler, "The composite pseudo Wigner distribution (CSWD): a computable and versatile approximation to the Wigner distribution (WD)," *Proceedings of IEEE ICASSP-83*, pp. 254-256, 1983.

[135] L. D. Jacobson and H. Wechsler, "Joint spatial/spatial-frequency representation," *Signal Processing*, vol. 14, pp. 37-66, January 1988.

[136] A. J. E. M. Janssen, "Gabor representation of generalized functions," *J. Math. Anal. Appl.*, vol. 38, pp. 377-394, 1981.

[137] A. J. E. M. Janssen, "Positively properties of phase-plane distribution functions," *J. Math. Phys.*, vol. 25, pp. 2240-2252, July 1984.

[138] A. J. E. M. Janssen and T. A. C. M. Claassen, "On positively of time-frequency distributions," *IEEE Trans. Acoustics, Speech, Signal Processing*, vol. 33, no. 4, pp. 1029-1032, 1985.

[139] A. J. E. M. Janssen, "A note on positive time-frequency distributions," *IEEE Trans. Acoustics, Speech, Signal Processing*, vol. 35, no. 5, pp. 701-703, May 1987.

[140] A. J. E. M. Janssen, "On the locus and spread of pseudo-density functions in the time-frequency plane," *Phillips J. Res.*, vol. 37, no. 3, pp. 79-110, 1982.

[141] A. J. E. M. Janssen, "The Zak transform: a signal transform for sampled time-continuous signals," *Phillips J. Res.*, vol. 43, pp. 23-69, 1988.

[142] A. J. E. M. Janssen, "Positively of time-frequency distribution functions," *Signal Processing*, vol. 14, pp. 243-252, 1988.

[143] A. J. E. M. Janssen, "Optimality property of the Gaussian window spectrogram," *IEEE Trans. Signal Processing*, vol. 39, no. 1, pp. 202-204, January 1991.

[144] A. J. E. M. Janssen, "Duality and biorthogonality for Weyl-Heisenberg frame," to appear in *J. Fourier Anal. Appl.*

[145] A. J. E. M. Janssen, "On rationally oversampling Weyl-Heisenberg frame," submitted to *Signal Processing.*

[146] J. Jeong and W. Williams, "Alias-free generalized discrete-time time-frequency distributions," *IEEE Trans. Signal Processing*, vol. 40, no. 11, pp. 2757-2765, November 1992.

[147] J. Jeong and W. Williams, "Kernel design for reduced interference distributions," *IEEE Trans. Signal Processing*, vol. 40, pp. 402-412, 1992.

[148] D. J. Jones and R. G. Baraniuk, "A simple scheme for adaptive time-frequency representations," *IEEE Trans. Signal Processing*, vol. 42, pp. 3530-3535, December 1994.

[149] D. J. Jones and R. G. Baraniuk, "An adaptive optimal-kernel time-frequency representation," *IEEE Trans. Signal Processing*, vol. 43, pp. 2361-2371, October 1995.

[150] L. K. Jones, "On a conjecture of Huber concerning the convergence of PP-regression," *Ann. Statist.*, vol. 15, pp. 880-882, 1987.

[151] S. M. Kay and S. L. Marple, "Spectrum analysis-a modern perspective," *Proceedings of the IEEE*, vol. 69, pp. 1380, 1981.

[152] J. G. Kirwood, "Quantum statistics of almost classical ensembles," *Phys. Rev.*, vol. 44, pp. 31-37, 1933.

[153] K. Kodera, C. de Villedary, and R. Gendrin, "A new method for numerical analysis of non-stationary signals," *Phys. Earth and Plan. Int.*, vol.12, pp. 142-150, 1976.

[154] K. Kodera, R. Gendrin, and C. de Villedary, "Analysis of time-varying signals with small BT values," *IEEE Trans. Acoustic, Speech, and Signal Processing*, vol. 26, pp. 64-76, 1978.

[155] P. J. Kootsookos, B. C. Lovell, and B. Boashash, "A unified approach to the STFT, TFD's, and instantaneous frequency," *IEEE Trans. Signal Processing*, vol. 40, no. 8, pp. 1971-1982, August 1992.

[156] M. Kostadinov, F. Yamazaki, and K. Sudo, "Comparative study on liquefaction detection methods using strong motion records," *Proceedings of 25th JSCE Earthquake Engineering Symposium*, pp. 409-412, 1999.

[157] W. Kozek and F. Hlawatsch, "A comparative study of linear and nonlinear time-frequency filters," *Proceedings of IEEE-SP Int. Symp. Time-Frequency Time-Scale Analysis*, pp. 163-166, Victoria, B.C., October 1992.

[158] M. L. Kramer and D. L. Jones, "Improved time-frequency filtering using an STFT analysis-modification-synthesis method," *Proceedings of Symp. Time-Frequency and Time-Scale Analysis*, pp. 264-267, Philadelphia, Pennsylvania, 1994.

[159] W. Krattenthaler and F. Hlawatsch, "Improved signal synthesis from pseudo Wigner distribution," *IEEE Trans. Signal Processing*, vol. 39, no. 2, pp. 506-509, February 1991.

[160] W. Krattenthaler and F. Hlawatsch, "Time-frequency design and processing of signals via smoothed Wigner distribution," *IEEE Trans. Signal Processing*, vol. 41, no. 1, pp. 278-287, January 1993.

[161] A.J. van Leest and M.J. Bastiaans, "Gabor's discrete signal expansion and the discrete Gabor transform on a non-separable lattice," *Proceedings of ICASSP-00*, vol. 1, pp. 101-104, June 2000.

[162] Shidong Li, "The theory of frame multiresolution analysis and its applications," Ph.D dissertation, University of Maryland, May 1993.

[163] Shidong Li, "A general theory of discrete Gabor expansion," *Proceedings of SPIE'94 Mathematical Imaging: Wavelet Applications*, San Diego, July 1994.

[164] Shidong Li, "General frame decompositions, pseudo-duals and its application to Weyl-Heisenberg frames," to appear in *Numerical Functional Analysis and Optimization*.

[165] Shidong Li, "Non-separable 2-D discrete Gabor expansion for image processing," submitted to Special Issue of *Multidimensional Signal Processing*.

[166] S. Li and D. M. Healy Jr., "A parametric class of discrete Gabor expansions," *IEEE Trans. Signal Processing*, vol. 44, no. 2, pp. 201-211, February 1996.

[167] S. Li and S. Qian, "A complement to a derivation of discrete Gabor expansion," *IEEE Signal Processing* Lett., vol. 2, no. 2, pp. 31-33, February 1995.

[168] Weiping Li, "Wigner distribution method equivalent to dewdrop method for detecting a chirp signal," *IEEE Trans. Acoustics, Speech, Signal Processing*, vol. 35, no. 8, August 1988.

[169] P. Louphlin, J. Pitton, and L. Atlas, "Construction of positive time-frequency distributions," *IEEE Trans. Signal Processing*, vol. 42, pp. 2697-2705, 1994.

[170] Y. Lu and J. M. Morris, "On equivalent frame conditions in the Gabor expansion," *Signal Processing*, vol. 49, no. 2, pp. 97-100, March 1996.

[171] Y. Lu and J. M. Morris, "Gabor expansion for adaptive echo cancellation," *IEEE Signal Processing Magazine*, vol. 16, no. 2, pp. 68-80, March 1999.

[172] S. Mallat and Z. Zhang, "Matching pursuit with time-frequency dictionaries," *IEEE Trans. Signal Processing*, vol. 41, no. 12, pp. 3397-3415, December 1993.

[173]H. Margenau and R. L. Hill, "Correlation between measurements in quantum theory," *Prog. Theoret, Phys.*, pp. 722-738, 1961.

[174] J. M. Morris and D. Wu, "Some results on joint time-frequency representation via Wigner-Ville distribution decomposition," *Proceedings of 1992 Conference Information Science & System,* vol. 1, pp. 6-10, 1992.

[175] J. M. Morris and D. Wu, "On alias-free formulation of discrete-time Cohen's class of distributions," *IEEE Trans. Signal Processing*, vol. 44, pp. 1355 -1364, June 1996.

[176] J. M. Morris and Y. Lu, "On biorthogonal-like sequences for generalized discrete Gabor expansion of discrete-time signals," *IEEE Trans. Signal Processing*, vol. 44, pp. 1378-1391, 1996.

[177] S. H. Nawab and T. F. Quatieri, "Short-time Fourier transform," in J.S. Lim and A.V. Oppenheim, editors, *Advanced Topics in Signal Processing*, Prentice Hall, Upper Saddle River, New Jersey, 1988.

[178] E. Niedermayer and F. Lopes Da Silva, "*Electroencephalography*: *basic principles,"* in *Clinical Applications and Related Fields*, Fourth edition, pp. 1154, Williams & Wilkins, Philadelphia, Pennsylvania, 1954.

[179] A. H. Nuttall, "Wigner distribution function: relation to short-term spectral estimation, smoothing, and performance in noise," *Technical Report 8225*, Naval Underwater Systems Center, 1988.

[180] A. H. Nuttall, "The Wigner distribution function with minimum spread," *Technical Report 8317*, Naval Underwater Systems Center, 1988.

[181] A. H. Nuttall, "Alias-free Wigner distribution function and complex ambiguity function for discrete signals," *Technical Report 8533*, Naval Underwater Systems Center, 1989.

[182] X. Ouyang and M. G. Amin, "Short-time Fourier transform receiver for nonstationary interference excision in direct sequence spread spectrum communications," *IEEE Trans. Signal Processing*, vol. 49, no. 4, pp. 851-863, April 2001.

[183] R. S. Orr, "The order of computation of finite discrete Gabor transforms," *IEEE Trans. Signal Processing*, vol. 41, no. 1, pp. 122-130, January 1993.

[184] R. S. Orr, "Derivation of Gabor transform relations using Bessel's equality," *Signal Processing*, vol. 30, pp. 257-262, 1993.

[185] C. H. Page, "Instantaneous power spectrum," *J. Appl. Phys.*, vol. 23, pp. 103-106, 1952.

[186] A. Papandreou and G. F. Boudreaux-Bartels, "A generalization of the Choi-Williams and the Butterworth time-frequency distributions," *IEEE Trans. Signal Processing*, vol. 41, pp. 463-472, 1993.

[187] M. Pasquier, P. Goncalves, and R. Baraniuk, "Hybrid linear/bilinear time-scale analysis," *IEEE Trans. Signal Processing*, vol. 47, pp. 254-259, January 1999.

[188] S. C. Pei and T. Y. Wang, "The Wigner distribution of linear time-variant systems," *IEEE Trans. Acoustics, Speech, Signal Processing*, vol. 36, pp. 1681-1684, 1988.

[189] Flemming Pedersen, "A Gabor expansion-based positive time-dependent power spectrum," *IEEE Trans. Signal Processing*, vol. 47 no. 2, pp. 587-590, February 1999.

[190] T. Posch, "Wavelet transform and time-frequency distributions," *Proceedings of SPIE, Int. Soc. Optical Engineering*, vol. 1152, pp. 477-482, 1988.

[191] M. R. Protnoff, "Time-frequency representation of digital signal and systems based on short-time Fourier analysis," *IEEE Trans. Acoustics, Speech, Signal Processing,* vol. 28, no. 1, pp. 55-69, February 1980.

[192] S. Qian and D. Chen, "Signal representation in adaptive Gaussian functions and adaptive spectrogram," *Proceedings of the Twenty-seventh Annual Conference on Information Sciences and Systems*, pp. 59-65, Department of Electrical and Computer Engineering, The Johns Hopkins University, Baltimore, Maryland, March 24-26, 1993.

[193] S. Qian and J. M. Morris, "Wigner distribution decomposition and cross-term deleted representation," *Signal Processing*, vol. 25, no. 2, pp. 125-144, May 1992.

[194] S. Qian, K. Chen, and S. Li, "Optimal biorthogonal sequence for finite discrete-time Gabor expansion," *Signal Processing*, vol. 27, no. 2, pp. 177-185, May 1992.

[195] S. Qian and D. Chen, "Discrete Gabor transform," *IEEE Trans. Signal Processing*, vol. 41, no. 7, pp. 2429-2439, July 1993.

[196] S. Qian and D. Chen, "Optimal biorthogonal analysis window function for discrete Gabor transform," *IEEE Trans. Signal Processing*, vol. 42, no. 3, pp. 687-694, March 1994.

[197] S. Qian and D. Chen, "Signal representation using adaptive normalized Gaussian functions," *Signal Processing*, vol. 36, no. 1, pp. 1-11, March 1994.

[198] S. Qian and D. Chen, "Decomposition of the Wigner-Ville distribution and time-frequency distribution series" *IEEE Trans. Signal Processing*, vol. 42, no. 10, pp. 2836-2841, October 1994.

[199] S. Qian, M. E. Dunham, and M. J. Freeman, "Trans-ionospheric signal recognition by joint time-frequency representation," *Radio Science*, vol. 30, no. 6, pp. 1817-1829, November-December 1995.

[200] S. Qian and D. Chen, "Joint time-frequency analysis," *IEEE Signal Processing Magazine*, vol. 16, no. 2, pp. 52-67, March 1999.

[201] S. Qiu and H. G. Feichtinger, "Discrete Gabor structure and optimal representation," *IEEE Trans. Signal Processing*, vol. 43, no. 10, pp. 2258-2268, October 1995.

[202] S. Qiu, "Block-circulant Gabor-matrix structure and discrete Gabor transforms," *Optical Engineering*, vol. 34, no. 10, pp. 2872-2878, October 1995.

[203] A. Rechtschaffen and A. Kales, Eds., "A Manual of Standarized Terminology," *Techniques and Scoring System for Sleep Stages in Human Subjects*: U.S. Gov. Printing Office, 1968.

[204] W. Rihaczek, "Signal energy distribution in time and frequency," *IEEE Trans. Inst. Radio Engineers (IRC)*, vol. IT-14, pp. 369-374, 1968.

[205] O. Rioul, "Wigner-Ville representations of signals adapted to shifts and dilations," *Tech. Memo. 112277-880422-03-TM*, AT&T Bell Labs, 1988.

[206] B. E. Saleh and N. S. Subotic, "Time-variant filtering of signals in the mixed time-frequency domain," *IEEE Trans. Acoustics, Speech, Signal Processing*, vol. 33, no. 3, pp. 1479-1485, 1985.

[207] R. J. Sclabassi and R. M. Harper, "Laboratory computers in neurophysiology," *Proceedings of the IEEE*, vol. 61, no. 11, pp. 1602-1614, 1973.

[208] R. J. Sclabassi, M. Sun, D. N. Krieger, P. Jasiukaitis, and M. S. Scher, "Time-frequency analysis of the EEG signal," *Proceedings of ISSP 90, Signal Processing, Theories, Implementations and Applications*, pp. 935-942, Gold Coast, Australia, 1990.

[209] R. J. Sclabassi, M. Sun, D. N. Krieger, P. Jasiukaitis, and M. S. Scher, "Time-frequency domain problems in the neurosciences," *Time-Frequency Signal Analysis: Methods and Applications*, edited by B. Boashash, pp. 498-519, Wiley Halsted Press, Longman-Cheshire, 1992.

[210] M. Sun, C. C. Li, L. N. Sekhar, and R.J. Sclabassi, "Efficient computation of the discrete pseudo-Wigner distribution," *IEEE Trans. Acoustics, Speech, Signal Processing,* vol. 37, pp. 1735-1742, 1989.

[211] M. Sun and R. J. Sclabassi, "Discrete instantaneous frequency and its computation," *IEEE Trans. Signal Processing*, vol. 41, pp. 1867-1880, 1993.

[212] M. Sun, S. Qian, X. Yan, S. B. Baumann, X-G. Xia, R. E. Dahl, D. N. Ryan, and R. J. Sclabassi, "Localizing functional activity in the brain through time-frequency analysis and synthesis," *Proceedings of the IEEE*, vol. 84, no. 9, pp. 1302-1312, September 1996.

[213] S. M. Sussman, "Least squares synthesis of radar ambiguity functions," *Trans. Inst. Radio Engineers (IRE)*, vol. 8, pp. 246-254, 1962.

[214] H. H. Szu and J. A. Blodgett, "Wigner distribution and ambiguity function," in L. M. Narducci, editor, *Optics in Four Dimensions,* pp. 355-381, American Institute of Physics, New York, 1981.

[215] H. H. Szu, "Two dimensional optical processing of one-dimensional acoustic data," *Optical Engineering,* vol. 21, pp. 804-813, 1982.

[216] H. H. Szu, "Signal processing using bilinear and nonlinear time-frequency joint-distributions," in Y. S. Kim and W. W. Zachary, editor, *The Physics of Phase Space*, pp. 179-199, Springer Verlag, New York, 1987.

[217] R. Tolimieri and R. S. Orr, "Poisson summation, the ambiguity function and the theory of Weyl-Heisenberg frames," *IEEE Trans. Inst. Radio Engineers (IRC)*, 1995.

[218] L. C. Trintinalia and H. Ling, "Extraction of waveguide scattering features using joint time-frequency ISAR," *IEEE Microwave Guided Wave Lett.*, vol. 6, pp. 10-12, January 1996.

[219] G. T. Venkatesan and M. G. Amin, "Time-frequency distribution kernels using FIR filter design techniques," *IEEE Trans. Signal Processing*, vol. 45, no. 6, pp. 1645-1650, June 1997.

[220] J. Ville, "Théorie et applications de la notion de signal analylique," *Câbles et Transmission*, 2A, pp. 61-74, 1948.

[221] H. Vold and J. Leuridan, "Order tracking at extreme slew rates using Kalman tracking filters," *SAE Paper Number 931288*, 1993.

[222] C. Wang and M. G. Amin, "Performance analysis of instantaneous frequency-based interference excision techniques in spread spectrum communications," *IEEE Trans. Signal Processing*, vol. 46, no. 1, pp. 70 -82, January 1998.

[223] J. Wang and J. Zhou, "Aseismic design based on artificial simulations," *IEEE Signal Processing Magazine*, vol. 16, no. 2, pp. 94-99, March 1999.

[224] D. Wei and A.C. Bovik, "On the instantaneous frequency of multicomponent AM-FM signals," *IEEE Signal Processing Letter*, vol. 5, no. 4, pp. 84-87, April 1998.

[225] J. Wexler and S. Raz, "Discrete Gabor expansions," *Signal Processing*, vol. 21, no. 3, pp. 207-221, November 1990.

[226] E. P. Wigner, "On the quantum correction for thermodynamic equilibrium," *Phys. Rev.*, vol. 40, pp. 749, 1932.

[227] E. P. Wigner, "Quantum-mechanical distribution functions revisited," in W. Yourgrau and A. van der Merwe, editors, *Perspectives in Quantum Theory*, pp. 25-36, MIT Press, Cambridge, Massachusetts, 1971.

[228] W. Williams, H. P. Zaveri, and J. C. Sackellares, "Time-frequency analysis of electrophysiology signals in epilepsy," *IEEE Trans. Engr. Med. Biol.,* pp. 133-143, March/April 1995.

[229] J. Whittaker, "Interpolaroty Function Theory," *Cambridge Tracts in Math. and Math. Physics*, vol. 33, 1935.

[230] D. Wu and J. M. Morris, "Time-frequency representations using a radial Butterworth kernel," *Proceedings of IEEE-SP Int. Symp. on Time-Frequency and Time-Scale Analysis*, pp. 60-63, Philadelphia, Pennsylvania, October 25-28, 1994.

[231] X-G. Xia and S. Qian, "Convergence of an iterative time-variant filtering based on discrete Gabor transform," *IEEE Trans. Signal Processing*, vol. 47, no. 10, pp. 2894-2899, October 1999.

[232] J. Yao, "Complete Gabor transformation for signal representation," *IEEE Trans. Image Processing*, vol. 2, pp. 152-159, April 1993.

[233] Qinye Yin, Shie Qian, and Aigang Feng, "A fast algorithm for adaptive chirplet decomposition," submitted to *IEEE Trans. Signal Processing* for publication.

[234] B. Zhang and S. Sato, "A time-frequency distribution of Cohen's class with a compound kernel and its application to speech signal processing," *IEEE Trans. Signal Processing,* vol. 42, no. 4, pp. 54-64, January 1994.

[235] Y. Zhang and M. G. Amin, "Spatial averaging of time-frequency distributions for signal recovery in uniform linear arrays," *IEEE Trans. Signal Processing*, vol. 48, no. 10, pp. 2892 -2902, October 2000.

[236] Y. Zhao, L. E. Atlas, and R. J. Marks, "The use of cone-shaped kernels for generalized time-frequency representations of nonstationary signals," *IEEE Trans. Acoustics, Speech, Signal Processing*, vol. 38, no. 7, pp. 1084-1091, July 1990.

[237] M. Zibulski and Y. Y. Zeevi, "Oversampling in the Gabor scheme," *IEEE Trans. Signal Processing*, vol. 41, no. 8, August 1993.

[238] M. Zibulski and Y. Y. Zeevi, "Frame analysis of the discrete Gabor-scheme," *IEEE Trans. Signal Processing*, vol. 42, no. 4, pp. 942-943, April 1994.

Papers for Wavelet Analysis and Filter Banks

[239] B. Alpert, "A class of bases in l^2 for the sparce representation of integral operators," *SIAM J. Math. Analysis*, 1993.

[240] F. Argoul, A. Arneodo, J. Elezgaray, and G. Grasseau, "Wavelet transform of fractal aggregates," *Physics Letters A.*, 135:327-336, March 1989.

[241] F. Bao and N. Erdol, "On the discrete wavelet transform and shiftability," *Proceedings of the Asilormar Conference on Signals, Systems and Computers*, pp. 1442-1445, Pacific Grove, California, November 1993.

[242] F. Bao and N. Erdol, "The optimal wavelet transform and translation invariance," *Proceedings of IEEE ICASSP*-94, III: 13-16, Adelaide, May 1994.

[243] S. Basu and C. Chiang, "Complete parameterization of two dimensional orthonormal wavelets," *Proceedings of IEEE-SP Symposium on Time-Frequency and Time-Scale Methods '92*, Victoria, BC, 1992.

[244] A. Benveniste, R. Nikoukhah, and A. S. Willsky, "Multiscale system theory," *IEEE Trans. Circuits and Systems* I, 41(1):2-15, January 1994.

[245] Jonathan Berger, Ronald R. Coifman, and Maxim J. Goldberg, "Removing noise from music using local trigonometric bases and wavelet packets," *Journal of the Audio Engineering Society*, 42(10):808-817, October 1994.

[246] Albert P. Berg and Wasfy B. Mikhael, "An efficient structure and algorithm for the mixed transform representation of signals," *Proceedings of the 29th Asilomar Conference on Signals, Systems, and Computers*, Pacific Grove, California, November 1995.

[247] G. Beylkin, R. R. Coifman, and V. Rokhlin, "Fast wavelet transforms and numerical algorithms I," *Communications on Pure and Applied Mathematics*, 44:141-183, 1991.

[248] J. N. Bradley, C. M. Brislawn, and T. Hopper, "The FBI wavelet/scalar quantization standard for gray-scale fingerprint image compression," *Visual Info. Process. II, SPIE*, Orlando, Florida, April 1993.

[249] Andrew Brice, David Donoho, and Hong-Ye Gao, "Wavelet analysis," *IEEE Spectrum*, 33(10):26-35, October 1996.

[250] C. M. Brislawn, J. N. Bradley, R. Onyshczak, and T. Hopper, "The FBI compression standard for digitized fingerprint images," *Proceedings of the SPIE Conference 2847, Applications of Digital Image Processing XIX*, 1996.

[251] A. G. Bruce, D. L. Donoho, H.-Y. Gao, and R. D. Martin, "Denoising and robust nonlinear wavelet analysis," *Proceedings of Conference on Wavelet Applications*, pp. 325-336, *SPIE*, Orlando, Florida, April 1994.

[252] C. S. Burrus and J. E. Odegard, "Wavelet systems and zero moments," *IEEE Trans. Signal Processing*, submitted, November 1 1996.

[253] C. Sidney Burrus and Jan E. Odegard, "Generalized coiflet systems," *Proceedings of the International Conference on Digital Signal Processing*, Santorini, Greece, July 1997.

[254] Shaobing Chen and David L. Donoho, "Basis pursuit," *Proceedings of the 28th Asiomlar Conference on Signals, Systems, and Computers*, pp. 41-44, Pacific Grove, California, November 1994.

[255] Charles K. Chui and J. Lian, "A study of orthonormal multi-wavelets," *Applied Numerical Mathematics*, 20(3):273-298, March 1996.

[256] A. Cohen, I. Daubechies, and J. C. Feauveau, "Biorthogonal bases of compactly supported wavelets," *Communications on Pure and Applied Mathematics*, 45:485-560, 1992.

[257] Albert Cohen, Ingrid Daubechies, and Pierre Vial, "Wavelets on the interval and fast wavelet transforms," *Applied and Computational Harmonic Analysis*, 1(1):54-81, December 1993.

[258] A. Cohen and Q. Sun, "An arithmetic characterization of the conjugate quadrature filters associated to orthonormal wavelet bases," *SIAM Journal of Mathematical Analysis*, 24(5):1355-1360, 1993.

[259] Ronald. R. Coifman, "Wavelet analysis and signal processing," in Louis Auslander, Tom Kailath, and Sanjoy K. Mitter, editors, *Signal Processing, Part I: Signal Processing Theory*, pp. 59-68, Springer-Verlag, New York, 1990.

[260] Ronald R. Coifman, Y. Meyer, S. Quake, and M. V. Wickerhauser, "Signal processing and compression with wave packets," in Y. Meyer, editor, *Proceedings of the International Conference on Wavelets*, Paris, 1992.

[261] Ronald R. Coifman and M. V. Wickerhauser, "Entropy-based algorithms for best basis selection," *IEEE Transaction on Information Theory*, 38(2):713-718, March 1992.

[262] Ronald R. Coifman and D. L. Donoho, "Translation-invariant de-noising," in Anestis Antoniadis, editor, *Wavelets and Statistics*, Springer-Verlag, 1995.

[263] T. Cooklev, A. Nishihara, M. Kato, and M. Sablatash, "Two-channel multifilter banks and multiwavelets," *Proceedings of IEEE ICASSP-96*, pp. 2769-2772, 1996.

[264] B. N. Cuffin, "A method for localizing EEG sources in realistic head models," *IEEE Trans. Biomed. Engr.*, vol. 42, pp. 68-71, 1995.

[265] Ingrid Daubechies, "Time-frequency localization operators: a geometric phase space approach," *IEEE Trans. Information Theory*, 34(4):605-612, July 1988.

[266] Ingrid Daubechies, "Orthonormal bases of compactly supported wavelets," *Comm. Pure Appl. Math.*, vol. 4, pp. 909-996, November 1988.

[267] Ingrid Daubechies, "The wavelet transform, time-frequency localization and signal analysis," *IEEE Trans. Information Theory*, pp. 961-1005, September 1990.

[268] Ingrid Daubechies and Jeffrey C. Lagarias, "Two-scale difference equations, part I. existence and global regularity of solutions," *SIAM Journal of Mathematical Analysis*, 22:1388-1410, 1991.

[269] Ingrid Daubechies and Jeffrey C. Lagarias, "Two-scale difference equations, part II. local regularity, infinite products of matrices and fractals," *SIAM Journal of Mathematical Analysis*, 23:1031-1079, July 1992.

[270] Ingrid Daubechies. "Orthonormal bases of compactly supported wavelets II, variations ant theme," *SIAM Journal of Mathematical Analysis*, 24(2):499-519, March 1993.

[271] Ingrid Daubechies, "Where do wavelets come from? — a personal point of view," *Proceeding of the IEEE*, 84(4):510-513, April 1996.

[272] I. Daubechies, H. Landau, and Z. Landau, "Gabor time-frequency lattices and the Wexler-Raz identity," submitted for publication.

[273] Z. Doganata, P. P. Vaidyanathan, and T. Q. Nguyen, "General synthesis procedures for FIR lossless transfer matrices, for perfect-reconstruction multirate filter bank applications," *IEEE Trans. Acoustics, Speech, and Signal Processing*, 36(10):1561-1574, October 1988.

[274] David L. Donoho, "Nonlinear wavelet methods for recovery of signals, densities, and spectra from indirect and noisy data," in Ingrid Daubechies, editor, *Different Perspectives on Wavelets, I*, pp. 173-205, American Mathematical Society, Providence, 1993.

[275] David L. Donoho, "Unconditional bases are optimal bases for data compression and for statistical estimation," *Applied and Computational Harmonic Analysis*, 1(1):100-115, December 1993.

[276] David L. Donoho and Lain M. Johnstone, "Ideal denoising in an orthonormal basis chosen from a library of bases," *C. R. Acad. Sci.*, series I, p. 319, Paris, 1994.

[277] David L. Donoho and Lain M. Johnstone, "Ideal spatial adaptation via wavelet shrinkage," *Biometrika*, 81:425-455, 1994.

[278] David L. Donoho, Lain M. Johnstone, Gerard Kerkyacharian, and Dominique Picaxd, "Wavelet shrinkage: asymptopia?" *Journal Royal Statistical Society B.*, 57(2):301-337, 1995.

[279] David L. Donoho, Lain M. Johnstone, Gerard Kerkyacharian, and Dominique Picard, "Discussion of Wavelet Shrinkage: Asymptopia?" *Journal Royal Statist. Soc. Ser B.*, 57(2):337-369, 1995.

[280] David L. Donoho and Lain M. Johnstone, "Adapting to unknown smoothness via wavelet shrinkage," *Journal of American Statist. Assn.*, 1995. Also Stanford Statistic Dept. Report TR-425, June 1993.

[281] David L. Donoho, "De-noising by soft-thresholding," *IEEE Trans. Information Theory*, 41(3):613-627, May 1995.

[282] T. Eirola, "Sobolev characterization of solutions of dilation equations," *SIAM Journal of Mathematical Analysis*, vol. 23, no. 4, pp. 1015-1030, July 1992.

[283] J. S. Geronimo, D. P. Hardin, and P. R. Massopust, "Fractal functions and wavelet expansions based on several scaling functions," *Journal of Approximation Theory*, 78:373-401, 1994.

[284] R. Glowinski, W. Lawton, M. Ravachol, and E. Tenenbaum, "Wavelet solution of linear and nonlinear elliptic, parabolic and hyperbolic problems in one dimension," *Proceedings of the Ninth SIAM International Conference on Computing Methods in Applied Sciences and Engineering*, Philadelphia, 1990.

[285] T. N. T. Goodman, S. L. Lee, and W. S. Tang, "Wavelets in wandering subspaces," *Than. American Math. Society*, 338(2):639-654, August 1993.

[286] T. N. T. Goodman and S. L. Lee, "Wavelets of multiplicity *r*," *Than. American Math. Society*, 342(1):307-324, March 1994.

[287] R. A. Gopinath and C. S. Burrus, "Efficient computation of the wavelet transforms," *Proceedings of IEEE ICASSP-90*, pp. 1599-1602, Albuquerque, New Mexico, April 1990.

[288] R. A. Gopinath and C. S. Burrus, "On the moments of the scaling function ψ_0, *Proceeding of the IEEE International Symposium on Circuits and Systems*, ISCAS-92, pp. 963-966, San Diego, California, May 1992.

[289] R. A. Gopinath, 3. E. Odegard, and C. S. Burrus, "On the correlation structure of multiplicity M scaling functions and wavelets," *Proceedings of the IEEE International Symposium on Circuits and Systems*, ISCAS-92, pp. 959-962, San Diego, California, May 1992.

[290] R. A. Gopinath and C. S. Burrus, "Theory of modulated filter banks and modulated wavelet tight frames," *Proceedings of IEEE ICASSP-93*, III:169-172, Minneapolis, April 1993.

[291] R. A. Gopinath, E. Odegard, and C. S. Burrus, "Optimal wavelet representation of signals and the wavelet sampling theorem," *IEEE Trans. Circuits and Systems II*, 41(4):262-277, April 1994.

[292] R. A. Gopinath, "Modulated filter banks and wavelets, a general unified theory," *Proceedings of IEEE ICASSP-96*, pp. 1585-1588, Atlanta, May 7-10, 1996.

[293] P. Groupillaud, A. Grossman, and J. Morlet, "Cyclo-octave and related transforms in seismic signal analysis," *SIAM J. Math. Anal.*, 15:723-736, 1984.

[294] P. Goupillaud, A. Grossmann, and J. Morlet, "Cycle-octave and related transformation in seismic signal analysis," *Geoexploration*, vol. 23, pp. 85-102, 1984.

[295] J. Götze, J. E. Odegard, P. Rieder, and C. S. Burrus, "Approximate moments and regularity of efficiently implemented orthogonal wavelet transforms," *Proceedings of IEEE ICASSP-96*, II: 405-408, Atlanta, May 12-14, 1996.

[296] Gustaf Gripenberg, "Unconditional bases of wavelets for Sobelov spaces," *SIAM Journal of Mathematical Analysis*, 24(4):1030-1042, July 1993.

[297] A. Grossman and J. Morlet, "Decomposition of hardy functions into square integrable wavelets of constant shape," *SIAM J. Math. Anal.*, vol. 15, pp. 723-736, 1984.

[298] H. Guo, J. E. Odegard, M. Lang, R. A. Gopinath, I. Selesnick, and C. S. Burrus, "Speckle reduction via wavelet soft-thresholding with application to SAR based ATD/R," *Proceedings of SPIE Conference 2260*, San Diego, July 1994.

[299] H. Guo, J. E. Odegard, M. Lang, R. A. Gopinath, I. W. Selesnick, and C. S. Burrus, "Wavelet based speckle reduction with application to SAR based ATD/R," *Proceedings of the IEEE International Conference on Image Processing*, I: 75-79, Austin, Texas, November 13-16, 1994.

[300] H. Guo, M. Lang, J. E. Odegard, and C. S. Burrus, "Nonlinear processing of a shift-invariant DWT for noise reduction and compression," *Proceedings of the International Conference on Digital Signal Processing*, pp. 332-337, Limassol, Cyprus, June 26-28, 1995.

[301] Haitao Guo and C. Sidney Burrus, "Convolution using the discrete wavelet transform," *Proceedings of IEEE ICASSP-96*, III: 1291-1294, Atlanta, May 7-10, 1996.

[302] Haitao Guo and C. Sidney Burrus, "Approximate FFT via the discrete wavelet transform," *Proceedings of SPIE Conference 2825*, Denver, August 6-9 1996.

[303] Haitao Guo and C. Sidney Burrus, "Waveform and image compression with the Burrows Wheeler transform and the wavelet transform," *Proceedings of the IEEE ICIP-97*, Santa Barbara, October 26-29 1997.

[304] Haitao Guo and C. Sidney Burrus, "Wavelet transform based fast approximate Fourier transform," *Proceedings of IEEE ICASSP-97*, pp. 111:1973-1976, Munich, April 21-24 1997.

[305] A. Haar, "Zur Theorie der Orthogonalen Funktionenysystem," *Math. Annal.*, Vol. 69, pp. 331-371, 1910.

[306] C. Heil and D. Walnut, "Continuous and discrete wavelet transforms," *SIAM Rev.*, vol. 31, pp. 628-666, 1989.

[307] C. Heil, G. Strang, and V. Strela, "Approximation by translates of refinable functions," *Numerische Mathernatik*, 73(1)75-94, March 1996.

[308] Peter N. Heller, "Rank m wavelet matrices with n vanishing moments," *SIAM Journal on Matrix Analysis*, 16:502-518, 1995. Also as technical report AD940123, Aware, Inc., 1994.

[309] P. N. Heller, V. Strela, G. Strang, P. Topiwala, C. Heil, and L. S. Hills, "Multiwavelet filter banks for data compression," *IEEE Proceedings of the International Symposium on Circuits and Systems*, pp. 1796-1799, 1995.

[310] Cormac Herley, Jelena Kovacevic, Kannan Ramchandran, and Martin Vetterli, "Time-varying orthonormal tilings of the time-frequency plane," *Proceedings of the IEEE Signal Processing Society's International Symposium on Time-Frequency and Time-Scale Analysis*, pp. 11-14, Victoria, BC, Canada, October 4-6, 1992.

[311] Cormac Herley, Jelena Kovacevic, Kannan Ramchandran, and Martin Vetterli, "Tilings of the time-frequency plane: construction of arbitrary orthogonal bases and fast tiling algorithms," *IEEE Trans. Signal Processing*, 41(12):3341-3359, December 1993.

[312] O. Herrmann, "On the approximation problem in nonrecursive digital filter design," *IEEE Transactions on Circuit Theory*, 18:411-413, May 1971. Reprinted in DSP reprints, IEEE Press, 1972, page 202.

[313] A. N. Hossen, U. Heute, O. V. Shentov, and S. K. Mitra, "Subband DFT — part II: accuracy, complexity, and applications," *Signal Processing*, 41:279-295, 1995.

[314] Plamen C. Ivanov, Michael G. Rosenblum, C.-K. Peng, Joseph Mietus, Shlomo Havlin, H. Eugene Stanley, and Ary L. Goldberger, "Scaling behavior of heartbeat intervals obtained by wavelet-based time-series analysis," *Nature*, 383:323-327, September 26, 1996.

[315] Björn Jawerth and Wim Sweldens, "An overview of wavelet based multiresolution analyses," *SIAM Review*, 36:377-412, 1994.

[316] R. Q. Jia, "Subdivision schemes in L_p spaces," *Advances in Computational Mathematics*, 3:309-341, 1995.

[317] D. L. Jones and R. G. Barniuk, "Efficient approximation of continuous wavelet transforms," *Electronics Letters*, 27(9):748-750, 1991.

[318] H. W. Johnson and C. S. Burrus, "The design of optimal DFT algorithms using dynamic programming," *Proceedings of IEEE ICASSP-82*, pp. 20-23, Paris, May 1982.

[319] R. L. Josho, V. J. Crump, and T. R. Fischer, "Image subband coding using arithmetic coded trellis coded quantization," *IEEE Trans. Circuits and Systems*, pp. 515-523, December 1995.

[320] A. A. A. C. Kalker and Imran Shah, "Ladder structures for multidimensional linear phase perfect reconstruction filter banks and wavelets," *Proceedings of SPIE Conference 1818 on Visual Communications and Image Processing*, 1992.

[321] Hyeongdong Kim and Hao Ling, "Wavelet analysis of radar echo from finite-size targets," *IEEE Trans. Antennas Propagation*, vol. 41, no. 2, pp. 200-207, February 1993.

[322] R. D. Koilpillai and P. P. Vaidyanathan, "Cosine modulated FIR filter banks satisfying perfect reconstruction," *IEEE Trans. Signal Processing*, 40(4):770-783, April 1992.

[323] H. Krim, S. Mallat, D. Donoho, and A. Willsky, "Best basis algorithm for signal enhancement," *Proceedings of IEEE ICASSP-95*, pp. 1561-1564, Detroit, May 1995.

[324] M. Lang, H. Guo, J. E. Odegard, C. S. Burrus, and R. O. Wells, Jr., "Nonlinear processing of a shift-invariant DWT for noise reduction," in Harold H. Szu, editor, *Proceedings of SPIE Conference 2491, Wavelet Applications II*, pp. 640-651, Orlando, April 17-21, 1995.

[325] M. Lang, H. Guo, J. E. Odegard, C. S. Burrus, and R. O. Wells, Jr., "Noise reduction using an undecimated discrete wavelet transform," *IEEE Signal Processing Letters*, 3(1):10-12, January 1996.

[326] Markus Lang and Peter N. Heller, "The design of maximally smooth wavelets," *Proceedings of IEEE ICASSP-96*, pp. 1463-1466, Atlanta, May 1996.

[327] Wayne M. Lawton, "Tight frames of compactly supported affine wavelets," *Journal of Mathematical Physics*, 31(8):1898-1901, August 1990.

[328] Wayne M. Lawton, "Multiresolution properties of the wavelet Galerkin operator," *Journal of Mathematical Physics*, 32(6): 1440-1443, June 1991.

[329] Wayne M. Lawton, "Necessary and sufficient conditions for constructing orthonormal wavelet bases," *Journal of Mathematical Physics*, 32(1):57-61, January 1991.

[330] R. E. Learned, H. Krim, B. Claus, A. S. Willsky, and W. C. Karl, "Wavelet-packet-based multiple access communications," *Proceedings of SPJE Conference, Wavelet Applications in Signal and Image Processing*, Vol. 2303, pp. 246-259, San Diego, July 1994.

[331] James M. Lewis, *The continuous wavelet transform: a discrete approximation*, M.S. thesis, Rice University, 1998.

[332] Jie Liang and Thomas W. Parks, "A two-dimensional translation invariant wavelet representation and its applications," *Proceedings of the IEEE International Conference on Image Processing*, 1:66-70, Austin, November 1994.

[333] K-C. Lian, J. Li, and C-C. J. Kuo, "Image compression with embedded multiwavelet coding," *Proceedings of SPIE, Wavelet Application III*, pp. 165-176, Orlando, Florida, April 1996.

[334] Y. Lin and P. P. Vaidyanathan, "Linear phase cosine-modulated filter banks," *IEEE Trans. Signal Processing*, vol. 43, 1995.

[335] S. M. LoPresto, K. Ramchandran, and M. T. Orchard, "Image coding based on mixture modeling of wavelet coefficients and a fast estimation-quantization framework," *Proceedings of DCC*, March 1997.

[336] F. Low, "Complete sets of wave packets," in *A Passion for Physics — Essays in Honor of Geoffrey Chew*, World Scientific, Singapore, 1985.

[337] Stéphane Mallat, "A theory for multiresolution signal decomposition: the wavelet representation," *IEEE Trans. Pattern Anal., Machine Intell.*, vol. 11, pp. 674-693, July 1989.

[338] Stéphane Mallat, "Multifrequency channel decompositions of images and wavelet models," *IEEE Trans. Acoustics, Speech, Signal Processing*, vol. 37, pp. 2091-2110, 1989.

[339] Stéphane Mallat, "Multiresolution approximations and wavelet orthonormal bases of $L^2(R)$," *Trans. of Amer. Math. Soc.*, vol. 315, pp. 69-87, September 1989.

[340] Stéphane Mallat, "A theory for multiresolution signal decomposition: the wavelet representation," *IEEE Transactions on Pattern Recognition and Machine Intelligence*, 11(7):674-693, July 1989.

[341] Stéphane Mallat, "Zero-crossings of a wavelet transform," *IEEE Trans. Information Theory*, 37(4): 1019-1033, July 1991.

[342] Stéphane Mallat and Frédéric Falzon, "Understanding image transform codes," *Proceedings of SPIE Conference*, Aerosense, Orlando, April 1997.

[343] Stephen Del Marco and John Weiss, "Improved transient signal detection using a wavepacket-based detector with an extended translation-invariant wavelet transform," *IEEE Trans. Signal Processing*, vol. 43, 1994.

[344] Stephen Del Marco and John Weiss, "M-band wavepacket-based transient signal detector using a translation-invariant wavelet transform," *Optical Engineering*, 33(7):2175-2182, July 1994.

[345] Stephen Del Marco, John Weiss, and Karl Jagler, "Wavepacket-based transient signal detector using a translation invariant wavelet transform," *Proceedings of SPIE Conference on Wavelet Applications*, pp. 792-802, Orlando, Florida, April 1994.

[346] T. G. Marshall, Jr., "Predictive and ladder realizations of subband coders," *Proceedings of IEEE Workshop on Visual Signal Processing and Communication*, Raleigh, NC, 1992.

[347] T. G. Marshall, Jr., "A fast wavelet transform based on the eucledean algorithm," *Proceedings of Conference on Information Sciences and Systems*, Johns Hopkins University, 1993.

[348] J. Mau, "Perfect reconstruction modulated filter banks," *Proceedings of IEEE ICASSP-92*, IV:273, San Francisco, California, 1992.

[349] Y. Meyer, "Ondelettes et functions splines," *Seminaire EDS*, Ecole Polytechnique, Paris, 1986.

[350] Y. Meyer, "Orthonormal wavelets," in J. M. Combes, A. Grossmann, and P. Tchamitchian, editors, *Wavelets, Time-Frequency Methods and Phase Space*, Springer-Verlag, Berlin, 1989.

[351] J. E. Odegard, R. A. Gopinath, and C. S. Burrus, "Optimal wavelets for signal decomposition and the existence of scale limited signals," *Proceedings of IEEE ICASSP-92*, IV: 597-600, San Francisco, California, March 1992.

[352] Mohammed Nafie, Murtaza Au, and Ahmed Tewfik, "Optimal subset selection for adaptive signal representation," *Proceedings of IEEE ICASSP-96*, pp. 2511-2514, Atlanta, May 1996.

[353] T. Q. Nguyen and P. P. Vaidyanathan, "Maximally decimated perfect-reconstruction FIR filter banks with pairwise mirror-image analysis and synthesis frequency responses," *IEEE Trans. Acoustics, Speech, and Signal Processing*, 36(5):693-706, 1988.

[354] Truong Q. Nguyen, "A class of generalized cosine-modulated filter banks," *Proceedings of IEEE ISCAS*, pp. 943-946, San Diego, California, 1992.

[355] T. Q. Nguyen and R. D. Koilpillai, "The design of arbitrary length cosine-modulated filter banks and wavelets satisfying perfect reconstruction," *Proceedings of IEEE-SF Symposium on Time-Frequency and Time-Scale Methods '92*, pp. 299-302, Victoria, BC, 1992.

[356] Truong Q. Nguyen, "Near perfect reconstruction pseudo QMF banks," *IEEE Trans. Signal Processing*, 42(1):65-76, January 1994.

[357] Truong Q. Nguyen, "Digital filter banks design quadratic constrained formulation," *IEEE Trans. Signal Processing*, 43(9):2103-2108, September 1995.

[358] Truong Q. Nguyen and Peter N. Heller, "Biorthogonal cosine-modulated filter band," *Proceedings of IEEE ICASSP-96*, pp. 1471-1474, Atlanta, May 1996.

[359] J. E. Odegard, H. Guo, M. Lang, C. S. Burrus, R. O. Wells, Jr., L. M. Novak, and M. Hiett, "Wavelet based SAR speckle reduction and image compression," *Proceedings of SPIE Conference 2487, Algorithms for SAR Imagery II*, Orlando, April 17-21 1995.

[360] Jan E. Odegard and C. Sidney Burrus, "Toward a new measure of smoothness for the design of wavelet basis," *Proceedings of IEEE ICASSP-96*, III:1467-1470, Atlanta, May 7-10, 1996.

[361] Jan E. Odegard and C. Sidney Burrus, "New class of wavelets for signal approximation," *Proceedings of IEEE ISCAS-96*, II:189- 192, Atlanta, May 12-15, 1996.

[362] J. C. Pesquet, H. Krim, and H. Carfantan, "Time-invariant orthonormal wavelet representations," *IEEE Trans. Signal Processing*, 44(8):1964-1970, August 1996.

[363] See-May Phoong and P. P. Vaidyanathan, "A polyphase approach to time-varying filter banks," *Proceedings of IEEE ICASSP-96*, pp. 1554-1557, Atlanta, 1996.

[364] G. Plonka, "Approximation properties of multi-scaling functions: a fourier approach," *Rostock. Math. Kolloq.* vol. 49, pp. 115-126, *1995*.

[365] K. Ramchandran and M. Vetterli, "Best wavelet packet bases in a rate-distortion sense," *IEEE Trans. Image Processing*, 2(2):160-175, 1993.

[366] T. A. Ramstad and J. P. Tanem, "Cosine modulated analysis synthesis filter bank with critical sampling and perfect reconstruction," *Proceedings of IEEE ICASSP-91*, pp. 1789-1792, 1991.

[367] P. Rieder and J. A. Nossek, "Smooth multiwavelets based on two scaling functions," *Proceedings of IEEE Conf. on Time-Frequency and Time-Scale Analysis*, pp. 309-312, 1996.

[368] Olivier Rioul, "Fast computation of the continuous wavelet transform," *Proceedings of IEEE ICASSP-91*, Toronto, Canada, March 1991.

[369] O. Rioul and M. Vetterli, "Wavelets and signal processing," *IEEE Signal Processing Magazine*, pp. 14-39, October 1991.

[370] O. Rioul and P. Duhamel, "Fast algorithms for discrete and continuous wavelet transform," *IEEE Trans. Inform. Theory*, vol. 38, pp. 569-586, March 1992.

[371] O. Rioul and P. Flandrin, "Time-scale energy distributions: a general class extending wavelet transforms," *IEEE Trans. Signal Processing*, vol. 40, pp. 1746-1757, 1992.

[372] Olivier Rioul, "Simple regularity criteria for subdivision schemes," *SIAM J. Math. Anal.*, 23(6): 1544-1576, November 1992.

[373] Olivier Rioul, "A discrete-time multiresolution theory," *IEEE Trans. Signal Processing*, 41(8):2591-2606, August 1993.

[374] Olivier Rioul, "Regular wavelets: a discrete-time approach," *IEEE Trans. Signal Processing*, 41(12):3572-3579, December 1993.

[375] Olivier Rioul and Pierre Duhamel, "A Remez exchange algorithm for orthonormal wavelets," *IEEE Trans. Circuits and Systems II*, 41(8):550-560, August 1994.

[376] J. O. A. Robertsson, J. O. Blanch, W. W. Symes, and C. S. Burrus, "Galerkin-wavelet modeling of wave propagation: optimal finite-difference stencil design," *Mathematical and Computer Modeling*, 19(1):31-38, January 1994.

[377] M. Sablatash and J. H. Lodge, "The design of filter banks with specified minimum stopband attenuation for wavelet packet-based multiple access communications," *Proceedings of 18th Biennial Symposium on Communications*, Queen's University, Kingston, ON, Canada, June 1996.

[378] A. Said and W. A. Pearlman, "A new, fast, and efficient image codec based on set partitioning in hierarchical trees," *IEEE Trans. Cir. Syst. Video Tech.*, 6(3):243-250, June 1996.

[379] A. Said and W. A. Penman, "An image multiresolution representation for lossless and lossy image compression," *IEEE Trans. Image Processing*, 5:1303-1310, September 1996.

[380] Ivan W. Selesnick, Markus Lang, and C. Sidney Burrus, "Magnitude squared design of recursive filters with the Chebyshev norm using a constrained rational Remez algorithm," *IEEE Trans. Signal Processing*, 1997.

[381] J. M. Shapiro, "Embedded image coding using zerotrees of wavelet coefficients," *IEEE Trans. Signal Processing*, 41(12):3445-3462, December 1993.

[382] M. J. Shensa, "The discrete wavelet transform: wedding the a trous and Mallat algorithms," *IEEE Trans. Information Theory*, 40:2464-2482, 1992.

[383] O. V. Shentov, S. K. Mitra, U. Heute, and A. N. Hossen, "Subband DFT — Part I: definition, interpretation and extensions," *Signal Processing*, 41:261-278, 1995.

[384] M. J. Smith and T. P. Barnwell, "Exact reconstruction techniques for tree-structured subband coders," *IEEE Trans. Acoustics, Speech, and Signal Processing*, 34:434-441, June 1986.

[385] M. J. Smith and T. P. Barnwell, "A new filter bank theory for time-frequency representation," *IEEE Trans. Acoustics, Speech, and Signal Processing*, 35:314-327, March 1987.

[386] W. So and J. Wang, "Estimating the support of a scaling vector," *SIAM J. Matrix Anal. Appl.*, 18(1):66-73, January 1997.

[387] H. V. Sorensen and C. S. Burrus, "Efficient computation of the DFT with only a subset of input or output points," *IEEE Trans. Signal Processing*, 41(3):1184-1200, March 1993.

[388] A. K. Soman and P. P. Vaidyanathan, "On orthonormal wavelets and paraunitary filter banks," *IEEE Trans. Signal Processing*, 41(3):1170-1183, March 1993.

[389] A. K. Soman, P. P. Vaidyanathan, and T. Q. Nguyen, "Linear phase paraunitary filter banks: theory, factorizations and designs," *IEEE Trans. Signal Processing*, 41(12):3480- 3496, December 1993.

[390] P. Steffen, P. N. Heller, R. A. Gopinath, and C. S. Burrus, "Theory of regular M-band wavelet bases," *IEEE Trans. Signal Processing*, 41(12):3497-3511, December 1993.

[391] V. Strela and G. Strang, "Finite element multiwavelets," *Proceedings of SPIE, Wavelet Applications in Signal and Image Processing II*, pp. 202-213, San Diego, CA, July 1994.

[392] Gilbert Strang, "Wavelets and dilation equations: a brief introduction," *SIAM Rev.*, vol. 31, pp. 614-627, December 1989.

[393] Gilbert Strang, "Wavelets," *American Scientist*, 82(3):250-255, May 1994.

[394] G. Strang and V. Strela, "Short wavelets and matrix dilation equations," *IEEE Trans. Signal Processing* 43(1):108-115, January 1995.

[395] M. Sun, F-C. Tsui, and R. J. Sclabassi, "Partially reconstructible wavelet decomposition of evoked potentials for dipole source localization," *Proceedings of 15th Annual Int. Conf., IEEE Engr. in Medicine and Biology Soc.*, pp. 332-333, San Diego, 1993.

[396] M. Sun, F-C. Tsui, and R. J. Sclabassi, "Multiresolution EEG source localization using the wavelet transform," *Proceedings of the IEEE 19th Northeast Biomedical Engineering Conf.*, pp. 88-91, Newark, New Jersey, March 1993.

[397] Wim Sweldens, "The lifting scheme: a custom-design construction of biorthogonal wavelets," *Applied and Computational Harmonic Analysis*, 3(2):186-200, 1996.

[398] Hai Tao and R. J. Moorhead, "Lossless progressive transmission of scientific data using biorthogonal wavelet transform," *Proceedings of the IEEE Conference on Image Processing*, ICIP-94, Austin, November 1994.

[399] Hai Tao and R. J. Moorhead, "Progressive transmission of scientific data using biorthogonal wavelet transform," *Proceedings of the IEEE Conference on Visualization*, Washington, October 1994.

[400] Jun Tian, "The mathematical theory and applications of biorthogonal coifman wavelet systems," Ph.D. dissertation, Rice University, February 1996.

[401] J. Tian and R. O. Wells, "Image compression by reduction of indices of wavelet transform coefficients," *Proceedings of DCC*, April 1996.

[402] Michael Unser, "Approximation power of biorthogonal wavelet expansions," *IEEE Trans. Signal Processing*, 44(3):519-527, March 1996.

[403] P. Vaidyanathan, "Quadrature mirror filter banks, M-band extensions and perfect-reconstruction techniques," *IEEE Acoustics, Speech, and Signal Processing Magazine,* 4(3):4-20, July 1987.

[404] P. P. Vaidyanathan, "Theory and design of M-channel maximally decimated quadrature mirror filters with arbitrary M, having perfect reconstruction properties," *IEEE Trans. Acoustics, Speech, and Signal Processing*, 35(4):476-492, April 1987.

[405] P. P. Vaidyanathan and Phuong-Quan Hoang, "Lattice structures for optimal design and robust implementation of two-channel perfect reconstruction QMF banks," *IEEE Trans. Acoustics, Speech, and Signal Processing*, 36(1):81-93, January 1988.

[406] P. P. Vaidyanathan and S. K. Mitra, "Polyphase networks, block digital filtering, LPTV systems, and alias-free QMF banks: a unified approach based on pseudocirculants," *IEEE Trans. Acoustics, Speech, and Signal Processing*, 36:381-391, March 1988.

[407] P. P. Vaidyanathan, T. Q. Nguyen, Z. Doganata, and T. Saramäki, "Improved technique for design of perfect reconstruction FIR QMF banks with lossless polyphase matrices," *IEEE Trans. Acoustics, Speech, and Signal Processing*, 37(7):1042-1056, July 1989.

[408] J. D. Villasenor, B. Belzer, and J. Liao, "Wavelet filter evaluation for image compression," *IEEE Trans. Image Processing*, vol. 4, August 1995.

[409] M. Vetterli, "Filter banks allowing perfect reconstruction," *Signal Processing*, vol. 10, no. 3, pp. 219-244, April, 1986.

[410] Martin Vetterli, "A theory of multirate filter banks," *IEEE Trans. Acoustics, Speech, and Signal Processing*, 35(3):356-372, March 1987.

[411] M. Vetterli and D. Le Gall, "Perfect reconstruction FIR filter banks: some properties and factorizations," *IEEE Trans. Acoustics, Speech, and Signal Processing*, 37(7):1057-1071, July 1989.

[412] M. Vetterli and C. Herley, "Wavelets and filter banks: theory and design," *IEEE Trans. Acoustics, Speech, and Signal Processing*, pp. 2207-2232, September 1992.

[413] Hans Volkmer, "On the regularity of wavelets," *IEEE Trans. Information Theory*, 38(2):872-876, March 1992.

[414] M. J. Vrhel, C. Lee, and M. Unser, "Fast continuous wavelet transform: a least-squares formulation," *Signal Processing*, 57(2):103-120, March 1997.

[415] D. Wei, J. E. Odegard, H. Guo, M. Lang, and C. S. Burrus, "SAR data compression using best-adapted wavelet packet basis and hybrid subband coding," in Harold H. Szu, editor, *Proceedings of SPIE Conference 2491, Wavelet Applications II*, pp. 1131-1141, Orlando, April 17-21, 1995.

[416] D. Wei and C. S. Burrus, "Optimal soft-thresholding for wavelet transform coding," *Proceedings of IEEE International Conference on Image Processing*, I:610-613, Washington, DC, October 1995.

[417] D. Wei, J. E. Odegard, H. Guo, M. Lang, and C. S. Burrus, "Simultaneous noise reduction and SAR image data compression using best wavelet packet basis," *Proceedings of IEEE International Conference on Image Processing*, III:200-203, Washington, DC, October 1995.

[418] Dong Wei, Jun Tian, Raymond O. Wells, Jr., and C. Sidney Burrus, "A new class of biorthogonal wavelet systems for image transform coding," *IEEE Trans. Image Processing*, 1997.

[419] Dong Wei and Alan C. Bovik, "On generalized coiflets: construction, near-symmetry, and optimization," *IEEE Trans. Circuits and Systems II*, 1998.

[420] Dong Wei and Alan C. Bovik, "Sampling approximation by generalized coiflets," *IEEE Trans. Signal Processing*, 1999.

[421] R. O. Wells, Jr., "Parameterizing smooth compactly supported wavelets," *Transactions of the American Mathematical Society*, 338(2):919-931, 1993.

[422] I. Witten, R. Neal, and J. Cleary, "Arithmetic coding for data compression," *Communications of the ACM*, 30:520-540, June 1987.

[423] J. Wu, K. M. Wong, and Q. Jin, "Multiplexing based on wavelet packets," *Proceedings of SPIE Conference*, Aerosense, Orlando, April 1995.

[424] G. Wornell and A. V. Oppenheim, "Estimation of fractal signals from noisy measurements using wavelets," *IEEE Trans. Acoustics, Speech, and Signal Processing*, 40(3):611- 623, March 1992.

[425] G. W. Wornell and A. V. Oppenheim, "Wavelet-based representations for a class of self-similar signals with application to fractal modulation," *IEEE Trans. Information Theory*, 38(2):785-800, March 1992.

[426] X-G. Xia, "Topics in wavelet transforms," Ph.D. dissertation, University of Southern California, 1992.

[427] X-G. Xia, C-C. Jay Kuo, and Z. Zhang, "Wavelet coefficient computation with optimal prefiltering," *IEEE Trans. Signal Processing*, vol. 42, pp. 2191-2197, 1994.

[428] X-G. Xia, J. S. Geronimo, D. P. Hardin, and B. W. Suter, "Design of prefilters for discrete multiwavelet transforms," *IEEE Trans. Signal Processing*, 44(1): 25-35, January 1996.

[429] X-G. Xia and B. W. Suter, "Vector-valued wavelets and vector filter banks," *IEEE Trans. Signal Processing*, 44(3):508-518, March 1996.

[430] Z. Xiong, C. Herley, K. Ramchandran, and M. T. Orcgard, "Space-frequency quantization for a space-varying wavelet packet image coder," *Proceedings of Int. Conf. Image Processing*, I:614-617, October 1995.

[431] H. Zou and A. H. Tewfik, "Design and parameterization of M-band orthonormal wavelets," *Proceedings of the IEEE International Symposium on Circuits and Systems*, pp. 983-986, San Diego, 1992.

[432] H. Zou and A. H. Tewfik, "Discrete orthogonal M-band wavelet decompositions," *Proceedings of IEEE ICASSP-92*, pp. IV-605-608, San Francisco, California, 1992.

Index

E

F

G

H

I

J

L

M

N

O

P

Q

R

S

T

U

W

Z

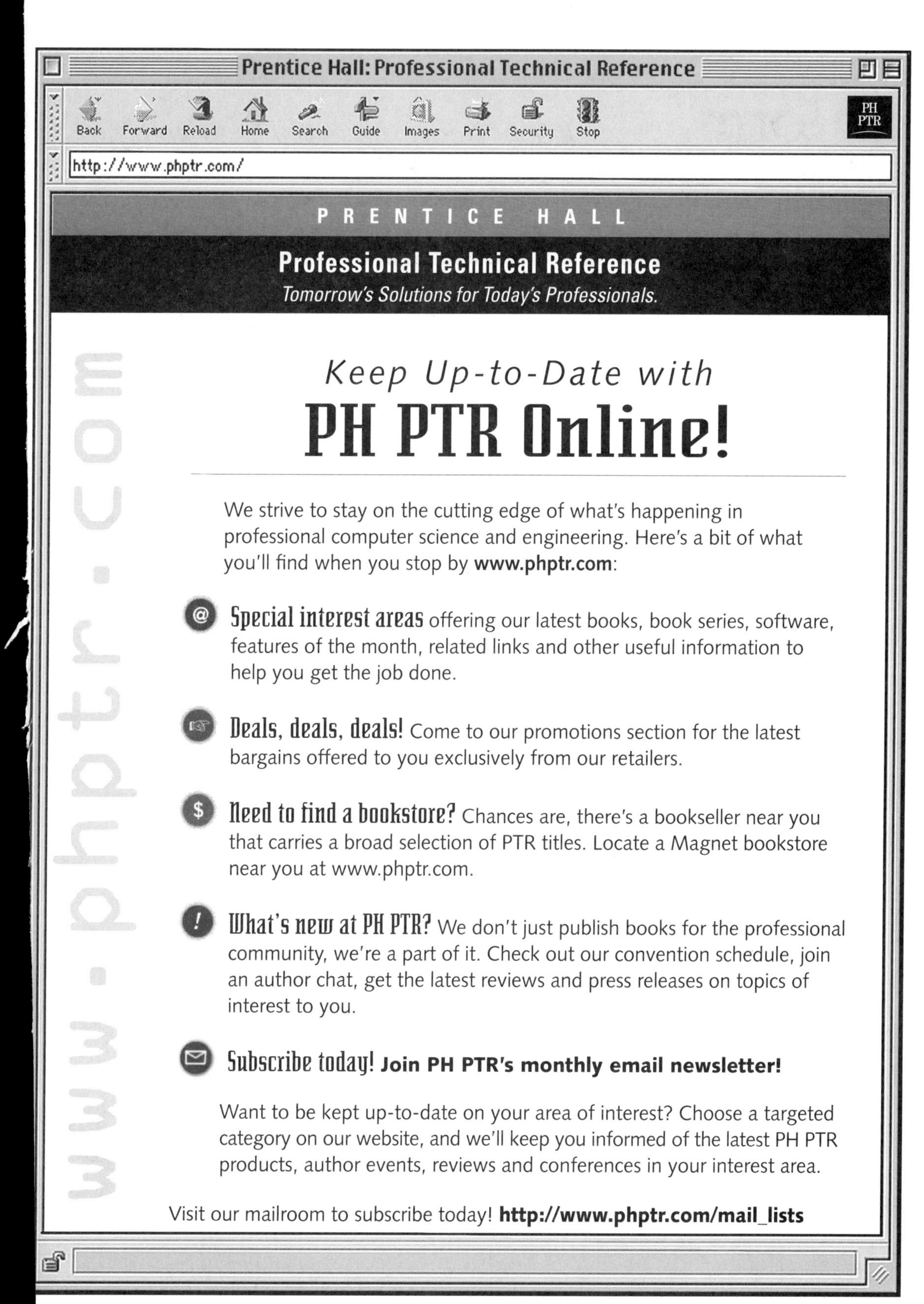
Prentice Hall: Professional Technical Reference
Back
Forward
Reload
Home
Search
Guide
Images
Print
Security
Stop
PH PTR
http://www.phptr.com/
PRENTICE HALL
Professional Technical Reference
Tomorrow's Solutions for Today's Professionals.
www.phptr.com
Keep Up-to-Date with
PH PTR Online!
We strive to stay on the cutting edge of what's happening in professional computer science and engineering. Here's a bit of what you'll find when you stop by www.phptr.com:
Special interest areas offering our latest books, book series, software, features of the month, related links and other useful information to help you get the job done.
Deals, deals, deals! Come to our promotions section for the latest bargains offered to you exclusively from our retailers.
Need to find a bookstore? Chances are, there's a bookseller near you that carries a broad selection of PTR titles. Locate a Magnet bookstore near you at www.phptr.com.
What's new at PH PTR? We don't just publish books for the professional community, we're a part of it. Check out our convention schedule, join an author chat, get the latest reviews and press releases on topics of interest to you.
Subscribe today! Join PH PTR's monthly email newsletter!
Want to be kept up-to-date on your area of interest? Choose a targeted category on our website, and we'll keep you informed of the latest PH PTR products, author events, reviews and conferences in your interest area.
Visit our mailroom to subscribe today! http://www.phptr.com/mail_lists

Solutions from experts you know and trust.

Articles | Free Library | eBooks | Expert Q & A | Training | Career Center | Downloads | MyInformIT

Login Register About InformIT

Topics

Operating Systems

Web Development

Programming

Networking

Certification

and more...

www.informit.com

- Free, in-depth articles and supplements
- Master the skills you need, when you need them
- Choose from industry leading books, ebooks, and training products
- Get answers when you need them - from live experts or InformIT's comprehensive library
- Achieve industry certification and advance your career

Expert Access

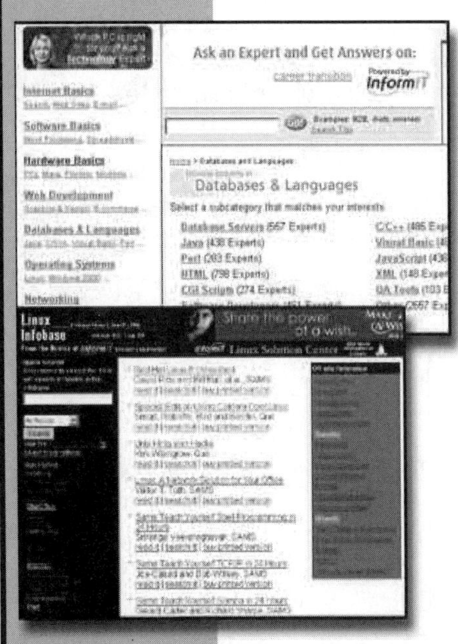

Free Content

Visit InformIT today and get great content from PH PTR

Prentice Hall and InformIT are trademarks of Pearson plc / Copyright © 2000 Pearson